Springer Texts in Business and Economics

Springer Texts in Business and Economics (STBE) delivers high-quality instructional content for undergraduates and graduates in all areas of Business/Management Science and Economics. The series is comprised of self-contained books with a broad and comprehensive coverage that are suitable for class as well as for individual self-study. All texts are authored by established experts in their fields and offer a solid methodological background, often accompanied by problems and exercises.

More information about this series at http://www.springer.com/series/10099

Joachim Weimann
Jeannette Brosig-Koch

Methods in Experimental Economics

An Introduction

Springer

Joachim Weimann
Otto-von-Guericke University
Magdeburg
Magdeburg, Germany

Jeannette Brosig-Koch
University of Duisburg-Essen
Essen, Germany

Original German edition published by Springer Fachmedien Wiesbaden
GmbH, Wiesbaden, Germany, 2019

ISSN 2192-4333 ISSN 2192-4341 (electronic)
Springer Texts in Business and Economics
ISBN 978-3-319-93362-7 ISBN 978-3-319-93363-4 (eBook)
https://doi.org/10.1007/978-3-319-93363-4

This Springer imprint is published by the registered company Springer Nature Switzerland AG
The registered company address is: Gewerbestrasse 11, 6330 Cham, Switzerland

To our families and in memory of Thomas

About This Book

Experimental economic research has become an integral part of modern economic research. Laboratory experiments, the central subject of this book, are now being used to address a wide range of economic issues in all areas of economics. They can be found in business economics research, industrial economics, finance, capital market research, macroeconomics, health economics, and many other fields. The boom in experimental research has been accompanied by the development of an increasingly sophisticated methodology that has contributed greatly to the fact that the quality of laboratory experiments has continued to rise.

A consequence of this is that it has become increasingly important to know very precisely how a research question can be investigated in the laboratory in a methodologically sound way. This textbook aims to help researchers who want to use the laboratory to do just that. We have set ourselves the goal of writing a book that will help both scientists who have already gained experience in the laboratory and those who are starting to work with this method. No special prior knowledge is required to use this book.

Economic research is, of course, the focus of this book, but we hope that colleagues from other related disciplines in which laboratory experiments are starting to be used can also benefit from it. In any case, we have made an effort to assume as little knowledge of economic theory as possible. This is reflected, for example, in the fact that we have written two appendices in which, on the one hand, important game theoretical terms are explained and, on the other hand, important basic experiments that are used in economics are introduced.

The book is divided into four chapters (plus the appendices). The first chapter seeks to place the experimental method in the context of economic research. This seemed necessary and sensible to us because economics was traditionally not an experimental discipline and, in its beginnings, it clearly distinguished itself from psychology. We therefore need an explanation of how normative theory, with its axiomatic models, together with experiments and the increasingly important field of behavioral economics came to dominate this scientific field today. We endeavor to provide such an explanation in the first chapter.

The second chapter of the book is in a sense its core, for it is devoted to the methodological foundations. We have sought to address what we consider to be the most important methodological questions. Of course, we do not claim it to be exhaustive and it naturally remains a subjective selection. We were assisted in this by regularly reading the newsletter of the Economic Science Association (esa-announce@googlegroups.com), which provided us with many valuable ideas. We would like to take this opportunity to thank the ESA Community.

The third chapter of the book deals with the practical implementation of experiments. This is important in view of the fact that both economists and social scientists are generally not accustomed to getting up from their desks and doing practical work in a laboratory. For this reason, it seemed important to us not only to give some useful advice on how to organize the work there but also to point out the most serious traps that lurk in laboratory work.

The fourth chapter of the book deals with the statistical analysis of the data generated in the laboratory. It was of importance to us to point out that this analysis should not begin only when the experiment is completed. On the contrary, it is advisable to already give some thought to the ensuing statistical analysis when designing the experiment. Errors made during the design of the experiment cannot be rectified by statistics. The fourth chapter posed the greatest challenge for us when it came to selecting material since the literature on the statistical methods that can be used for laboratory data is very extensive and the number of possible methods is exceedingly large. We therefore had to carefully consider what we would include and how far we would go into the details of a procedure. We refer the reader to more specialized textbooks in many places because it would have gone beyond the scope of this book to report on the methods in even greater depth.

Sönke Hoffmann supported us in our work on chapter four. We cannot emphasize enough how important his contribution is and would have liked Sönke to be a coauthor of this book. His contribution certainly justifies it. We would also like to thank the Springer Verlag staff, namely, Barbara Fess and Isabella Hanser, for their support and, above all, for their patience with us. Finally, we would like to express our appreciation and thanks to Brian Browne for translating this book.

We very much hope that our book will be put to use in teaching and that it will help those who conduct research in the laboratory to meet the ever-higher methodological standards that research demands. If everything goes well, there will be more editions of this book, and we would be very happy if those who read and use it could let us know if they feel something is missing or see things differently from the way we have presented them. And of course, we look forward to receiving any positive feedback from readers who like the book. We can be reached at:

Joachim.Weimann@ovgu.de and
Jeannette.Brosig-Koch@ibes.uni-due.de

Contents

Supplementary Information

About the Authors

Joachim Weimann

was born in Düsseldorf. He studied economics at the University of Bielefeld. He received his doctorate and habilitation from the University of Dortmund. After a first call to the Ruhr-University Bochum, he got the call to the Otto-von-Guericke-University Magdeburg in 1994. There he still holds the Chair of Economic Policy. He is the author of numerous publications in international scientific journals and of seven monographs, including three textbooks. He is chairman of the German Society for Experimental Economics and executive director of the MaXLab (Magdeburg Laboratory for Experimental Economic Research), member of the Senate and Approval Committee for Research Training Groups of the German Research Foundation, chairman of the ISSM (Institute for Location Research and Tax Policy Magdeburg), and member of ACATECH (Academy of Engineering Sciences). He was dean of the Faculty of Economics at the University of Magdeburg from 1998 to 2008 and a member of the Scientific Senate of the University of Magdeburg from 1998 to 2011. In addition to experimental economic research, his scientific interests include labor market research, happiness research, and environmental economics. Prof. Weimann also frequently speaks in public about labor market and environmental policy issues. The FAZ listed him several times in the list of the 50 most influential economists in Germany.

Jeannette Brosig-Koch

holds a Chair for Quantitative Economic Policy at the University of Duisburg-Essen, Germany, and is the founding director of the Essen Laboratory for Experimental Economics (elfe). Her main research interests are in the fields of experimental health economics and market design. She obtained her doctoral degree in economics in 2003 and her habilitation in 2008, both from the University of Magdeburg, Germany. During this period, she spent several months as a research fellow at Pennsylvania State University, USA, and served as managing director of the Magdeburg Laboratory for Experimental Economics (MaXLab). From 2006 to 2008, she was interim professor at the University of Cologne, Germany. Since April 2008, Jeannette Brosig-Koch has held a full professorship for quantitative economic policy at the University of Duisburg-Essen. Jeannette Brosig-Koch is a member of the review board of the German Research Foundation (DFG), chair of the Socio-scientific Committee of the German Economic Association (VfS), and general secretary of the German Health Economics Association (dggö). Moreover, she serves as a member of the management board of the Behavioural Experiments in Health Network which is a European network for experimental and behavioral research in health economics. The network aims to foster the use of experimental methods and behavioral insights in health economics, policy, and management.

The Study of Behavior

J. Weimann, J. Brosig-Koch, *Methods in Experimental Economics*, Springer Texts in Business and Economics, https://doi.org/10.1007/978-3-319-93363-4_1

1

Overview

In this first chapter of the book we are, in a sense, preparing the stage for what comes later. Experiments have only been part of the economic toolbox for a relatively short time and it is important to understand how this instrument fits into the economic toolbox. It is also important to provide an idea of where it fits in the big picture of the discipline. The explanations on the history of the subject are helpful, but can be skipped by readers who are only interested in the methodological aspects.

The excursions into the history of the subject, in ► Sects. 1.2 and 1.3, serve to explain how behavioral economics could emerge despite the long dominance of normative theory and why both should be understood as complementary parts.

► Section 1.5 deals with the external validity of experiments. This refers to the question of whether experimental findings can be transferred to the real world or not. At the end of the first chapter, the reader should be able, on the one hand, to put experimental economics in the context of economics research, and on the other hand, to understand how economic experiments have been integrated into the division of labor between the various disciplines. In addition, the reader should know that the issue of external validity is an important question that cannot be answered generally, but this does not mean that experimental economics research will fail on account of this. In the first chapter (as in the other chapters) there will be "questions" and summaries ("important") and we have built in one or the other "box" in which interesting background information is conveyed.

1.1 Introduction

It is no longer possible to imagine the economic sciences without experimental research. It has become a well-established method and is used in virtually all branches of economics. Today it is a matter of course that experimental work is published in nearly all international economic journals and are regularly featured in the top journals. This has not always been the case. Only 30 years ago, experimental work was an absolute rarity in important journals and experimental research on a wider scale has only existed since the 1970s. Although economic experiments have been conducted since the 1930s, at the end of the 1960s it was still possible to present the whole body of literature on experimental research in a single survey paper. Even well into the 1980s, it was relatively easy to study economics without once learning in a lecture that economists also carry out experiments.

Experiments are used to study people's behavior under controlled conditions. This can be done in the laboratory or in the field. These may be experimental setups designed by scientists (in the laboratory or the field); they may also be experimental designs that arise naturally. A good example of such a natural experiment is provided by a classical orchestra in Bremen, Germany, which decided 1 day to make as its home a school in an area populated by low-income families. Life within the orchestra became, to an extent, interwoven with life within the school. As a result, the number of applicants for a place at the school soared. The school authority did not know what else to do but draw lots for the few available places. Thus a wonderful experimental design was created since the decision as to who was allowed to be at the school was made purely by chance. In this way, one could be sure that no systematic selection took place. It would, therefore, now be possible to investigate the causal effects of upbringing at an "orchestra school",[1] although so far the school authority has unfortunately not been able to bring itself to take this step.

Economics is a "behavioral science" since it focuses on people's *decisions* on the use of scarce resources. Since about the beginning of the twentieth century economics has stood apart from other disciplines dealing with human behavior in its broadest sense due to the fact that it uses formal mathematical models and abstract theories based on clearly defined assumptions. Its approach is deductive: by means of mathematical reasoning, consequences are deduced from the model's assumptions. This method has very considerable advantages for experimentation. To put it more precisely, the existence of formal theories allows a perfect interplay between theory and experiment.

Economic research based on models allows the assumptions on which a scientific statement is based to be precisely stated in a mathematical sense, thus enabling us to very precisely specify the situation to which the theory is to be applied. The decisive factor here is that the formal method allows the conclusions that economists draw to be derived and proven from the assumptions with exactly the same precision. Economists can therefore make "if-then" statements whose clarity and precision would be unimaginable without formal methods.

At first sight, economists pay a high price for the clarity of these statements. A real-world test of the theory is scarcely possible because the assumptions used in the models economists construct have a high degree of abstraction from the real conditions of the everyday economic world and because they use ideal-typical behavioral models. And what good is a theory that makes unambiguous and mathematically elegant statements if it is not possible to know whether these statements have any significance for economic reality? This "empirical weakness" is transformed into a strength when the possibility of verifying the theory experimentally under laboratory conditions is factored in. It is precisely because economic theory provides such detailed information about the assumptions it uses that these assumptions can be created in the laboratory. And since the mathematical formulation generates clear "if-then" statements, clear hypotheses as to what is to be observed in the laboratory also result if the conditions theory requires are created there. A high degree of *internal validity* can be achieved with experiments, i.e. in the laboratory the experimenter can create a representation of what the theoretician thought up at his desk. In this way, theory becomes verifiable in an ideal-typical manner and the contradiction between formal rigor and empirical lack of substance is resolved.

Definition

Internal validity: This refers to how successfully a model or theory can be represented in the laboratory. An experiment that is internally valid tests exactly the model for which it was designed to test.

1 We thank Nora Szech, who told us this story.

1

Definition

External validity: This refers to the capacity of an experiment to make statements about reality. An externally valid experiment yields observations that can be transferred to reality.

It is mainly thanks to two important characteristics that experiments are so well suited to testing formal theory. First, experiments make it possible to vary the conditions under which decisions are made in a targeted and controlled way. To give a simple example, the question of which information a decision-maker has and does not have available can be answered in every conceivable way in an experiment. The experimenter is in control of what information he provides the subject, and can thus systematically investigate what influence the extent to which the decision-maker is informed has on the decision. This possibility to adjust the variables that are important for behavior in a controlled way represents a second very important feature of experiments. It allows the researcher to pose the question that specifically interests him and to gather the data that is specifically relevant to this question. He is not reliant on economic reality to provide the data he requires to investigate a particular issue. Rather, he is in a position to generate the data to virtually every question that can be asked.

This means, though, that the opportunities offered by experimental research go beyond merely verifying theories. It can also be used to search for stylized facts, regularities in behavior that have not, at least not yet, been described by theories. By this means, explorative experiments yield observations that could provide valuable information as to how successful behavioral theories can be descriptively formulated. The range of application of the experimental method is therefore not limited to those areas determined by existing theories. Roth (1995) once aptly described this by attributing three main functions to experiments. They can "speak to theoreticians" by testing theories and helping to find new theories, they can "search for facts" by uncovering stylized facts, and finally they can "whisper in the ear of princes", i.e. they can be used to provide policy advice.

Question

Given the advantages of experiments as described in the last two paragraphs, how would you characterize research based on non-experimental field data?

If the experimental method is so universally applicable and so well in harmony with formal theory, why did it take so long for experimental research to become established? Considering all the advantages, isn't making use of experiments an obvious thing to do? In order to understand what was preventing economists from investigating the behavior of real people for such a long time, it is necessary to spend a little time on examining the history of economics and to throw some light on the relationship between experimental research and the rest of the discipline. This is very useful and helpful in order to gain an understanding of the experimental method, although not absolutely essential, The reader who is solely interested in the "techniques" of experimental research can therefore feel free to skip the following comments on the history of the research and go straight to ▶ Sect. 1.5, which deals with external validity, for it is a topic of major importance for anyone carrying out experiments.

1.2 Normative Theory and Behavioral Economics

If we were to take a bird's-eye view of the methodological conventions of economics, we would be struck by a curious division. On one side, we have the empirical methods and, among them, the experimental method, with which actual observable behavior is studied. On the other side, we see the methods applied to the development of theories, without any apparent connection to the empirical parts of the discipline at all. Someone who, for example, deals with general equilibrium theory or works with game-theoretical models to address industrial-economic issues uses assumptions about human behavior that are not explicitly based on empirical findings. We will use the term "normative theory" to describe this part of economic research. This does not mean these theories contain normative statements in the sense of "should-be statements", but that they make assumptions that are normative in the sense that they are not empirically based. For example, such models can show us what findings can be expected in situations where the assumptions used are fulfilled. They therefore provide a clear reference point for further analysis. This does not, however, alter the fact that normative theory exists and is applied quite independently of any empirical analysis. The connection to empiricism is not always made, and if so, then usually *ex post,* i.e. after the theory has been developed.

A major feature of neoclassical normative theory is the use of the "rational choice model". This means that decisions are seen as a rational choice from a set of alternatives on the basis of a well-defined preference ordering that possesses particular characteristics. This implies that neoclassical theory models decisions without recourse to psychological insights. The rational choice model uses a priori assumptions about preferences and otherwise uses only mathematics as its instrument of analysis. Psychology does not play a role in the formulation of assumptions. We will give some thought in the next section as to how it could come about that neoclassical theory could deal with decision behavior so successfully for almost a century without even taking notice of psychology.

Alongside neoclassical economic theory, *behavioral economics* has managed to establish itself over the past 30 years. We will also briefly describe how this came about in the next section. It differs from neoclassical economics in that it is focused on replacing assumptions made in the rational choice model with empirically based behavioral assumptions. The roots of behavioral economics are to be found in psychology, but economists have used these roots to cultivate the growth of things that are very similar to those in neoclassical economics in one respect. Like neoclassical models, modern behavioral economics utilizes formal models to describe human decision-making behavior. This is a method that is quite foreign to psychologists, but it has the advantage that it affords the rigorous modeling of psychologically motivated hypotheses and the derivation of testable predictions.

Question

Do you have a hypothesis as to why psychology largely abandoned the development of formal models of decision-making behavior? If you do, it might be worth checking it in the light of the explanations in ▶ Sects. 1.2 and 1.5.

It should be made clear at this point that behavioral economics cannot be equated with experimental research. It is true that behavioral economists very frequently use experiments to gain information on human decision-making behavior. However, experiments may well have the result that the rational choice model of neoclassical economics delivers the best explanation for the data. This has, in fact, been observed in a number of studies. The experimental method is in the first instance nothing more and nothing less than an instrument with which human behavior can be observed under controlled conditions. And this textbook is about how one must proceed in order to obtain reliable results. In a way, the experimental method is a kind of bridge between neoclassical economic theory and behavioral economics.

Normative theory still dominates the development of economic theory, but behavioral science approaches are clearly gaining ground.[2] As a consequence, economics can be divided into two "camps". While in one camp an effort is made to understand how people actually behave and actually reach decisions, in the other little attention is given to this question and consistent use is made of the assumptions that people behave strictly rationally and have well-defined, stable preferences. In order to avoid any misunderstanding at this stage, it should be stressed that these are not hostile camps. On the contrary, not only is a peaceful co-existence possible between them, they can by all means support each other and can be regarded as complementary in the economist's toolbox.

Question

What do you understand by the terms

- normative theory?
- the rational choice model?

At first glance, however, they appear irreconcilable and contradictory. This is not necessarily worth worrying about. As long as the methods of one or the other camp (or both) are mastered, every researcher can be happy and make a successful career in the scientific community of economists. Whoever would like to resolve this contradiction, however, should take a closer look at the history of experimental research and the history of the entire discipline.

That is precisely what we will do in the following. First, we will have a look at how experimental research developed from its beginnings, and then we will attempt to place this in the "big picture" of the subject.

Important

In economics there are different approaches to the object of enquiry. Normative theory uses the rational choice model, which *assumes* that people make error-free choices on the basis of a given preference ordering. Behavioral economics, on the other hand, tries to develop theories that describe the actual observable behavior of humans and that use assumptions that deviate from the rational choice model.

---

2 See DellaVigna (2009) for a survey.

1.3 The History of Economic Experiments

Where does the history of a scientific method begin? When it was first used? Or when it first left its mark? Roth (1995) answered these questions by noting that it is not important when something is first discovered, but when it is last discovered. Someone, at some time or another, carrying out an experiment for the first time is inconsequential for the scientific community if only few people ever find out about it. A new method will only become a true discovery once it actually has some influence on the methodology of the whole profession. This makes it somewhat of a challenge to date the beginning of the history of experimental research since it was presumably only the combination of several pioneering achievements that led to economists as a group taking notice of the experimental method.

Even if it is difficult to characterize a clear chronology in its historical development, it is fair to say that before 1960 there was only little evidence for the existence of experimental research in economics. That does not mean that this anecdotal evidence is without significance – on the contrary, some of it is extremely important.

It is probably no exaggeration to say that the course of the history of experimental research would have been totally different had there not been another methodological innovation whose birth year can be dated quite precisely. In 1944, "A Theory of Games and Economic Behavior" was published. With this book, the authors John von Neumann and Oskar Morgenstern laid the foundation for game theory and, in the process, for the analysis of strategic interactions. At the same time, with the expected utility theory, they created a basis for the analysis of individual choices under uncertainty (von Neumann and Morgenstern 1944). It was above all John Nash, Reinhard Selten and John Harsanyi who extended this foundation in such a way that it became a solid basis for a gigantic research program, now equipping economists with a powerful instrument with which strategic interactions can be precisely described and analyzed: non-cooperative game theory. What is needed for this is first and foremost a formal description of the rules according to which participants of the game interact with each other. Put in very simple terms, these rules provide details on who is playing, which alternatives are open to each player (which moves they can make), what information is available to that player and what consequences every possible combination of moves has for each player.[3]

Definition

Two players A and B are in a *strategic interaction* when the payoffs of A depend on which action B chooses and at the same time the payoffs of B depend on which action A chooses. *Non-cooperative game theory* deals with the analysis of such strategic interactions.

With a little effort and imagination, any possible description of a game can be read as the instructions on how to run an experiment. In other words, game-theoretical models virtually cry out to be tested experimentally because the games created on paper by

3 At the end of the book is an appendix in which, among other things, the most important concepts of game theory are briefly explained.

the theoretician can as a rule actually be conducted in the laboratory. It is only there that they can be carried out since the laboratory allows us to create the exact boundary conditions and incentives that are present in the model. Experiments make it possible to produce a one-to-one copy, so to speak, of the theory from paper to the laboratory. For example, if a theoretician is using a *ceteris paribus* clause in a model and is only investigating the role of a few endogenous variables, this can be accurately reproduced in the laboratory. Therefore, in the laboratory it is possible to achieve what cannot be achieved in the real world: to test a theory under the conditions which the theory itself formulates for its application and which systematically differ from the conditions that are met in the real world. By using monetary incentives and with the assistance of the "induced value method" – which we will discuss in more detail in ▶ Sect. 2.2.1 of the second chapter – not only can the constraints and assumptions of a model be applied in the laboratory, but the preferences assumed in the model can also be induced in the experimental subjects.

With game theory, a methodology that facilitates the interplay between theory and experiment in an almost ideal way entered the stage of economic research. The game-theoretical method in a way forces the theoretician to formulate and formalize explicitly all that is necessary to create a well-defined decision-making situation in the laboratory. Creating this situation is sometimes significantly more difficult in models that employ other methods and do not model any strategic interactions. It is, however, possible and was done early on, i.e. before 1960.

Important

Economic experiments are closely related to game theory. Game-theoretical models contain all the information needed to accurately recreate the model in the laboratory. This has made game theory the ideal basis for experimental research.

As early as 1931 Louis Leon Thurstone addressed the question of whether it was experimentally possible to test or represent a central concept of neoclassical economics that today is still as dominant as it was then (Thurstone 1931). He attempted to derive indifference curves experimentally by offering subjects a (hypothetical) choice between alternative bundles of goods. Although the subject of this experiment concerned a core element of the economic rational choice model, Thurstone himself was not an economist, but a psychologist, and the experimental methods he used were more consistent with those of experimental psychology and less with those used by experimental economists today. The criticism that no less than Milton Friedman and Wilson Allen Wallis directed at this experiment in 1942 anticipated the classic criticism that economists voice about psychological experiments (Wallis and Friedman 1942). This is essentially that nothing can be learned from hypothetical questions because the subjects are not provided with the right incentives. This criticism was not actually leveled at experimental economics – which did not yet exist at that time – but at the attempts of psychologists to use their methods to investigate economic questions.

Of a somewhat different nature is the experiment published in 1948 by Edward Hastings Chamberlin (Chamberlin 1948). Chamberlin, along with Joan Robinson, is regarded as the founder of the theory of imperfect competition. He created the concept "monopolistic competition" and was the first to attempt to create markets in the laboratory, applying methods that are still also used today in an enhanced form to conduct market experiments. The aim of his experiment was to prove that it could not at all be expected that market equilibria come about in markets. Chamberlin himself regarded the results he obtained as supporting his skepticism concerning this. The further development of his own method by such researchers as Vernon Smith, however, later led to a convincing confirmation in the laboratory that even under difficult conditions (e.g., in the case of very limited information) markets are indeed able to generate equilibrium prices.

No matter how Chamberlin's findings are interpreted, his experiments are characterized by three things: first, they were carried out by an economist; second, their objective was to experimentally test an equilibrium concept central to economic theory; and, third, Chamberlin paid his subjects on the basis of their behavior in the experiment. It is presumably these three things that led many economists to view Chamberlin's experiments as the birth of experimental economics. The name "experimental economics", however, was coined by someone else: Heinz Sauermann, who together with his student Reinhard Selten made significant contributions to experimental research on oligopolistic markets (Sauermann and Selten 1959). Sauermann and Selten, two important German economists, undoubtedly rank among the pioneers of experimental research. In 1977 Sauermann founded the *Gesellschaft für experimentelle Wirtschaftsforschung* (German Society for Experimental Economics), which is still today the scientific association for German-speaking experimental economists and can claim to be the oldest scientific association of experimental economists in the world.[4] It is also thanks to Sauermann and Selten that the contributions of German-speaking experimental economists to the development of this still young methodology gained international attention in the 1960s and 1970s.

In contrast to Chamberlin's market experiments, the oligopoly experiments of Sauermann and Selten were characterized by game-theoretical analysis. Important though the first experiments of psychologists and economists on economic issues may have been, it is game theory that gave the experimental method a decisive boost. Two experiments in particular provide key examples of this.

In 1950 Melvin Dresher and Merrill Flood devised a game that has since then had a remarkable career in economics: the prisoner's dilemma (see Flood 1952, 1958). These two authors thought up this game with the intention of subjecting the concept of the Nash equilibrium to a particularly hard test (today this would be called a stress test). The payoffs in the original game developed by Dresher and Flood were as follows (◘ Table 1.1):

The row player and the column player can simultaneously choose between the first and second rows or columns. It is clear that the Nash equilibrium is (2, 1) since the dominant strategy for the row player is to select option 2, while the column player's dominant strategy is to choose option 1. At equilibrium both players collect a payoff that

4 Also see the anthology published on the occasion of the 30th anniversary of the German Society for Experimental Economics (Sadrieh and Weimann 2008) with which Reinhard Tietz, another German pioneer of experimental research, was honored.

Table 1.1 Prisoner's dilemma payoffs (the first number is the payoff received by the column player, the second number is the payoff received by the row player, see Flood 1952, 1958)

	Column player Option 1	Column player Option 2
Row player Option 1	−1, 2	½, 1
Row player Option 2	0, ½	1, −1

is ½ lower than the payoff obtainable if they chose (1, 2). The equilibrium is thus highly inefficient, leading to the question of whether real people, in light of this, still select the rational solution and play their dominant strategies. Therein lies the stress test.

Definition

A **dominant strategy** is a strategy that is always the best response (i.e., always maximizes the payoff for this player) a player can provide to what other players do. Having a dominant strategy frees a player from the need to form expectations about what other players will do. No matter what they do, the dominant strategy is always the best response.

Dresher and Flood conducted this game as part of an experiment with (only) one pair of subjects, repeating the game 100 times, with the payoff being in US pennies.[5] This means that, at equilibrium, the column player earns half a dollar and the row player comes out empty-handed. If both players act efficiently and choose (1, 2), each will gain half a dollar more than they do at equilibrium. The results of this experiment showed that the subjects neither played the Nash equilibrium, nor were capable of cooperating to obtain an efficient solution. It is not the result itself that lends significance to this experiment, but rather the fact that for the first time a game-theoretical prediction (rational players choosing dominant strategies) and at the same time a game-theoretical equilibrium concept were put under the microscope in an experiment. Since then, the prisoner's dilemma and the closely related theory of public goods have been subjected to this countless times. With their experiment, Dresher and Flood opened up a whole world of research.

A similar lasting impact was achieved by an experiment that examined not an equilibrium concept, but expected utility theory, thus singling out another central building block of game theory and modern economic theory as a whole. Maurice Allais was not only awarded the Nobel Prize for his experimental work, but his name is inextricably linked to an experimental finding that stands in clear contradiction to the expected utility theory of von Neumann and Morgenstern. The Allais paradox

5 The choice of the optimal sample size will be discussed in ▶ Sect. 4.5.2.

describes a choice between lotteries that systematically and reproducibly leads to results indicating that the decision-makers do not maximize their expected utility. This finding is just as momentous as that of Dresher and Flood since expected utility theory is still today the central theory used to describe choices under uncertainty, thus making it possible for economists to model how people deal with risk. By today's standards, Allais's experiment was methodologically inadequate owing to its use of hypothetical payoffs. But this is not critical for the significance of the experiment. What is more important is that, similar to the case of the prisoner's dilemma, the Allais paradox provided a path that has since been taken by many experimental and theoretical economists. It is achievements such as these that are honored with the Nobel Prize.

Important

Research questions derived from early experiments to examine predictions of expected utility theory and game theory are still the subject of economic experiments today.

The closeness of the relationship between game theory and experimental research is also evident in the fact that a number of outstanding game theoreticians are among the early experimentalists. In addition to Reinhard Selten, whom we mentioned earlier, there are also John Nash and Thomas Schelling, for example. Although Nash made only a relatively short foray into the world of experiment, Schelling was intensively involved with experiments on coordination games as early as 1957 (Schelling 1957). As close as the connection between game theory and experiment may be, the building of a bridge from theory to experiment is by no means a foregone conclusion. The reason for this lies in the difficulties for theory that arise from many experimental findings. Expected utility theory and, as a consequence, also game theory use the notion of optimizing players who act strictly rationally as a basic premise. Beginning with the early experiments on the prisoner's dilemma of Dresher and Flood and the experiments that tested expected utility theory, the history of experimental research has also time and again been a history of findings that are at odds with the assumption of rationality. This does not mean that experiments always show non-rational behavior, but it occurs relatively frequently. Theoreticians, who took it for granted that the agents in their models act hyper-rationally, were faced with results that showed that people systematically display bounded rational behavior. That does not make it easy to take up the path to experimentation. The fact that bridge building is nevertheless possible and very productive, despite the clashes and contradictions, is illustrated by an anecdote of an incident that took place at an annual conference of the *Verein für Socialpolitik* (German Economic Association).

Reinhard Selten held a plenary lecture at this conference. After the lecture, which dealt with the findings of experimental studies, the chairman of the plenary session, Hans-Werner Sinn, asked him whether he could not be accused of being, in a certain sense, schizophrenic. After all, he had received the Nobel Prize for studies in which he had pushed the rationality assumption of game theory to the limit, as it were; on the other hand, he was now searching for a theory of bounded rational behavior. Reinhard Selten responded by saying that he was not schizophrenic, but a *methodological dualist*. This was because it simply made sense, on the one hand, to investigate where perfect rationality would lead or what characterized a strictly rational decision, only to concede, on the other hand, that people were not capable of this perfect rationality and then to set out to find a theory that could describe what real people actually do when they make a decision. It is difficult to argue with this assessment. It makes it clear that despite contradictions it is possible for normative theory and experimental research interested in real behavior to exist side by side.

The last two to three decades have shown that even more is possible. Behavioral economics, using the methods of normative theory, attempts to develop models of human behavior that are consistent with the experimental findings of economists and psychologists. The breakthrough in behavioral economics can be dated relatively accurately. It came in 1979 with the publication of the article *Prospect Theory: An Analysis of Decision under Risk* by Daniel Kahneman and Amos Tversky in the renowned economics journal *Econometrica* (Kahneman and Tversky 1979). Both authors were psychologists and Kahnemann was awarded the 2002 Nobel Prize for this article, amongst others.[6] Unlike neoclassical expected utility theory, prospect theory provides an explanation for decisions under risk based on psychological findings and not on a system of axioms that is not empirically verified, such as that of von Neumann and Morgenstern.

Without prospect theory, the dynamism with which experimental research established itself in economics would probably have been substantially less and the development of behavioral economics would have progressed at a much slower pace. Prospect theory to an extent paved the way for studies deviating from the strict assumptions of the rational choice model and made them respectable in economics. Thus the number of publications based on experimental work grew dramatically and a culture of research geared to the special needs of experimental and behavioral economics research evolved. The current role of behavioral economics in the economic sciences is clearly demonstrated by the Nobel Prize once again being awarded to a behavioral economist, Richard Thaler, in 2017.

In order to better understand this research culture, it is important to bear in mind that the nature of the knowledge gained in economic experiments is fundamentally different to that generated by theoretical papers. An experiment can only ever represent one individual observation made under particular conditions at a particular place at a particular time. Further experiments under at least similar conditions in other locations at some other time are necessary before the observations become a finding that can claim to possess a certain degree of generality.

This means that progress in experimental economics is rather slow. It simply takes time to carry out all the experiments needed to produce such things as stylized facts. It also means there needs to be some degree of coordination between those who conduct experimental research. It is necessary to reach agreement on which phenomena will be investigated in order to find out which observations are reproducible patterns of behavior and which are merely artifacts of a particular experimental design. This type of coordination has become increasingly successful since the beginning of the 1970s. Series of experiments have emerged from this, i.e. many experiments are conducted on

---

6 He shared it with Vernon Smith. Amos Tversky would probably have been awarded the Nobel Prize as well had he not passed away in 1996.

one and the same "basic issue", each with variations that can be exploited to separate the wheat from the chaff amongst the findings. It has also seen the emergence of specific issues that are almost solely handled in cooperation between experimentalists and behavioral economists. Evidence for the existence of social behavior and its characterization is one such issue, parts of auction design can be seen in a similar light, and the basic assumptions of prospect theory are still today the subject of experimental studies in the laboratory and in the field.

Box 1.1 Prospect Theory

Prospect theory assumes that the valuation of lottery payoffs is based on a reference point. The reference point is not fixed, but can change. Based on the reference point payoffs are evaluated as gains or losses. The value function is concave in the gain area and convex in the loss area. In addition, it is steeper in the loss area than in the gain area. This is an expression of so-called loss aversion: losses carry more weight than equivalent gains. Note that, besides the valuation of payoffs, prospect theory also assumes a weighting of probabilities.

Today, no one seriously doubts the value of experimental research in economics. Laboratory experiments are an important addition to the toolbox of economists. They allow us to ask and answer questions that could not be answered solely on the basis of theoretical models and classical empiricism (i.e. statistics and econometrics based on field data). The development of experimental research has become possible thanks to expected utility theory and game theory, which, like no other sub-disciplines of economics, provided the basis for experimental work and which to a certain extent had already created experiment in their own methodology. Without the assistance of expected utility theory and game theory at its beginning, experimental economics could never have developed into what it is today. This insight provides us with the key to answering the question we posed at the beginning and which we have not yet been able to answer: how did it come about that economics became a discipline in which it is not unusual to make use of experiments?

Question

Food for thought: Do you tend to see similarities or differences between experimental research in economics and experimental physics?

1.4 The History of the Neoclassical Rational Choice Model and the Return of Psychology

At the beginning of the 20th century, the transition from classical to neoclassical economics began to take shape. The objective theory of value of classical economics was superseded by the subjective theory of value, which ascribed the value of objects to their scarcity and their utility for people. During this phase of its history, in search of basic principles, economics was only just beginning to become established as an independent discipline. Its connections to philosophy and, in particular, psychology were still very strong. We know from Pareto's correspondence that it was his express goal to make economics an independent discipline possessing its own object of enquiry and methodology that clearly distinguished it from its neighboring disciplines.[7] It was precisely the subjective theory of value, however, that at first stood in the way of such independence. Bruni and Sugden (2007) point out that there were some early exponents of neoclassical economics who even wanted to intensify the relationship to psychology. Francis Ysidro Edgeworth and William Stanley Jevons, for example, regarded "utility", which constituted the heart of the subjective theory of value, as a psychological category. It was Edgeworth, in particular, who wanted to make this more objective by applying the *Weber-Fechner Law*, which describes the logarithmic relationship between the intensity of a stimulus and its resulting subjective perception, to the measurement of utility. The notion that utility could be measured and compared intersubjectively represented an obstacle, so to speak, to the independence of economics. Edgeworth's research program could only have been feasible if it had been possible to examine the nature of utility. It would have been necessary to investigate what utility actually is, what it depends on and how it can be measured, compared and evaluated. An intensive psychologization of economic research would have been needed to manage all these tasks. However, the methodology of psychology research would then have become the most important instrument of economics, and this would have had consequences that many neoclassical economists would have found difficult to live with.

The claim to the uniqueness of economics would have practically been abandoned. Economic research would have become some kind of specialized area of psychology, which above all else dealt with the question of how people perceive pleasure and pain and how they seek one but avoid the other. Not only would its independence have receded into the distant future, but its claim to being scientific would also have been severely jeopardized by a close proximity to psychology. Despite the methodological advances made in psychology at the beginning of the twentieth century thanks to the work of Ernst-Heinrich Weber (1795–1878) and Gustav Theodor Fechner (1801–1887),[8] there was always some controversy associated with the issue of whether psychology was scientific. For example, although quantitative methods were used, introspection remained an important methodological instrument despite it obviously not being suitable for generating objective, intersubjectively comparable data (Bruni and Sugden 2007, pp. 150). Psychology's preoccupation with pleasure and pain therefore led to the suspicion, and this was not completely unjustified, that a certain degree of arbitrariness was permitted, resulting in the accusation that psychology was not a true science.

The alternative to a psychology-based research program consisted of the attempt to liberate economics completely from metaphysics and to align it with the natural sciences with respect to objectivity and scientificity. It was precisely this that Pareto wanted to achieve with his research program. But how are we to eliminate psychology if we want

7 See the very interesting essay by Bruni and Sugden (2007), which strongly inspired our explanations in this section.

8 Fechner is regarded as the founder of modern psychology. 1860 saw the publication, in two volumes, of his main work "Elements of Psychophysics", containing the description of methods which made it possible to make quantitative statements on feelings and sensations and which were based on the earlier work of Weber (Fechner 1860).

to analyze people's actions and decisions? At first glance, it might seem somewhat far-fetched that psychology should not play a role in this, but it is actually possible. The trick is that Pareto's economics is not concerned with the essence of things – for example, the question of the nature of utility and its measurement – but with secondary principles that can be derived from the essence of things. Specifically, this means economic analysis is not based on a metaphysical consideration of the nature of utility, but only on objectively observable decisions that individuals base on their own subjective calculation of utility. The idea is remarkably simple. Let us suppose that every person is precisely aware of how much utility he will gain from a certain level of consumption or a certain choice. Let us also suppose that the way people behave depends on their perceived utility, i.e. in a choice between two alternatives they select the one providing them with greater benefit. Based on these assumptions, it must then be possible to infer the underlying utility concept from observing people making a choice. The question is, however, when can we make these assumptions? When can we conclude from people's observable decisions that they make these decisions precisely because they maximize their utility?

Pareto used this to create the concept of rational actions. Essentially, it is based on the idea that people act as if they are constantly optimizing. They use a utility function that assigns a value to every available alternative in order to measure the benefit they could realize: the higher the value, the better the alternative. The optimization problem consists of choosing precisely the combination of goods for which the utility function takes on the highest possible value at the given prices and income. But when can we conclude from observing people making a choice that their actions are indeed based on a utility maximizing calculus? A substantial part of the Paretian research program consisted of the question "When may it be assumed that people have a preference ordering that can be represented by means of a utility function?" This question is referred to as the *integrability problem* and it intensely occupied some of the best economists of the last century (Bruni and Sugden 2007, p. 159).[9]

It is important to note that if the integrability problem is solved, economics can dispense altogether with psychology. If it is permissible to assume that people make choices as if they maximize some utility function, any decision can be described as the result of an optimization calculation, and mathematics takes the place of psychology. It is a matter of discussion whether the integrability problem has actually been solved in all its generality and sufficiently comprehensively or not. What is important is the fact that the economics profession views this problem as having been adequately solved for more than 70 years and does not attach any particular importance to any questions that may still be open. The solution consists essentially of the revealed preference theory, which was decisively influenced by Paul Samuelson (1938). This is not a suitable place to go into details and analyze the differences between the various axioms of revealed preference theory. Explaining the basic principle will suffice.

The Weak Axiom of Revealed Preference can be described in a rather loose way as follows. Suppose you want to buy yourself or a friend a new tie. The sales assistant offers you 20 different ties and you choose the one with red stripes on a white background. The

9 The two last paragraphs have been adapted from the manuscript of the book "Measuring Happiness. The Economics of Well-Being" by Andreas Knabe, Ronnie Schöb and Joachim Weimann (Weimann et al. 2015).

weak axiom requires that you also make this choice if you are not offered 20, but only 18 or 15 ties. As long as the one with the red stripes is included, you will choose it and thus fulfill the requirements of the weak axiom. Samuelson showed that it is then possible to describe your decision as the result of utility maximization and it can be assumed that you have constructed a system of ranking the ties that allows you to indicate which tie you prefer whenever you compare any two of the ties.

In this way, an important part of Pareto's program was realized. Since we only need to assume that, to apply the rational choice model, an ordinal ranking of the alternatives (the ties) exists, and it is the fulfillment of the weak axiom that allows its existence to be assumed, it is no longer necessary to have any kind of psychology to ask the question of why of all the ties it is the red-and-white striped tie that generates the highest level of utility. Moreover, the economist is permitted to describe the decision by maximizing a utility function whose concrete functional form is left to his own discretion, since many functions are suited to describing ordinal preference rankings. Metaphysics was thus completely eliminated from the description of the individual act of choosing and economics' claim to being scientific was secured since the basis of all analyses was now the objective observation of people making a choice.

Box 1.2 Rational Choice Model

The rational choice model is based on the assumption that people make decisions on the basis of a preference ordering of all available alternatives. In general, it is assumed that this order is complete, reflexive and transitive. The preference ordering permits the decision-maker to specify which alternatives are preferred or whether the decision-maker is indifferent between the alternatives. Formally, this preference ordering can be represented by a utility function that assigns a number to each alternative, whereby the greater the number, the higher up the alternative is in the preference ordering. The rational choice model assumes that all the acts of choosing made on the basis of this preference ordering are carried out without error, i.e. we always choose the alternative which is "as high as possible" in the preference ordering, or which has the highest utility index of all achievable alternatives.

This did not mean, however, that the full implementation of the Paretian research program had been completed. This is because economic research is concerned not only with the individual act of choosing, but also with the resolution of conflicts arising from this act. In the final analysis, economists are interested in the question of how to allocate resources in the face of a fundamental scarcity problem that prevents the demands of all individuals from being met at the same time. Is it not necessary to bring psychology back on board if we want to solve this problem? Assuming there are resources available that can be used for, say, the construction of a home for the aged, a street, a kindergarten or a bowling center, shouldn't we weigh up the advantages of each use against the other and at the same time determine the utility each individual measure generates? This would in fact mean we were back to the "essence of things", having to deal with "utility" in the way Pareto wanted to avoid.

The completion of the Paretian research program only came about with the concept of *Pareto efficiency*. This requires that resources be allocated in such a way that it is no longer possible for an individual to become better off without at the same time another individual becoming worse off. As long as we fail to achieve Pareto efficiency, we are clearly wasting resources.

Box 1.3 Pareto Criterion
Among other things, Vilfredo Pareto was an engineer and he used the criterion named after him for the construction of machines: a machine is not yet optimally designed if it is possible to improve a performance parameter without compromising another parameter value.

Question
Why does the Pareto criterion include a value judgment? What other value judgments could be used in its place when it comes to the question of how to use scarce resources?

The Pareto criterion interacts perfectly with the rational choice model, which is based on the concept of ordinal utility. Together they provide neoclassical economics with a tremendously powerful methodological tool, allowing whole areas of research to be opened up and explored. The massive success of neoclassical economics over the last 70 years can be attributed to its methods being largely value-free, the universal applicability of mathematics as a "behavioral model" and the flexibility that has so far enabled neoclassical economics either to nullify all criticism or to assimilate the critics. This flexibility arises from the straightforwardness of revealed preference theory, which says nothing about *what* choices are available and still less about what is *preferable*. Neoclassical economics does not specify what people prefer and so pointing out, for instance, that people are not only self-interested (as is assumed in many neoclassical models) cannot undermine neoclassical economic theory. It allows people to have a preference for giving something to others (Samuelson 2005). The "neoclassical repair shop", which has incorporated many critical movements in the neoclassical edifice, is so successful because neoclassical economists ultimately do not commit themselves to very much.

The fact that many people nevertheless gain the impression that the neoclassical "homo oeconomicus" is "completely selfish" (Levitt and List 2008, p. 909) can be attributed to the fact that neoclassical economics has no other choice than to make very concrete assumptions about preferences. If this were not the case, the theory would more or less be devoid of content. On account of this, the assumption of "more is better than less" is made after all and used to assert that people always prefer a higher income to a lower income. The reason may well be that a higher income provides more opportunities for consumption and we observe that people have a certain preference for having more opportunities. Thus both are correct: neoclassical economics is, in principle, open for any assumption concerning what generates benefits for people, but the premise that people selfishly strive for higher income is, in fact, the dominant assumption in neoclassical economic theory. This conflicting nature is very nicely illustrated by the avowed neoclassicists Binmore and Shaked (2010). At first they report that experimental findings do not put neoclassical economics at risk of being refuted since it makes no assumptions concerning people's preferences (p. 88). Later, however, they only speak about money maximization as the behavioral hypothesis that has proven superior to inequity aversion models. This does not represent a contradictory argumentation, but rather an expression of the flexibility of neoclassical economics. *In principle*, it assumes that people behave selfishly in a material sense. However, in situations in which this is clearly not the case, it is by all means open to accepting that people can also be moved by other motives. This flexibility regarding assumptions on the content of preference ordering preserves the very essence of neoclassical theory: the assumption that *given their preferences* people behave rationally.

Important

The core of neoclassical economics is the assumption that the rational choice model is capable of representing human behavior. In contrast, the assumption that humans behave selfishly is not essential for neoclassical economics, although it is used in many neoclassical models. In fact, neoclassical economics can be reconciled with many assumptions concerning the content of individual preferences. This gives it considerable flexibility, because by varying the assumptions of preference, it is possible to rationalize behavior that would not be rational under the condition of strictly self-interested preferences.

In view of the success of neoclassical economics and its "ability to put up a fight", the question naturally arises as to how the return of psychology to the world of economics could come about. It should be noted here that for all the flexibility of the neoclassical method, it succeeded in keeping psychology out for a long time. This was achieved, on the one hand, by incorporating into the rational choice model the possibility of people having social preferences. In this way, it was possible to integrate into the neoclassical edifice a considerable share of the experiments yielding intractable deviations from the standard model without having to make compromises to the rationality assumption. On the other hand, the demarcation from psychology was also achieved by experimental economics taking *methodological* approaches different to those used in (the older) experimental *psychology*. Two major differences are worth noting.

In general, real incentives are provided in economic experiments, i.e. these experiments are always about money.[10] Psychologists, in contrast, do not generally use monetary incentives. The second difference is that experimental economists utterly disapprove of manipulating experimental subjects by not informing them of the complete truth about the experiment. The reason for this is that the reputation of the experimenters and, thus, the control over the decision environment would otherwise be jeopardized. The subjects would not believe the instructions placed before them and their behavior would no longer be open to meaningful interpretation because it would not be clear what game they had actually played (or they thought they had played). Psychologists are usually less sensitive in this respect and lie to their subjects at first, only to explain during the "debriefing" afterwards what really happened in the experiment.[11]

These methodological differences made it possible to clearly define the boundaries between experimental economics and psychology. In so doing, experimental research could be incorporated into neoclassical economics without it being necessary for it to open up to psychology in the process. Experiments in economics still deal to a great extend with rational behavior and attempt to establish what the preferences of subjects look like. As in the past, the analysis of both the psychological basis of the perception of

10 We will go into this in more detail in ► Sect. 2.2.

11 This point will also be covered in more detail in ► Sect. 2.3.

utility and the choices derived from it are still not the object of neoclassical economics. So how did psychology manage to return to economics?

The initial push unmistakably came from the side of psychology. As mentioned earlier, in 1979 an article by Daniel Kahneman und Amos Tversky, "*Prospect Theory: An Analysis of Decision under Risk*", was published in *Econometrica*. With this article, behavioral economics took the academic stage, as it were. What was special about Kahneman and Tversky's work was that psychologists had presented a formal theory that was able to explain a number of observations that the standard economic theory, expected utility theory, could not explain. This was achieved not with a priori assumptions on the rationality of the actors as a starting point of the modeling, but rather with experimentally substantiated findings in psychology on how real people deal with uncertainty. This was an approach that differed fundamentally from that of neoclassical economics since it cast doubt on the premise that people always behave rationally. Psychology pointed out that there were findings indicating that people could deviate systematically from that which would be considered rational in the sense of neoclassical economic theory, with these results for the most part being obtained by experimental research. This brought together two things that combined to promote the development of behavioral economics on a massive scale. On the one hand, the objection of psychologists, which could not be dismissed out of hand, and on the other hand, experimental economics, which made experiments in economics acceptable to economists, thus building the bridge that enabled psychological findings to cross to economics.

From that time on, it has been necessary to address the question of in which cases the rational choice model of neoclassical economics is applicable and in which cases it is not. The answer to this question is by no means trivial. It is not enough to have observed once in an experiment that people are not behaving as the rational choice model predicts. It is, in fact, also necessary to clarify the nature of the deviation and whether it can be considered systematic.

In order to identify deviations, it is first necessary to reach agreement on what is meant by the rational choice model. Two variants are worth considering: on the one hand, models that solely assume people behave rationally when pursuing given but arbitrary goals and, on the other hand, models that additionally employ assumptions that people behave strictly selfishly, i.e. they will always prefer a higher income to a lower income. In the following, we will take the first variant as our starting point since the assumption of strict self-interest is not essential for the rationality of behavior. The second variant corresponds to standard neoclassical economic theory where actors are generally assumed to be self-interested.

Given this interpretation, deviations from the rational choice model are reduced to cases in which people systematically make decisions that are not consistent with the goals they are pursuing – no matter what those goals are. This means that deviations from standard neoclassical economic theory that can be resolved by replacing the assumption of self-interested motives with some other assumption of preferences are *not* to be interpreted as deviations from the rational choice model. For example, many cases in which experimental observations contradict standard theory can be reconciled with the rational model of choice by assuming people have "social preferences".[12]

12 The literature on social preferences is in the meantime very extensive. The term social preferences basically means that people not only consider their own well-being, but also that of others when making decisions.

A significant proportion of the experimental economics literature deals with the question of what kind of preferences are suitable for organizing observations in such a way that they can be reconciled with the assumption of rational behavior. For instance, the inequity aversion models of Bolton and Ockenfels (2000) and Fehr and Schmidt (1999) suggest that people have a preference not only for their own absolute payoff, but also for their own relative payoff. Another prominent thesis is that people behave reciprocally, i.e. they are willing to do something good for people who were "nice" to them before and punish people who treated them unfairly.[13] Charness and Rabin (2002) assume people have both an inequity aversion and a respect for efficiency. This branch of literature has with some justification also been called the "neoclassical repair shop" because the core of the neoclassical model – the assumption of rational behavior – can be protected by suitably modifying the assumptions concerning the underlying preferences. However, in complex decision-making situations it becomes apparent that the deviation from the assumptions of strictly selfish behavior alone is not sufficient to reflect the observed behavior (e.g., Bolton and Brosig-Koch 2012).

Therefore, the neoclassical repair shop cannot integrate all the behavior observed under controlled conditions in the laboratory. A whole host of deviations remain for which the only possible explanation is that people do not actually act rationally. Behavioral economics refers to *heuristics* and *biases* in this context: people use heuristics to simplify complicated decisions, and they are subject to biases, i.e. they take decisions that are not truly in their interest. The causes and manifestations of such biases are diverse. Chetty (2015) and DellaVigna (2009) propose a formal characterization that basically identifies three areas in which deviations from the rational choice model can occur. First, the preference ordering does not exhibit the standard properties that the neoclassical model requires. For example, the dependence of preferences on a reference point, as assumed in prospect theory, violates the standard assumptions of neoclassical economics. Second, people do not form rational expectations. For instance, they systematically overestimate their own abilities, which, for example, can lead to time-inconsistent decisions. The inability to form Bayesian expectations leads to phenomena such as the "gambler's fallacy". Third, in the case of given expectations and preferences, it is not possible for the decision-making process itself to take place rationally. People, for example, allow themselves to be influenced by the manner in which decisions are presented (this is termed "framing effects"), or they do not, or do not sufficiently, consider

Definition

Time-inconsistent behavior occurs when, at time t, a decision is made on the behavior at time $t + 1$ and the actual behavior at time $t + 1$ deviates from this decision. The gambler's fallacy is understood to mean that it is mistakenly assumed that random draws out of a hat occur without replacement although replacement does in fact take place. It is for this reason that a roulette player who is subject to the gambler's fallacy believes that the likelihood of the ball falling in red is greater than that of it falling in black after a long run of black previously.

13 See Camerer (2003), Fehr and Schmidt (2006) and Cooper and Kagel (2015) for a survey.

important factors in the decision because they do not pay attention to these or because they are influenced by social pressure.

Chetty (2015) proposes a simple characterization of the biases relevant to behavioral economics. Whereas in the neoclassical model the preference ordering of the decision-maker is accurately represented by a utility function, which is maximized to eventually provide the rational decision, in behavioral economics a distinction is made between *experienced utility* $u(c)$ and *decision utility* $v(c)$. The former is the utility that the decision-maker actually experiences when her decision takes effect. The latter is the utility that the decision-maker believes to be her true utility at the time of making her decision. While neoclassical economics assumes that $u(c) = v(c)$, behavioral economics permits the two to differ from one other. By analogy to an external effect, $e(c) = u(c) - v(c)$ can be interpreted as *internality*, a wedge that the decision-maker herself drives between her true utility and the utility she uses as the basis for her decision.[14]

The papers of Chetty (2015) and DellaVigna (2009) mentioned earlier provide a very good survey of the various forms that heuristics and biases can assume and demonstrate, moreover, that these are not phenomena that can only be observed in the laboratory, but that there are in the meantime a whole host of empirical findings indicating that these deviations from the rational choice model can be found in the field as well. This means that the findings of behavioral economics not only reveal contradictions to the rational choice model, but that they also possess a high degree of *external validity*, i.e. they can be also be detected outside the laboratory. Thus we come to a point that is of great significance for experimental economics research as a whole. For this reason, a separate section is devoted to it.

Important

Since economic experiments are methodologically different from those typically carried out in psychology, it was possible to include the experimental method in the toolbox of economics without endangering the autonomy of economics. With the publication of prospect theory and the resultant strengthening of behavioral economics, however, the rational choice model of neoclassical economic theory found itself being put to the test. Rationality means "consistent behavior in relation to given goals". The test concerns the consistency of the behavior. Inconsistency can generally be characterized as an *internality*. This refers to the difference between the decision utility (the utility assumed at the time of the decision) and the experienced utility (the utility that is actually experienced).

The decisive advantage of this criterion is that, as is the case with revealed preference theory, a comparison of "utility" is not needed. It may well be the case that this criterion does not assist in deciding between the kindergarten and the home for the aged, since both uses are efficient in the sense of Pareto, but economists can live with this due to the great advantages this efficiency criterion offers. It enables economics to make

14 Parts of this section have been taken from Weimann (2015).

statements concerning the preferability of social conditions without having to resort to value judgments that go beyond the acceptance of the Pareto criterion – and who would reject a call to avoid wasting resources?

1.5 External Validity

Internal validity deals with whether an experiment does in fact test the model or theory it is supposed to test. *External* validity concerns the question of whether what is observed in the laboratory can be translated to the real world outside the laboratory. There is a vitally important relationship between the internal and external validity of experiments that occasionally gives rise to confusion. Some researchers hold the view that external validity does not play a significant role, at least in classical experiments, since the purpose of experiments is to verify theories, and so the question of whether experimental results are also an indicator of what happens in the real world is irrelevant (Schram 2005). Even if one shares this view, there remains the question as to what the point then is of testing theories in the laboratory. There must be some connection to the real world; otherwise research (theory and experiment) would abandon its claim to being empirically relevant. Purely normative theories that quite deliberately construct counterfactual alternatives to reality are permitted to do this. But these need not be tested experimentally. The empirical testing of theories really only makes sense if the aim of the research is ultimately to explain real-world phenomena. This means, however, that issues of internal and external validity always occur together. Successful research requires that the experiment does in fact test the theory it wants to test, and that this results in observations that contribute to a better understanding of real phenomena. In the process, internal and external validity grapple with the same problem in the same place.

Economic theory is very careful to derive as general statements as possible. Specific assumptions concerning utility functions or production functions are therefore only made if statements that are even more general are not possible without them. This goal of modeling is very useful in its own right. It does also mean, however, that almost all economic models function without any context. They are not limited to particular conditions that have to be fulfilled in the "setting" of the phenomenon being studied, since this setting is considered irrelevant. Experimenters take advantage of this. If the context does not play a role, then a theory can also be tested in the artificial environment of a laboratory, since it claims to be valid there too. Should a theory be refuted in the laboratory, however, then the theoreticians are certain to counter by pointing out that they constructed the model for a real economic context and not for the laboratory. Bolton (2010) describes this problem using a very nice metaphor. Theories can be understood as maps that do not provide any details so as to highlight the generally valid abstract context. If you want to go from A to B, the context of the streets, the building development along the streets and the number of trees on the sides of the streets are all irrelevant as long as the streets you need are shown. The level of abstraction, i.e. the degree of generality of the model, depends on the context. Bolton refers to a subway map as an example. Such maps are well known, showing only straight lines but neither streets nor public squares. They are extremely helpful if you want to know which line to take to get from A to B and where you have to change lines. But they are only useful to the subway rider; they are totally unsuitable for pedestrians. So, if an experimenter comes up with the idea of testing the subway map on a pedestrian, he will come to the conclusion that the map is no good. This test, however, neglects the context in which the map should be seen. This is not a problem in the case of subway maps since information on the context for which they were designed is always given. Economic models do not have this information and, for this reason, it could happen that an experimenter sends a pedestrian, although it would only be reasonable to send a subway rider.

This problem actually puts experimenters in a comfortable position. After all, they can quite rightly point out that it is theory that makes a claim to generality and it is theory that should be measured against this claim. If theoreticians are going to claim that their theory is only valid for a particular context, then the context ought to be incorporated into the modeling. As long as this is not the case, experimenters are off the hook. They should not celebrate too soon, though, because they are faced with the same problem with external validity. If context plays a role in making decisions, then the laboratory context is relevant and observations made in the laboratory cannot be applied to the real world – not so easily at least.[15]

A biologist observing a rare species of animal in the wild does not need to worry about whether his observations are "externally valid". The situation is no longer so clear if the same scientist is observing animals kept in the laboratory. This is because the living conditions in the laboratory are simply quite different from those that prevail in the wild, and is not clear whether behavior displayed in the laboratory is also found under natural conditions. The situation in experimental economics is very similar. People in the laboratory are in an artificial environment and they have to make decisions in a way and under conditions they would probably never encounter in real life. Can we still assume that experiments are externally valid? Is it permissible to simply extrapolate findings obtained in the laboratory to real-world situations?

When dealing with this question, two methodological aspects must be clearly separated. The first concerns the opportunities and limitations of inductive conclusions, and the second pertains to the fact that the individual observations from which these conclusions are drawn have been made in a laboratory environment. A question that also arises outside the laboratory is: what possibilities does induction offer? The biologist carrying out field research by watching animals in their natural habitat gathers individual observations which in themselves do not yet allow any general statements about the typical behavior of a species to be made. Only repeated, independent observations of one and the same behavior allow the conclusion to be drawn that the behavior is highly likely to be species-specific. Such inductive conclusions cannot be drawn with a high degree of certainty. No matter how many white swans have been observed, this does *not* allow the conclusion to be drawn that all swans are white. This would be an example of invalid inductive reasoning. Despite the logical impossibility of deriving general statements from individual observations, induction is an indispensible method

15 The issue of the context dependency of experimental findings has recently been intensely discussed. We will return to this discussion several times. See, for example, Smith (2010).

without which many natural sciences, especially all the experimental disciplines, would be inconceivable.[16]

The experimental method is fundamentally dependent on the fact that its observations can be used to deduce general relationships (that hold at least with a high probability). This does not, however, apply to *single* observations made in a *single* experiment. The existence of general relationships can only be presumed if the observations are *reproducible* and prove to be robust to any changes in the experimental design. This applies to all types of experimental inquiry. It does not matter whether it is an experiment to test a model or to provide advice on policy or to gather facts about behavior. Generalized conclusions can only be drawn if a large number of independent observations displaying the same, or at least similar, relationships are available. What are sought are *stylized facts* of behavior that can be confirmed time and again and reproduced under a wide variety of conditions.

That is not to say that a sufficiently large number of independent observations must be made within one experiment. As undoubtedly necessary this is – we will look at this in more detail in our discussion on statistical methods in ▶ Chap. 4 of this book – it is, however, only a precondition for obtaining one single statistically significant observation. But this observation only applies to the subjects who participated in this particular experiment and it only applies to the place and time at which the experiment took place. It is, in the first instance, another question entirely as to whether it also applies to different places, at different times, and to other subjects. Only if there are identical findings from different experiments can we assume that there is a high probability that we are dealing with a pattern of human behavior.

Question

Make sure that you have understood the following terms and are able to classify them correctly:

- Induction or inductive reasoning
- Singularity of an observation
- Reproducibility of observations

The reproducibility of experiments is then of the utmost importance – which poses something of a problem. However necessary reproducing experiments may be, they are not popular amongst experimentalists. Repeating somebody else's experiment is boring and generally there is no particular promise of success in getting the work published since only few journals are prepared to publish results that can be found elsewhere. As a result of this, straight replications are exceedingly rare. As a general rule, they are "hidden" in published papers investigating a new aspect of an old problem. These papers usually require a "baseline treatment" to which the results of the new experimental design can be compared. Since these baseline treatments are frequently identical in many experiments, the necessary replications are obtained in passing, as

16 Francesco Guala's book "The Methodology of Experimental Economics" (Guala 2005) intensively explores induction as a scientific method, putting experimental economics in a greater methodological context in the process. We will not be dealing with the theoretical aspects of science in any greater depth and therefore refer the interested reader to Guala's book.

it were. Charness (2010), who makes the same point, quite rightly points out on this issue that the necessity for replications is a powerful argument for using experiments. In contrast to field experiments, experiments conducted under laboratory conditions have the advantage that the experimental design can be easily reproduced. To this end, the instructions the subjects receive are also published, thus effectively allowing experiments to be repeated one-to-one.

Important

- **The external validity of experimental results is important because experiments are as a rule always accompanied by a claim to be able to provide an empirical explanation.**
- **The context of a decision *may* play a role. This is a problem both for formal models (which generalize from the context) and for laboratory experiments, because they take place in a specific laboratory *context*.**
- **Experiments must be reproducible. Only then will the transition from single observations to stylized facts succeed.**

At this point, we should draw the reader's attention to an aspect concerning not only replication studies, but also "completely normal" experiments. A goal of experimental research is to identify effects that are sufficiently robust. It is therefore not necessarily desirable that all the experiments in a series of experiments are always identical in all respects. On the contrary, it makes complete sense to incorporate minor variations. For example, one might be of the opinion that it is necessary for all the sessions of an experiment to be conducted by the same person, because some aspect of the experimenter might have an influence on subjects' behavior. This could of course be the case. For instance, the gender of the experimenter could possibly be important. This raises the question of whether we should really be interested in effects whose occurrence depends on whether it is a man or a woman who conducts the experiment! This certainly could not be described as a robust effect.

But even if the replications are successful and the effect can indeed be described as a stylized fact that is robust, it still only applies to a laboratory environment. The biologist who has observed a thousand white swans can justifiably claim that the next swan is very likely to be a white swan, in the process making a statement about the real world. An experimental economist who has carried out a dozen labor market experiments and found his results to be in line with those of comparable studies of other researchers can nevertheless still not claim that what happens on real labor markets is the same as that observed in the laboratory. It certainly is valid to criticize experimenters that their observations come from an artificial environment and therefore cannot readily be extrapolated to the real world.

For a time, experimental economists provided quite a clever reply to this criticism (for example, Plott 1982, Falk and Heckman 2009). They pointed out that decisions made by people in the laboratory are not artificial at all, but that they are most definitely real! That is indeed true. Subjects in economic experiments are faced with "real" decisions involving "real" money that they receive as a real payoff. They are not just pretending to make decisions in the laboratory; they really are making decisions. To this extent experimentalists are right when they say they do their utmost to bring about real decisions in the laboratory. The fact that experiments in economics always operate with real incentives makes this effort particularly evident. This means that the subjects' decisions have very real consequences for them – due to the more or less generous payoff they can pocket at the end of the experiment.

But if we are honest, we must admit that using real monetary incentives primarily serves to establish *internal* validity and only has something to do with *external* validity as a secondary consequence. By paying money, we are ensuring that those incentives the model assumes are effective actually exist in the experiment. Only then can experiments claim to represent theoretical models. They cannot by any means claim that because they offer monetary incentives they also directly apply to real situations. The response from experimenters really is very clever, but it does not of course solve the problem. The question of whether the observation of *real* laboratory decisions permits statements to be made about *real* decisions outside the lab therefore continues to remain unanswered.

In the end, it may not be possible to answer the question of external validity with greater generality since it is ultimately an empirical issue. Ideally, it would even be possible to experimentally check the transferability of experimental results. A closed methodological chain is required to design such an ideal case. It begins with a model. As an example, let us take a model investigating how rational bidders behave under the rules of an eBay auction. For this purpose, the theoretician determines the Nash equilibrium bidding strategy and derives a prediction of the behavior in the auction. He takes this model to the experimenter, asking him to test his theory-based prediction. The difficulty with testing the model empirically is that although each bidder's willingness to pay is decisive for his bid, it cannot be observed directly. This problem can be solved quite easily in the laboratory. Prior to the auction, the experimenter informs each subject of the value the item being auctioned has for him or her, i.e. how much the payoff will be for the winner of the auction. The price determined by the auction will be subtracted from this payoff. The winning subject can keep the rest. When the auction is then carried out, the experimenter knows how high each bidder's maximum willingness to pay is – he has just "induced" it himself. Inducing preferences is only possible in a controlled laboratory experiment. Friedman (2010, p. 29) referred to this technique, which is clearly described by Smith (1976), as a cornerstone of experimental economics.

With the aid of the induced value method, experiment provides observations that make it possible to measure the quality of theoretical predictions. Let us assume that the experiment has confirmed the theory, i.e. the subjects have submitted bids corresponding to those expected in the Nash equilibrium of the game. The model has thus been confirmed in the laboratory with an internally valid experiment. The next step is to repeat the whole thing outside the laboratory, using a controlled field experiment. Let us suppose this form of auction is to be deployed on an Internet platform. This platform, with all the real properties it possesses, is used as the experimental environment in the field experiment. The subjects recruited by the experimenter will now be sitting comfortably at home, and not in front of a PC in the laboratory. This means they are in exactly the same situation as the "real" platform users. In all other respects, the field experiment proceeds just as it does in the laboratory, i.e. the subjects are informed of the value the item has for them and make a bid. Suppose that the theoretical model's predictions also appear to be correct in this environment. The last step is then the natural field experiment, in which the behavior of real users (who do not know they are participating in an experiment) is observed on the Internet platform. Since their willingness to pay is of course unknown, the method admittedly has to be varied somewhat. For example, two dissimilar auction designs that lead to different Nash equilibria could be used. If we assume the model then provides a qualitative prediction by saying which design leads to a higher price (with identical goods being auctioned), it would be possible to set up controlled forms of real auctions that permit us to test this qualitative prediction. If this test also leads to a positive result, it demonstrates that there is a high probability that the theoretical model correctly describes the actual behavior of real bidders in real auctions.

Sometimes methodological chains also work in reverse order. Bolton et al. (2013) report such a case. eBay offers its users the possibility to provide feedback on the transactions they have conducted. Buyers and sellers can relate their experiences with each other, thus enabling the parties involved on eBay to create a positive reputation for themselves by collecting as much positive feedback as possible. What is striking is that the proportion of negative feedback is very small, also when compared internationally – with one exception. In Brazil, the proportion of negative feedback is significantly higher than in any other country. Do Brazilians have a different mentality? Does cultural background play a role in this? Or is this due to the different design of the reputation mechanism used in Brazil? In Brazil, the various parties can conceal the negative ratings they submit by giving blind feedback. In other words, they are protected from retaliation (in the form of bad ratings they might also receive in return). Bolton et al. (2008) managed to answer this question by conducting an experiment in which the effects of the two reputation mechanism designs were investigated in isolation. The findings showed that blind ratings led to a significantly higher proportion of negative feedback.

The auction experiment by Brosig and Reiß (2007) is also motivated by empirical findings. For example, empirical studies on procurement auctions showed that companies that had not won in previous auctions were more likely to participate in later auctions than those that had won in previous auctions (Jofre-Bonet and Pesendorfer 2000, 2003). It also turned out that companies that had lost in the morning auctions bid more aggressively in the afternoon than the winners of the morning auctions (De Silva et al. 2002). It was suspected that capacity constrains were responsible for this behavior. This was tested by Brosig and Reiß (2007) under controlled laboratory conditions. In fact, they observed similar behavior in their procurement auctions, which in their laboratory experiment can clearly be traced back to the limited capacities.

The methodological chain permits a direct verification of *external validity*, as is the case above, because it lets us conduct controlled field experiments, thereby extending the chain into real life. This is possible in our example because electronic marketplaces possess a wonderful property for experimentalists. On the one hand, they can be recreated one-on-one in the laboratory and, on the other hand, they can be used as a laboratory in their natural setting. Unfortunately, this is an exception. Generally the methodological chain cannot be extended as far as this since the leap from the laboratory to the field does not succeed completely. For instance, if in the laboratory we set up a labor market in which the those who provide jobs and those who seek jobs meet to negotiate wages and decide how intensively they would like to work, we need to be aware of the fact that there are many factors that play a decisive role in determining wages and work effort in a real labor market that are not to be found in this laboratory labor market. For this reason, the *external validity* of experimental labor markets is very limited.

This does not mean, however, that the things observed in a labor market in the laboratory do not play a role in real labor markets. On the contrary, it may well be that experiment reveals patterns in the interaction between employers and jobseekers that are hidden from the eyes of observers of real markets. Whether this is indeed the case, or whether what is observed in the laboratory is an experimental artifact cannot, however, be deduced solely from the laboratory experiment. Further empirical investigations using real markets are necessary for this, and these are considerably more difficult than the field experiments used in the Internet auction. The great advantage that experimental analysis provides in this case is that it focuses the empiricists' attention, showing them precisely where they should be looking.

Important

- **Experiments should generate robust results. But even robust stylized facts are not necessarily externally valid.**
- **Monetary incentives ensure *internal* validity, but do not automatically create *external* validity.**
- **Ultimately, the question of the external validity of an experiment can only be answered empirically. Methodological chains that range from theory to field tests are of great assistance.**

Can't we simply assume external validity without any empirical confirmation? Why should people in the laboratory behave completely differently than in "real life"? After all, the subjects of the experiment do not leave their personality, their attitudes, values and preferences at the laboratory door. So why should what they do in the laboratory be different to what they do outside? The decision-makers in the laboratory are the same as those outside, but are the decision-making situations in front of and behind the laboratory door the same? Definitely not, and the decisive question is, therefore, how much the actual form of the decision problem influences the behavior of the individual actors. An example may make this point somewhat clearer.

The dictator game is used in experimental research to gain insights into how people behave when they have a choice between a payment they receive themselves and effecting a payment to another person. For this purpose, one subject (the dictator) receives a cash endowment (of, say, 10 euros) while being informed that there is a second subject who has received no money. The dictator can then decide whether he will give part of his endowment to the other subject or keep the 10 euros for himself. It is a stylized fact that the dictators in such experiments give, on average, a considerable share to the second subject. This is a astonishing result and differs substantially from what we observe outside the laboratory, where it virtually never happens that people give money to people who are complete strangers and about whom they know practically nothing. People donate money, but for charitable purposes, not just to anyone. People give presents, but they generally also have good reasons for doing so, which are mostly related to the receiver. All this is not the case in dictator games. The dictators do not know whether the other subject is male or female, needy or well off. They know neither the person's name nor anything else about him or her – and yet they give a gift to this unknown person. Obviously, this happens because the specific experimental situation the experimenter has put them in encourages this behavior. However, it is not possible simply to conclude that the same subjects who give away money in the experiment will also give money to strangers in the real world. We will discuss this point in more detail in ▸ Sect. 2.5 of this book.

This example makes it clear that external validity presumably depends on how similar the decision situations inside and outside the laboratory are. Cherry et al. (2002), for instance, show that the amounts given in dictator experiments decrease markedly when the dictators do not receive their initial endowment as a gift from the experimenter, but have to earn it. This laboratory situation thus approaches the real-world situation since the money that is donated is not generally received as a gift but earned through paid work. In this connection, Smith (2010, p.13) demands that experimenters should always try to imagine how the subjects would probably act if they had to play with their own money. Using one's own money creates a kind of parallelism between laboratory experiments and the real world. The sooner such a parallelism exists, the sooner we can assume that the experiments are externally valid. The importance of ensuring that parallelism is as conspicuous as possible becomes apparent in relation to the phenomenon of social preferences. DellaVigna (2009) notes that considerably more "social behavior" can be observed in the laboratory than in the real world. It is an important task of experimental economics to find out why.

The issue of the external validity of laboratory experiments has moved to center stage in the discipline in the last few years due to a discussion revolving around the question of whether there is a qualitative difference between laboratory experiments and field experiments resulting from the fact that field experiments possess a higher external validity than laboratory experiments. This claim was made mainly by Levitt and List (2007, 2008) and it led to a controversy that has in part been carried out at the highest scientific level (measured by the importance of the journals in which the relevant articles were published). For instance, "Science" published both a paper by Levitt and List (2007) critical of laboratory experiments and a vehement defense of laboratory experiments by Falk and Heckman (2009). In our view, this discussion is of great interest, not because there is actually a hierarchy of methods, but because it has very clearly revealed potential weaknesses of the experimental method. The remarks we made on the importance of methodological chains demonstrate that we assume a strictly complementary relationship between laboratory and field experiments. They are not alternatives and each has its own strengths and weaknesses. However, it is worth using the discussion that unfolded between "field" and "laboratory" to highlight some methodological points where caution needs to be exercised. We will discuss these points in detail in ▸ Chap. 2.

In their article from 2007, Levitt and List present a list of shortcomings they identify in laboratory experiments and which they claim field experiments do not have, or at least have in a less pronounced form. Falk and Heckman (2009) take this list and attempt to rebut the individual points of criticism (Croson and Gächter 2010). The first item on Levitt and List's list concerns the choice of subjects for the experiment. They contend that this involves a selection effect that distorts the results of the experiment. This selection bias is not confined to the fact that generally only students serve as subjects; it also includes the fact that it is only *particular* students who volunteer for the experiments.

It is, in fact, quite conceivable that certain personality traits or characteristics are more common among the experimental subjects than in the general student population. For example, before their initial visit to the laboratory the subjects do not know exactly what to expect. People who are eager to try out new things and prepared to take risks are likely to see this uncertainty as something positive and therefore participate in the experiment to find out what will happen in the laboratory, whereas risk-averse students are likely to be deterred by the uncertainty.

Falk and Heckman (2009) counter this objection by arguing that the influence of personality traits on laboratory behavior can be controlled. This can be done by having the subjects fill out a questionnaire that can assist in determining these characteristics. Some well-known examples are the Big Five Personality Test or the NEO-FFI, a further development of the Big Five (Simon 2006). In this way, the personality structures of the subjects become a further explanatory variable for the observed behavior. Furthermore, Falk and Heckman note that selection bias can, of course, also occur in field experiments. While this is true, it must also be acknowledged that field experiments in which the experimenter does not actively select the subjects struggle less with this problem. This discussion on possible selection effects in the choice of subjects clearly demonstrates that this process deserves special attention. We will therefore deal with this issue in some detail in ► Sect. 2.3. However, there is now experimental evidence that a systematic selection bias in the recruitment of student subjects cannot be established (Falk et al. 2013).

The next point on the list concerns subjects' behavior in the laboratory. Levitt and List suspect that the fact that subjects are being observed will change their behavior. There is indeed no denying that the laboratory situation differs from the real world on this point – even if there are definitely situations in real life in which people are also under observation. It may be the case, for example, that the subjects display the behavior they assume the experimenter expects from them. It could also be the case that they avoid egotistical decisions because they do not want to be observed behaving selfishly. This objection to the external validity of laboratory experiments can be countered, however, by using a double-blind design to "protect" the subjects from being directly observed by the experimenter. A double-blind design leads to a situation in which the experimenter knows how the subject with the number X behaves, but does not know who, among all the people in the laboratory, had the number X. Such a design can of course only have the desired effect if this has been made sufficiently clear to the subjects so that they can actually be certain nobody can find out how they behaved. In ► Sect. 2.5.2, we will return to this point and explain how transparent double-blind designs can be created.

The next point on the list of Levitt and List is in a sense old hat. It is the concern that the monetary amounts used in laboratory experiments might be too small and insignificant to in fact provide the required incentives. This objection has been raised against laboratory experiments from the very beginning and Falk and Heckman rightly point out that it can in fact be considered to have been dealt with. It is obviously quite straightforward to vary the payoffs in experiments in order to check whether choosing significantly higher payoffs than usual in experiments makes a difference. The experimental findings on the question of how higher payoffs affect laboratory behavior are mixed. For instance, Carpenter et al. (2005) could not find any difference between treatments with

$10 stakes or $100 stakes, whether in ultimatum experiments or dictator experiments. In the ultimatum experiments, however, they could only make statements about the proposer because all (but one) of the offers were accepted and nothing could therefore be said about the rejection rate of the responder (who accepts or rejects). Andersen et al. (2011) demonstrate that this rejection rate tends to decrease at higher payoffs, claiming that it would approach zero for sufficiently high stakes. Slonim and Roth (1998) observe that the variation in behavior decreases as the stakes increase. In contrast, Camerer and Hogarth (1999), using a meta-analysis, find no evidence that the size of the payoff has any noticeable impact, while Fehr et al. (2014) likewise cannot determine any difference between high and low stakes. Summing up, although it appears that the payoff amount is important to some extent, the comparatively low payments in experiments do not seem to represent a fundamental problem. We will also return to this point in more detail in ► Sect. 2.2.2.

Some criticize that subjects of laboratory experiments do not generally have sufficient opportunity to gain experience and to learn from it. This objection should be taken more seriously, since it is often actually the case. Laboratory experiments usually take no longer than an hour or two. This short duration obviously offers very limited opportunity to gather experience. Falk and Heckman point out that the influence of learning and experience on behavior can be tested experimentally. However, the fact is that this hardly ever happens. This is another point we will come back to.

In addition to the individual points Levitt and List present, they criticize that laboratory experiments are simply too far removed from the real world for it to be possible for their findings to be transferable to it. What is of utmost importance here is the context already mentioned earlier in which economic processes and decisions take place and which, according to the firm belief of many economists, still has a significant impact. If this context dependency is ignored, it is not possible to draw analogies between the laboratory setting and the real-world setting. However, the question arises as to why this point is often only raised in connection with laboratory experiments. If there is a context dependency, it has to be taken into account on all methodological levels, i.e. in the models and field experiments as well. In striving for maximum possible generality, economic theory has thoroughly banished all thoughts of context dependency from its thinking. The fact that context also hardly plays a role in the laboratory is a direct consequence of this since if experiment is to test theories that are free of context, then the experimental design must also be free of context (Alekseev et al. 2017). In the debate on external validity, focusing only on the issue of laboratory experiments being too far removed from the real world falls short of the mark. If at all, then this issue must be applied to all the methodological tools used by economists.

By means of the assumptions of his model, the theoretician developing a labor market model also creates a very special space within which actors, with attributes also created by assumptions, make decisions. It is not at all clear a priori whether this space and the assumptions of the model are suitable for deriving statements that can also claim validity outside the model, in the real world. In many cases it is even highly uncertain. For example, in order to determine Nash equilibria, game-theoretical models require very far-reaching assumptions about the behavior of individual players. Not only must they be assumed to behave completely and strictly rationally, but it must also be assumed that this rationality is "common knowledge".[17] Furthermore, the expectations of the players (their beliefs) must also be common knowledge. In many cases, it can definitely not be presumed that these assumptions are fulfilled outside the model. They can best be met in the laboratory - at least the beliefs of the players can be controlled and induced there.

Whether models say something about the real world or not is a question that is closely related to the question about the external validity of experiments. The easiest way to answer this question is to use methodical chains, as described above using the Internet auction as an example. Seen in this light, theoreticians and experimental economists are sitting in the same boat on this issue. Should theorists reject experiments because their external validity is not guaranteed, they overlook the need to demonstrate the external validity of the theory and overlook the fact that they can best do this together with the experimenters.[18]

Important

In order to ensure the external validity of laboratory tests, the parallelism between the situation in the laboratory and real decisions is very important. The more marked the parallelism, the more likely it is that externally valid results can be expected.

Experimental research has some specific problems that have been discussed intensively in literature: possible selection effects during the acquisition of experimental subjects, changes in behavior due to being observed, a lack of opportunities to gain and to learn from experience, as well as payoffs that are potentially too low. All these points are discussed in specific sections of the second chapter in this book.

The objection that experiments are too artificial or too abstract to provide externally valid findings must be directed against all instruments of economic research that work with simplifications and abstractions, including formal modeling. Whether it is justified or not is in any event an empirical question that needs to be answered.

In summary, it can be said that the question of external validity is so important for experimental research because economics is essentially an empirical discipline whose aim is to develop statements about what happens in real economies. If we try to investigate economically relevant decision-making situations in the laboratory, the question of what we can learn about the real world is obvious. And it really is a pressing question. That is why experimental researchers are repeatedly faced with this question – and rightly so. An answer that is generally valid is not possible, but can only be given on a case-by-case basis, with the external validity of different experiments varying considerably.

---

17 Everyone is rational, everyone knows that everyone is rational, everyone knows that everyone knows that everyone is rational, etc.

18 See Weimann (2015) or the discussion in the fourth chapter of the "Handbook of Experimental Methodology" (Fréchette and Schotter 2015).

Question

Does the question of external validity also arise in natural science experiments, for example in physics or chemistry?
Under which conditions can it safely be assumed that experiments in the natural sciences are externally valid?

1.6 Behavioral Research: An Interdisciplinary Issue

Human behavior is not only the subject of economic research. In fact, for a long time economists were not particularly interested in investigating the behavior of real people. Other disciplines have been much more active in this respect and therefore boast a long tradition and a large pool of knowledge. Most notably, psychology has of course from the very beginning been interested in what drives human behavior. Social psychology studies such behavior in social contexts and focuses not only on individuals, but also on groups and their behavioral dynamics. The neurosciences and genetics, admittedly much younger than psychology, have a more medically oriented view of behavior.

How does experimental economic research relate to these neighboring disciplines? What can economists learn from psychologists, neuroscientists and geneticists? Of particular interest for this book are, of course, the questions as to which methodological differences exist between the disciplines and how collaboration can be organized – possibly in the face of existing methodological differences.

Economics and Psychology

We have already seen that the neoclassical research program was accompanied by an attempt to completely decouple economics from psychology and to conduct "psychology-free research". With experimental economics research on the one hand and behavioral economics on the other, the connection to psychology has returned and, at first glance, this seems to have led to a broad rapprochement of these disciplines. The fact that *Daniel Kahneman*, a psychologist, was awarded the Nobel Prize for Economics is indeed a visible sign of this. Nevertheless, there are still considerable methodological differences between experimental economic research, including behavioral economics, in which many psychologists are employed, and large parts of experimental psychology.

Psychology has a much longer experimental "history" than economics. Therefore, one could imagine that experimental economic research is more or less an application of psychological methods to specific – namely economic – questions. However, economists have once again chosen to go their own way in this regard, and nowadays one has to admit that psychologists who deal with issues of behavioral economics tend to draw more on the methodology of economics rather than the other way around. What is the difference between experiments in economics and psychology?

Apart from a few minor details, it is mainly two fundamental methodological decisions that distinguish the two disciplines. The first is that economists generally only accept experiments in which the subjects are exposed to real incentives. In almost all cases, these incentives are money. The background to this principle of economic experiments is the theoretical basis on which research in experimental economics takes place.

Economic models usually describe human behavior as the outcome of optimizations in which material incentives play a very significant role. These incentives are formally represented by utility functions and can be incorporated into the experiment by means of payoff functions. In this way, the aforementioned close connection between experiment and theory is accomplished. This connection is inevitably looser in psychology because psychologists do not usually formulate their models using mathematical equations or other formal language that can be translated one-on-one into the laboratory. Therefore, the use of monetary incentives is not mandatory for psychological experiments to the extent they are for those of economics.

Two questions are repeatedly discussed in relation to whether monetary incentives should be used or not. The first is obvious: does it make a difference whether experimental subjects play for real money or will they behave in the same way with purely fictitious payoffs? This question has of course been investigated experimentally and the results can be summarized very easily: sometimes it does not make a difference, often it does.[19] Economists draw a simple conclusion from this: it is better to use real incentives, because you are then on the safe side.

The decision in favor of monetary incentives has a pleasant side effect. It facilitates the acquisition of subjects for experiments. This is a very important point, as a large number of subjects are sometimes required in order to obtain a sufficiently large number of independent observations. Students in particular generally appreciate the fact that by participating in experiments they can earn a little extra money. This effect should be taken into account when choosing the size of the payoff. A laboratory that has a reputation for paying well will have little difficulty in finding subjects. On the other hand, being too frugal in the use of experimental funds can quickly mean that a lot of money has been saved in the wrong place. Although there is still money left to carry out further experiments, there are unfortunately no more subjects for the experiments.

On the other hand, refraining from using real incentives makes recruiting subjects a real problem. In psychology, this is sometimes solved in a way that is capable of creating new problems. Psychology students are simply *obliged* to participate in a certain number of experiments during their studies. This naturally raises the question of what motives a subject who compulsorily participates in an experiment follows. It is possible to imagine many things. Anger over the time sacrificed could just as well play a role as the desire to please the experimenter. The point is, we do not know – in other words, we have no control over it. It is thus very difficult to interpret the behavior of people who are forced to participate in the experiment.

The second fundamental decision made by economists, and which clearly distinguishes them from their psychology colleagues outside behavioral economics, is that they never lie to the subjects of experiments. The manipulation of subjects is highly frowned upon in experimental economics – and for good reason. The only way to learn from the behavior of subjects in an experiment is to know the conditions under which they have made their decisions. And that is the advantage of experimentation. It is possible to control the circumstances in which decisions are made. However, this is no longer possible if we cannot assume that the subjects believe the experimenter.

19 See ▶ Sect. 2.2 and Lichters et al. (2016), where the importance of real incentives for experiments in market research is discussed.

The experimental community needs the reputation that it tells the subjects exactly what actually happens in the experiment. If it squanders this reputation, no experimenter will know what the subjects consider to be the true experimental setup. Editors and reviewers of scientific journals therefore ensure that the rules that were implicitly agreed upon are observed. We will deal with this point in more detail in ▶ Sect. 2.3.1.

The argument for dealing honestly with the subjects is in fact very convincing. But not all psychologists are prepared to accept it. Sometimes they argue that many interesting issues cannot be addressed if the truth always has to be told. Economists do not believe that. They are convinced that it is possible to conduct experiments on any question without lying. However, it may take a little longer and it might be more expensive than an experiment in which falsehoods are told. What is also not very convincing is the suggestion that psychologists are always honest in the end because after the experiment the subjects undergo a debriefing, i.e. at the conclusion of the experiment the subjects are informed about what *really* took place. Basically, this only makes things worse, because then word really does get around that lies are told at the beginning of the experiment.

▪ Neuroscience and Genetics

In recent years, the natural sciences have become significantly more important for research into human behavior. This development was mainly due to technological innovations in the field of imaging techniques in neuroscience and the considerable progress made in decoding the genetic code. Let us begin with neuroscience, which has already developed very extensive collaborations with experimental economics.

Today, neuroscience employs a variety of techniques to render processes in the brain visible. The EEG (electroencephalography), in which voltage variations on the scalp are recorded, has long been well known. These fluctuations in electrical potential are due to the voltage changes of individual brain cells. The cumulative changes can then be measured on the scalp. The disadvantage of this method is the relatively poor spatial resolution, which can be several centimeters. Furthermore, only voltage changes originating from superficial areas of the brain can be detected. Areas situated deeper within the brain are not captured by the EEG. The advantage of this method is the high temporal resolution, i.e. very little time passes between the brain activity and its recording.

Imaging techniques in a narrower sense have in the meantime acquired considerably greater significance for neuroscience, above all fMRI (functional magnetic resonance imaging). In simple terms, this procedure measures the blood flow in the brain. In doing so, it takes advantage of the fact that hemoglobin as an oxygen transporter possesses different magnetic properties in the oxygenated state than in the deoxygenated state. In layman's terms, this can be exploited to show where in the brain oxygen is consumed and therefore where there is increased activity of the brain cells, thus measuring what is referred to as the blood oxygenation level-dependent (BOLD) effect.

In order to be able to interpret fMRI images correctly, it is important to know that these images represent statistically significant differences between brains in different situations. Normally, the MRI scanner first scans the brain in its resting state. Subjects are then presented with a task and, while they are solving it, a scan is made again. The measured neural activation of the individual brain areas is compared with those at rest and the statistically significant deviations are depicted. This kind of representation is now well known, with an image showing a cross-section through the brain with red and yellow patches. Such representations lead to the assumption that only these areas are currently active. In fact, the entire brain is permanently in action and the "patches" indicate only those areas that are more active than in the resting state.

The key point is that in this way it is possible to identify which parts of the brain are active at the moment when the subject is making a decision or taking action. This allows conclusions to be drawn as to what kind of "thoughts" the subjects might have if it is known which functions the individual brain regions are responsible for. Neuroscience obtains such information from various sources. An important role is played by the examination of patients with lesions of individual brain areas. If a certain brain region fails, it is possible to observe which brain function is no longer available. Animal experiments (especially research on apes) provide further information, and fMRI examinations themselves help to understand which part of the brain is carrying out which task. Recently, there has been also an increase in the use of transcranial magnetic stimulation (TMS), in which strong magnetic fields can either stimulate or inhibit certain parts of the brain.

These imaging techniques are still relatively new. The first images of brain activity date back to 1992 and are thus only 27 years old. The possibilities and limitations of these techniques are therefore very difficult to assess at this stage. The decisive factors will be whether the resolution of the images can be further increased and whether the temporal resolution, which is significantly worse than that of the EEG, can be improved, and also which knowledge is generated with regard to the functions of brain areas and how they interact. Of course, future developments will also determine how well economic issues can be addressed with the help of imaging techniques. But *neuroeconomics*, i.e. the combination of classical experimental methods of economics with the imaging techniques of neuroscience, is already on its way to securing a firm place in the scientific world. Although a number of experimental economists are now using methods of neuroeconomics, very few neuroeconomics studies have been published in economic journals. This is probably due to a certain skepticism on the part of many economists.

One example at this point is Douglas Bernheim (2009), who put forward some powerful arguments against neuroeconomics. From economists' point of view, the brain is a black box. We can observe what goes in (information) and what comes out (decisions), but not what happens in the box. Bernheim asks whether we need to open this box for the analysis of economic issues. Is it not enough, for example, to know that people behave reciprocally in certain situations? Do we have to know why they are doing this? We don't study process engineering in order to pursue production theory. The point is we do not have to open every black box. Sometimes it is better to leave it closed because this saves resources that can be better put to use elsewhere.

From Bernheim's point of view, neuroeconomics only makes sense if it meets at least one of two conditions. First, it must make it possible to discriminate between alternative theories where this would not be achievable without the use of imaging techniques. Second, knowledge of neural processes must have economic consequences. These two demands make sense and that is why we want to pick up the thread that Bernheim was spinning with them.

The theory of revealed preferences is based on observations of acts of choosing and it is not possible to observe the motives behind these actions. However, imaging techniques may open up the possibility of observing such motives. For example, with the scanner it is possible to determine whether and to what extent certain emotions have played a role in a decision or whether the decision-maker has struggled with a conflict or not, and much more. If we take Bernheim's demands seriously and regard the observability of motives as the decisive innovation in neuroscience, it follows that neuroeconomic research makes sense if it is then possible to choose between alternative theories on the basis of more precise information on the motives underlying actions or when the awareness of motives has economic consequences.

In this way, we have a criterion that allows us to decide whether the use of neuroscientific tools is justifiable and worthwhile or not. One more question remains to be answered: why should neuroscientists be interested in economic questions? The fact that they are indeed interested in this is a crucial prerequisite for neuroeconomic work; without the cooperation of neuroscientists, the vital expertise needed for this work would be lacking. Fortunately, some neuroscientists are very interested in collaborating with economists because they provide something that the natural sciences lack: structured and theoretically developed concepts of how decisions are made and what kinds of decisions need to be identified. It is simply not enough just to collect data on the brain. It is also necessary to have a theoretical framework in order to interpret the data meaningfully. The following quote illustrates this point quite clearly.

> "Within neuroscience, for example, we are awash with data that in many cases lack a coherent theoretical understanding (a quick trip to the poster floor of the Society for Neuroscience meeting can be convincing on this point)." (Cohen and Blum 2002, p. 194)

In contrast to neuroscience, which in the meantime has entered into a well-running collaboration with experimental economics,[20] the connection between experimental economics and genetics is still in its relatively early stages. Here, too, the question naturally arises as to whether close cooperation makes sense from the point of view of economics. Is it of any use to economists to find out which genes are responsible for the fact that people are more or less cooperative or possess other traits that could be relevant for economic decisions? It is much more difficult to find a convincing answer to this question than to the analogous question in the neurosciences. In the field of genetics, the natural science and medical disciplines concerned with behavioral genetics benefit from the models and methodology of experimental economic research because this makes it easier to find experimental designs that can provide information on the genetic determinants of certain behaviors. What benefits this cooperation has for economic research still remains to be seen.

---

20 A summary of the current state of the art can be found in Glimcher et al. (2013).

1

Important
Collaboration between experimental economists and psychologists is well developed and unproblematic in the field of behavioral economics. In parts of psychology, however, experimental methods are used that economists cannot adopt. Psychology experiments, for example, frequently do not use monetary incentives and at times do not truthfully inform subjects about the experimental setup. Cooperation with the neurosciences is well advanced and has led to the emergence of neuroeconomics, whereas cooperation with geneticists is still in its infancy.

References

Alekseev, A., Charness, G., & Gneezy, U. (2017). Experimental methods: When and why contextual instructions are important. *Journal of Economic Behavior & Organization, 134*, 48–59.

Andersen, S., Ertaç, S., Gneezy, U., Hoffman, M., & List, J. A. (2011). Stakes matter in ultimatum games. *American Economic Review, 101*(7), 3427–3439.

Bernheim, B. D. (2009). The psychology and neurobiology of judgement and decision making: What's in it for economists? In P. W. Glimcher, C. F. Camerer, E. Fehr, & R. A. Poldrack (Eds.), *Neuroeconomics: Decision making and the brain* (pp. 115–126). San Diego: Academic Press.

Binmore, K., & Shaked, A. (2010). Experimental economics: Where next? *Journal of Economic Behavior & Organization, 73*(1), 87–100.

Bolton, G. E. (2010). Testing models and internalizing context: A comment on "theory and experiment: What are the questions?". *Journal of Economic Behavior & Organization, 73*(1), 16–20.

Bolton, G. E., & Brosig-Koch, J. (2012). How do coalitions get built? Evidence from an extensive form coalition game with and without communication. *International Journal of Game Theory, 41*, 623–649.

Bolton, G. E., & Ockenfels, A. (2000). A theory of equity, reciprocity and competition. *American Economic Review, 90*(1), 166–193.

Bolton, G. E., Loebbecke, C., & Ockenfels, A. (2008). Does competition promote trust and trustworthiness in online trading? An experimental study. *Journal of Management Information Systems, 25*(2), 145–170.

Bolton, G. E., Greiner, B., & Ockenfels, A. (2013). Engineering trust: Reciprocity in the production of reputation information. *Management Science, 59*(2), 265–285.

Brosig, J., & Reiß, J. P. (2007). Entry decisions and bidding behavior in sequential first-price procurement auctions: An experimental study. *Games and Economic Behavior, 58*, 50–74.

Bruni, L., & Sugden, R. (2007). The road not taken: How psychology was removed from economics, and how it might be brought back. *The Economic Journal, 117*, 146–173.

Camerer, C. (2003). Behavioral game theory: Experiments in strategic interaction. Princeton: Princeton University Press.

Camerer, C. F., & Hogarth, R. M. (1999). The effects of financial incentives in experiments: A review and capital-labor-production framework. *Journal of Risk and Uncertainty, 19*, 7–42.

Carpenter, J., Verhoogen, E., & Burks, S. (2005). The effect of stakes in distribution experiments. *Economics Letters, 86*, 393–398.

Chamberlin, E. (1948). An experimental imperfect market. *Journal of Political Economy, 56*, 95–108.

Charness, G. (2010). Laboratory experiments: Challenges and promise: A review of "theory and experiment: What are the questions?" by Vernon Smith. *Journal of Economic Behavior & Organization, 73*(1), 21–23.

Charness, G., & Rabin, M. (2002). Understanding social preferences with simple tests. *Quarterly Journal of Economics, 111*(3), 817–869.

Cherry, T., Frykblom, P., & Shogren, J. (2002). Hardnose the dictator. *American Economic Review, 92*(4), 1218–1221.

Chetty, R. (2015). Behavioral economics and public policy: A pragmatic perspective. *American Economic Review, 105*(5), 1–33.
Cohen, J., & Blum, K. (2002). Reward and decision. *Neuron, 36*(2), 193–198.
Cooper, D. J., & Kagel, J. H. (2015). Other-regarding preferences. In J. H. Kagel & A. E. Roth (Eds.), *The handbook of experimental economics* (Vol. 2, pp. 217–289). Princeton: Princeton University Press.
Croson, R., & Gächter, S. (2010). The science of experimental economics. *Journal of Economic Behavior & Organization, 73*(1), 122–131.
De Silva, D. G., Dunne, T., & Kosmopoulou, G. (2002). Sequential bidding in auctions of construction contracts. *Economics Letters, 76*(2), 239–244.
DellaVigna, S. (2009). Psychology and economics: Evidence from the field. *Journal of Economic Literature, 47*(2), 315–372.
Falk, A., & Heckman, J. J. (2009). Lab experiments are a major source of knowledge in the social sciences. *Science, 326*(5952), 535–538.
Falk, A., Meier, S., & Zehnder, C. (2013). Do lab experiments misrepresent social preferences? The case of self selected student samples. *Journal of the European Economic Association, 11*(4), 839–852.
Fechner, G. T. (1860). *Elemente der Psychophysik*. Leipzig: Breitkopf und Hartel.
Fehr, E., & Schmidt, K. M. (1999). A theory of fairness, competition and cooperation. *Quarterly Journal of Economics, 114*(3), 817–868.
Fehr, E., & Schmidt, K. M. (2006). The economics of fairness, reciprocity and altruism–experimental evidence and new theories. In S.-C. Kolm & J. M. Ythier (Eds.), *Handbook of the economics of giving, altruism and reciprocity* (Vol. 1, pp. 615–691). Amsterdam: Elsevier.
Fehr, E., Tougareva, E., & Fischbacher, U. (2014). Do high stakes and competition undermine fair behaviour? Evidence from Russia. *Journal of Economic Behavior & Organization, 108*, 354–363.
Flood, M. M. (1952). Some Experimental Games. Research Memorandum RM-789, RAND Corporation, June.
Flood, M. M. (1958). Some experimental games. *Management Science, 5*(1), 5–26.
Fréchette, G. R., & Schotter, A. (2015). *Handbook of experimental economic methodology*. Oxford: Oxford University Press.
Friedman, D. (2010). Preferences, beliefs and equilibrium: What have experiments taught us? *Journal of Economic Behavior & Organization, 73*, 29–33.
Glimcher, P. W., Camerer, C. F., Fehr, E., & Poldrack, R. A. (2013). *Neuroeconomics. Decision making and the brain*. Amsterdam et al.: Academic Press.
Guala, F. (2005). *The methodology of experimental economics*. Cambridge; New York: Cambridge University Press.
Jofre-Bonet, M., & Pesendorfer, M. (2000). Bidding behavior in a repeated procurement auction. *European Economic Review, 44*(4–6), 1006–1020.
Jofre-Bonet, M., & Pesendorfer, M. (2003). Estimation of a dynamic auction game. *Econometrica, 71*(5), 1443–1489.
Kahneman, D., & Tversky, A. (1979). Prospect theory: An analysis of decision under risk. *Econometrica, 47*(2), 263–291.
Levitt, S. D., & List, J. A. (2007). What do laboratory experiments measuring social preferences reveal about the real world? *Journal of Economic Perspectives, 21*, 153–174.
Levitt, S. D., & List, J. A. (2008). Homo economicus evolves. *Science, 31*, 909–910.
Lichters, M., Müller, H., Sarstedt, M., & Vogt, B. (2016). How durable are compromise effects? *Journal of Business Research, 69*(10), 4056–4064.
Plott, C. R. (1982). Industrial organization theory and experimental economics. *Journal of Economic Literature, 20*(4), 1485–1527.
Roth, A. E. (1995). Introduction. In J. H. Kagel & A. E. Roth (Eds.), *The handbook of experimental economics* (pp. 3–109). Princeton: Princeton University Press.
Sadrieh, K., & Weimann, J. (2008). *Experimental economics in Germany, Austria, and Switzerland. A collection of papers in honor of Reinhard Tietz*. Marburg: Metropolis-Verl.
Samuelson, P. A. (1938). A note on the pure theory of consumer's behaviour. *Economica, 5*(17), 61–71.
Samuelson, L. (2005). Economic theory and experimental economics. *Journal of Economic Literature, 43*(1), 65–107.

1

Sauermann, H., & Selten, R. (1959). Ein Oligopolexperiment. Zeitschrift für die ge-samte Staatswissenschaft. *Journal of Institutional and Theoretical Economics, 115*(3), 427–471.
Schelling, T. C. (1957). Bargaining, communication, and limited war. *Conflict Resolution, 1*(1), 19–36.
Schram, A. (2005). Artificiality: The tension between internal and external validity in economic experiments. *Journal of Economic Methodology, 12*, 225–237.
Simon, W. (2006). *Persönlichkeitsmodelle und Persönlichkeitstests*. Offenbach.
Slonim, R., & Roth, A. E. (1998). Learning in high stakes ultimatum games: An experiment in the Slovak Republic. *Econometrica, 66*(3), 569–596.
Smith, V. (1976). Experimental economics: Induced value theory. American economic review. *Papers and Proceedings, 66*(2), 274–279.
Smith, V. L. (2010). Experimental methods in economics. In S. N. Durlauf & L. E. Blume (Eds.), *Behavioural and experimental economics* (pp. 120–136). Basingstoke: Palgrave Macmillan.
Thurstone, L. L. (1931). The indifference function. *Journal of Social Psychology, 2*(2), 139–167.
Von Neumann, J., & Morgenstern, O. (1944). *Theory of games and economic behavior*. Princeton: Princeton University Press.
Wallis, W. A., & Friedman, M. (1942). The empirical derivation of indifference functions. In O. Lange, F. McYntire, & T. Yntema (Eds.), *Studies in mathematical economics and econometrics* (pp. 175–189). Chicago: University of Chicago Press.
Weimann, J. (2015). Die Rolle von Verhaltensökonomik und experimenteller For-schung in Wirtschaftswissenschaft und Politikberatung. *Perspektiven der Wirtschaftspolitik, 16*(3), 231–252.
Weimann, J., Knabe, A., & Schöb, R. (2015). *Measuring happiness. The Economics of wellbeing*. Cambridge, MA: MIT Press.

Methodological Foundations

J. Weimann, J. Brosig-Koch, *Methods in Experimental Economics*, Springer Texts in Business and Economics, https://doi.org/10.1007/978-3-319-93363-4_2

Overview

Now that we have prepared the stage for experimental economics in the first chapter, the second chapter deals with the methodological foundations. The times are long gone when economists "just did an experiment" to see what happens when you let subjects make specific decisions. In the meantime, methodological standards and procedures have evolved. Following these procedures is an important prerequisite for obtaining experimental results that can claim to meet the scientific standards of the economics profession.

We have put the most important methodological fundamentals into groups, each of which is based on one component of an experiment. As in ▶ Chap. 1, there will also be summaries and questions in ▶ Chap. 2, and we have included one or two boxes in which interesting (marginal) aspects will be explained in more detail.

2.1 Introduction

> However, there is an aspect of experimental design that is probably closer connected to theory than empirical work: As designers, we are responsible for the environment in which the data are generated. (Muriel Niederle 2015, p. 105)

This quote from the "Handbook of Experimental Economic Methodology" (Fréchette and Schotter 2015) explains why it is so important to deal with the methodological foundations of experimental research. It all comes down to the design of an experiment and how the experiment is carried out in practice. Therefore, in this second chapter we will first deal with the fundamental questions of the design of an experiment and in the third chapter we will present a description closely based on experimental practice of how a concrete experiment can proceed.

All these considerations naturally raise the question of the benchmark. What is actually a good experiment? In Muriel Niederle's article, from which the above quote is taken, it is argued that a good experiment succeeds in testing the most important effect a theory describes, while at the same time controlling for all plausible alternative explanations. This is based on the fact that there can often (if not always) be more than one potential explanation for an empirically observable regularity. Good experiments are undoubtedly those that, from the many alternative causal relationships, can find the one that actually explains the observed phenomenon.

It should be borne in mind, however, that this is a very ambitious goal that can only relatively rarely be achieved. This is simply because not all the alternatives are always known. Therefore, the design of an experiment should eliminate the possibility that the results are influenced by factors that are not connected to the assumed causal relationships. If a faulty design or the inept execution of an experiment is responsible for the observed regularity, then one has certainly not carried out a good experiment. For this reason, it is worth considering the way in which various elements of an experiment's design can influence the behavior of the subjects. This is the only way to be certain that the results obtained are an empirically significant regularity and not just an artifact of a bad experiment.

This is also the case when experiments are primarily used to look for stylized facts. Here, too, there are certain elements of the design that, directly or indirectly, are not involved in designing empirically relevant effects in the laboratory. It must be avoided that these factors impact on the behavior of the subjects. The following example illustrates this. When subjects do certain things to please the experimenter (often referred to in this context as experimenter demand effects), a design element that should not actually play a role is responsible for the result, thus preventing the experiment from making statements about an empirically relevant relationship. The following considerations are intended to protect against a "bad design" in this sense.

2.2 It's About Money

2.2.1 The Induced Value Method

At the heart of all experimental investigations is the ability to make observations under controlled conditions. For example, experiments with animals attempt to study their learning ability. For this purpose, certain signals are combined with rewards (in the form of food) and the researchers observe whether the laboratory animal is able to make a connection between the signal and the reward associated with it. By controlling the variation of signals and rewards, conclusions can be drawn about the animals' ability to learn. It is implicitly assumed in such experiments that the animals like the food offered as a reward and that they prefer to have more of it. Only then is it justifiable to assume that the animals are making an effort (learning is strenuous) to obtain the food. Now the assumption that apes, for example, like to eat sweet fruit and are particularly keen on certain "treats" is not too daring. It is easy to see that this is the case. The behavioral hypothesis for an ape is therefore "I prefer more bananas to fewer bananas".

But what about experiments with humans? Ultimately, economic theory describes above all else how people make decisions. Of central importance are the preferences that are attributed to the actors. Rational choices are always related to the goal being pursued. Thus, it is not possible to make a prediction about what a rational decision-maker will do if it is not known which goal this person is pursuing. Experiments that test theories and also experiments that are not based on theory would therefore be worthless without assumptions about the underlying preferences. The problem is that people's preferences are doubtlessly more differentiated than those of apes. In other words, it makes little sense to assume that people prefer more bananas to fewer bananas. Fortunately, however, the difference between us (humans) and our closest genetic relatives is not so great, thanks to the fact that there is a banana equivalent of sorts for us, and that is money.

Vernon Smith introduced this equivalent systematically into experimental methodology in 1976 and gave it a name: the induced value method (Smith 1976). The idea is very simple. It is assumed that the consumption of every good generates utility, for which there is a monetary equivalent – the willingness to pay for the good in question. If each utility value can be expressed in terms of money, then the utility function can also be replaced by a "money function", and by introducing this money function into the experiment as a "payoff function", one has *induced* from the outside the utility function that is used for the evaluation of options for action. The following example best illustrates this process.

Public goods are characterized by the fact that it is not possible to exclude people from their consumption. For example, all residents benefit from improved air quality in a given region, but nobody can be barred from consuming clean air. Since this is the case, standard neoclassical economic theory predicts that public goods cannot be offered privately. No one would be willing to pay a price for them, because consumption is also possible without having to pay. A serious problem with the research on and the provision of public goods is that the real utility people gain from the consumption of such goods is not known due to it being private information. Only the individual resident knows what benefits can be gained from the public good "better air quality" and what its real value is. How, then, should an experiment investigate whether and under what circumstances people are willing to participate in the provision of a public good? If someone makes a contribution, it is not possible to know whether or not this is to his advantage.

Using the induced value method, it is not a problem to create a public good in the laboratory where the experimenter has full control over the subjects' preferences. To this end, he employs what is known as the voluntary contribution mechanism, which was first introduced by Isaac et al. (1984). The subjects receive an initial endowment amounting to z_i. Any share of this can be contributed towards the public good. Each investment of one euro in the public good results in *all* N subjects receiving a payoff of a_i each. The money that is not invested in the public good can be kept by subject i. The payoff function is designed in such a way that $Na_i > 1$ and $a_i < 1$, which means that it is not worthwhile for the individual – self-interested and rational – player to invest in the public good. The payoff that results from this for the player is less than that received from keeping the money. If subjects follow this strategy, they all realize a payoff of z_i. At the same time, however, it is best for the group if all subjects invest their entire initial endowment in the public good, because each subject then receives a payoff of $Nz_ia_i > z_i$. The payoff function of the voluntary contribution mechanism thus ensures that precisely the dilemma arises, which in the real world also constitutes the problem of public goods: individual rational behavior leads to collectively non-rational choices.

This example should demonstrate the principle of the induced value method. The subjects do not consume "clean air" or any other public good, but rather money. The induced value method requires that people react to money in the same way as apes react to bananas – more of which is always better than less. This is also one of a total of three requirements that Smith (1976) specifies need to be met for the induced value method to be applied. First, the utility function must grow monotonically in terms of money. In slightly more technical terms, if a decision-maker can choose from two alternatives and one of them has a higher payoff than the other, then the decision-maker will always choose the alternative with the higher payoff (Smith 1976, p. 275).

Second, the payoffs have to be salient. The so-called salience requirement is understood as meaning that the decision to be taken by a subject in an experiment must also be payoff-relevant. It is worth considering a little more deeply what this implies. In an essay published in 1989 in the American Economic Review, Glenn Harrison launched a debate on what characteristics a payoff function must have in order to properly induce preferences. His point was that it is important that the payoff function is not too flat. If it is, taking different decisions has little impact on the resulting payoffs, which can result in subjects not putting much effort into actually making what for them is the best decision, because a mistake has little financial impact. Harrison's article resulted in an intense debate, into which we will go into more detail later (see ▶ Box 2.1).

The third requirement that Smith lists is the dominance of the payoffs. As an experimenter, one has to be aware of the fact that experimental subjects could also have other things on their minds than the money they can earn in an experiment. For example, people dislike getting bored or thinking that their precious time is being wasted. Boredom may lead to subjects having an incentive to make things more interesting by trying things out without paying too much attention to their payoffs. There are many other factors that might discourage subjects from focusing exclusively on maximizing their payoffs. They could form expectations about what the experimenter wants from them and behave accordingly (the experimenter demand effect mentioned earlier). They might also make social comparisons and try to outperform the other subjects by doing things that adversely affect them. They may even develop altruistic feelings or think about fairness. All this and much more is possible. What Smith means by "dominance" is that, despite all these distractions, the pursuit of the highest possible payoff still comes first, and in case of doubt, the alternative that ensures the highest payoff is chosen.

At this point, we would like to draw the reader's attention to the explanations in the first chapter. We referred there to the development of behavioral economics and its fundamental assumptions, which mainly stem from psychology. Any assessment of whether the induced value method works or not depends on the strength of the various motives that drive people in the laboratory. Economists tend to assume that the desire to earn more income overshadows everything else and is much stronger than the above-mentioned "distractions". Psychologists take a slightly different view. Behavioral economics deals with phenomena that lead people to make distorted decisions precisely because they allow themselves to be distracted from their actual goal. And psychologists know that people are probably much more susceptible to such distortions than economists generally believe.

Even if it is the case that the motive to earn money dominates the experiment, it may be that more money is not always better. Two examples are worth mentioning. First, it is possible that within the experiment there are non-monetary subjective costs which are not taken into account in the payoff function but which nevertheless influence the decision of the subjects. A simple example is the cost associated with reading and understanding instructions. How high these costs are depends on the complexity of the experimental design, on the one hand, and on the ability of the subjects to grasp the facts, on the other. The experimenter can control the former and design the experiment as skillfully as possible, while he is not able to observe the latter directly but can influence it very indirectly by recruiting from a suitable pool of potential subjects. If a subject finds it difficult to understand the task presented in the experiment, he will probably deviate from a payoff-maximizing decision due to the much greater effort required to determine it. A second example is the existence of altruistic preferences. If a decision-maker has such preferences, it is obvious that more own money need not always be better than less own money. However, this raises the question of whether the existence of altruistic or – more generally – social preferences is in fact a case in which the induced value method cannot be used meaningfully.

What would happen if an experiment were perfectly designed in every respect, i.e. if all the requirements of the induced value method were met in full, and if it could also be assumed that the subjects were able to make a rational decision – for example, because the experiment was so simple that it would be very easy to act rationally? Basically, there are then only two possibilities. Either the subjects would do exactly what the theory of rational choice would predict under the assumption of a utility function increasing monotonically in terms of their own monetary payoff, or the subjects would have other motives than maximizing their own income. This means that even a perfectly controlled experiment still has degrees of freedom. And that is a good thing, because if this were not the case, the experiment would not have to be carried out. However, this also means that a correctly applied induced value method is capable of revealing individual preferences that do not follow the standard assumption that more is always better than less. In other words, the discovery that people may have social preferences would hardly have been possible without the induced value method.

Important

In economic theory, the motives for individual action are modeled using a preference ordering, which in turn can be represented by a utility function. In game theory, this becomes a payoff function, with payoffs in the form of utility units. Experimental research replaces the utility payoffs with monetary payoffs and thus induces a utility function. This can only be achieved if three conditions are met:

1. **The actual perceived utility from the income must increase monotonically with the payoff.**
2. **The payoffs in the experiment must be *salient*, i.e. they must noticeably depend on the subject's decision.**
3. **The payoff must be *dominant*, i.e. the income motive should dominate other motives (such as avoiding boredom).**

Even if all of these conditions are met, subjects still have the freedom to follow motives other than that of maximizing payoffs. This opens up the possibility using experiments to reveal precisely these motives.

What can be done to achieve the best possible control of the subjects' preferences? There are a few things to watch out for:

The payoffs should be noticeable This means that the subjects can only be expected to pay attention to the payoffs if they are structured in such a way that it is worth paying attention to them. We will shortly be dealing a little more closely with the question of how high the payoffs should ideally be on average, but here we can already suggest as a guideline the opportunity costs incurred by the subjects by participating in the experiment. For students this could, for example, be the hourly rate that they would receive as student assistants for the duration of the experiment.

Subjective costs should be minimized This means that it should be made as easy as possible for the subjects to understand the task presented in the experiment and to make the best decision for them. This is one of the reasons why experiments should be simple. One of the pitfalls of confronting subjects with complex decision-making problems is that it is difficult to know whether they understand them at all, or even whether they want to understand them. For this reason, the design of the instructions is important. They should be as simple and easy to understand as possible. A very sensitive question in this context is how to deal with questions that individual subjects have about the instructions. Should they be answered out loud in front of the whole group or should they be answered privately? Since this is more a question of experimental practice, we will deal with it in ► Chap. 3.

Use neutral language The main aim is to avoid experimenter demand effects. This means that the subjects should not be given the impression that the experiment serves a specific purpose. Let us give a simple example. Several experiments were carried out to find out whether East German students behave differently from West German students (Ockenfels and Weimann 1999; Brosig-Koch et al. 2011). In such an experiment, it must of course be checked whether a subject belongs to one group or the other. When it comes to comparing the decisions, the overall structure of the experiment must not reveal to the subjects that an East-West comparison is involved. In other words, it should not be possible to deduce this information from the recruitment method or from the instructions. If the subjects were to assume that such a comparison was at issue, it would most likely result in a kind of competition between the two groups, and this would no longer allow any conclusions to be drawn about the question of interest.

Provide an opportunity to learn Experiments should be simple – as we have already pointed out. Even simple games should, however, be practiced by the subjects before the actual experiment takes place. It is quite possible that learning processes take place in the first few rounds of an experiment. If the experiment is not designed to observe these learning processes, then learning the game has no place in the actual experiment. The aim is to test whether the subjects who know and understand the game behave as predicted by the experimental hypothesis or not. Therefore, the learning process must take place before the experiment. This can be done, for example, by playing practice rounds against the computer. The advantage of this is that the subjects in the practice rounds all have the same experience. On the other hand, practice rounds involving the subjects playing with each other should be avoided. Experiments must be carried out with several groups in order to obtain independent observations. If the practice rounds are played within each group, it is possible that the subjects in different groups gain very different kinds of experience prior to the experiment. These can then serve as reference points in the experiment and influence the behavior there. This should be avoided.

A very controversial question is whether to conduct an experimental session only once or whether to provide the subjects with a repeated opportunity to experience the experimental situation – for instance, by carrying out the experiment again at intervals of 1 week. It is not a matter of repeating the experiment within a session (this is quite common), but of repeating the session as a whole. This is the case if observing "mature" behavior is the actual point of interest, i.e. behavior that is not influenced by the fact that the decision situation is new and unfamiliar. In addition, repetition tends to increase external validity, since most of the decision problems that are explored in experiments are in fact not one-off occurrences in the real world, but recur at irregular intervals. The disadvantage of repeating the session is that the experimenter has no control over what happens between the repetitions. It may well be that the subjects in the meantime gain experience which has a strong influence on their behavior. The problem is that there is no way of knowing what experience this is. It is presumably this methodological problem that has hitherto prevented experimenters from repeating sessions to any large extent.[1] We will return to this issue at various points in this book.

2.2.2 The Size of Payoffs

It's about money. But how much money we are actually talking about is a question we have sidestepped a little so far. Do monetary incentives have an effect and if so, to what extent does it depend on the size of the payoffs? This is a question that is frequently directed at experimental economic research. Two extreme positions are conceivable. One of them assumes that it makes no difference at all whether monetary incentives are used or not. Should this hypothesis prove to be correct, it would mean that in the last decades quite a considerable amount of money has been wasted due to it then having been used unnecessarily in economic experiments as payoffs to subjects. The other extreme position is that all the deviations from the model of rational choice that can be observed in experiments disappear if the payoffs are set at a sufficiently high level. If this thesis were correct, the implications would not be so clear. In principle, it would mean that economic rational choice theory could only be applied to those decision-making situations in which large sums of money are at stake. However, most of the decisions that people make every day are not of this kind. Consequently, research dealing with deviations from the rational choice model would still have a very broad field of application.

In ▶ Sect. 1.4, we already referred to some experimental findings on the question of how the payoff level impacts on subjects' behavior. The majority of studies conclude that the effect is not substantial. Nevertheless, it is worth taking a closer look at the findings. Meta-studies that have been carried out on this question are fruitful because they evaluate a large number of experiments (not only economic ones).[2] In their study, Camerer and Hogarth (1999) analysed 74 papers in which the effect of different payoff levels was investigated. The most important message from their study is that the two extreme positions described above are wrong. Monetary incentives are not ineffective (i.e. decisions should not be elicited hypothetically, but provided with appropriate monetary consequences) and deviations from the rational choice model do not disappear at higher payoffs. The last point can be expressed a little more precisely. The authors have not found a single study where a deviation from the rational choice model observed at low payoffs disappears when the payoff is increased.

Nevertheless, the work of Camerer and Hogarth also shows that the effect of incentives is not always the same. It may well depend on the special circumstances of the experiment. For instance, an increase in payoffs has an impact if the payoff a subject receives at the end of the experiment depends on the effort involved. A good example is provided by experiments testing memory ability. Here it is profitable for the subjects to be more attentive and the more they can earn, the more attentive they actually are.

1 The exceptions are Volk et al. (2012), Carlsson et al. (2014), Sass and Weimann (2015), Brosig-Koch et al. (2017), and Sass et al. (2018).
2 For a survey see Feltovich (2011).

2

It is also evident that the allocations in the dictator game are certainly sensitive to the size of the payoffs. The more money allocated to the dictator, the smaller the proportion that is given to the receiver. One possible explanation for this could be that the dictators assume that the "other player" had to incur transaction costs to participate in the experiment and are willing to compensate him. As the transaction costs are independent of the amount initially allocated, this would explain the negative correlation between the size of the payoff and the relative share that is given away.

The size of the payoff, on the other hand, has no influence in experiments in which the subjects already have a sufficiently high level of intrinsic motivation[3] or in which any additional effort is not worthwhile because the payoff function is flat. Such experiments show, however, that the variance decreases, i.e. the average amount given remains the same, but there are fewer deviations upwards or downwards. The size of the payoff also has little influence on the behavior under risk. At best, there is a slight tendency towards more risk-averse behavior.

The results of Camerer and Hogarth are more or less clearly confirmed in some other smaller meta-studies, so that the following conclusion can be drawn. Incentives are important (who is to believe this if economists do not) and they have an impact when it is possible for the subjects to earn more money through greater effort. This is not surprising. On the other hand, the size of the payoff has less influence in many economic experiments in which the level of effort does not matter. To the extent that we are concerned with such experiments, the following rule of thumb should suffice: it is necessary to set noticeable but not exorbitantly high incentives. As a rule, the size of the payoff should be based on the opportunity costs of the subjects in the experiment.

Box 2.1 Flat Payoff Functions

In 1989, an article by Glenn W. Harrison entitled "Theory and Misbehavior of First-Price Auctions" was published in the American Economic Review (Harrison 1989). Contrary to what the title suggests, the paper is not primarily concerned with auctions and their analysis, but rather with a methodological point of general interest. This is probably the reason why, in 1992 alone, five contributions were published in the AER on the subject (including Harrison's reply to his critics).[4] It is quite entertaining to look at these contributions. One example is when Harrison points out that individual aspects of the other authors' works are as urgent as studying the breakfast menu on the Titanic (Harrison 1992, p. 1437). Nonetheless, in essence it involves an important methodological question.

The starting point for the discussion was an observation that Cox et al. (1982) made concerning experiments in the first-price auction. Bidders place higher bids in these auctions than the Nash equilibrium assuming risk-neutral behavior predicts. An explanation for this deviation from rational choice theory could be that bidders are not risk-neutral but risk-averse. Harrison (1989) mistrusts the findings made in the various experiments for the following reason. It is fundamentally necessary to distinguish between the message space and the payoff space of an experiment. The message space comprises the strategies or actions that the subjects can choose.

---

3 However, there is also evidence that monetary incentives can destroy intrinsic motivation (see, for example, Frey and Oberholzer-Gee 1997, Gneezy and Rustichini 2000, Mellström and Johannesson 2008, Fryer 2013).

4 Cox et al. (1992), Friedman (1992), Harrison (1992), Kagel and Roth (1992) and Merlo and Schotter (1992).

In an auction experiment, these are the bids they can make. The payoff space, on the other hand, is defined by the monetary payoffs resulting from the bids of all the subjects. The deviation from the Nash equilibrium was measured by Cox et al. (1982) in the message space, where it was very clear. Harrison points out that the deviations in the payoff space are extremely small. In one example, he calculates that a significant deviation of 13 cents in the message space leads to deviations in the payoff space, which (depending on the induced valuation of the good being auctioned) are between 0.1 and 0.6 cents. Why, Harrison concludes, should subjects worry about what the optimal bid is if they can only gain a fraction of a cent? This is a legitimate question and the methodological consequence of accepting Harrison's criticism is that payoff functions must not be flat. Strong deviations from the prediction must not lead to minimal differences in payoffs.

Ultimately, the main issue in the discussion is how important the criterion of dominance is. Harrison criticizes that it was violated in the experiments of Cox et al. (1982). From the subjects' point of view, the monetary consequence of deviating was not dominant enough to have a decisive influence on the decision. Therefore, according to Harrison, one cannot draw the conclusion from the experiments that people in first-price auctions make excessively high bids. Whether they are risk-averse or risk neutral in such auctions is therefore at best a secondary question (just as important as the breakfast menu on the Titanic after the collision with the iceberg).

Not all researchers were completely able to accept this line of argumentation, resulting in the already mentioned discussion in the AER. However, this was only to a lesser extent concerned with the methodological core of Harrison's criticism and to a greater extent with the interpretation of first-price auction experiments. For example, Friedman (1992) criticizes Harrison for not being able to explain the excessively high bids in the first-price auction with the metric he proposed (the deviations from the payoff in the Nash equilibrium), to which Harrison (1992, p. 1436) replies that was not what he wanted to do anyway. Kagel and Roth agree in principle with Harrison's methodological criticism and Merlo and Schotter (1992) also refer to Harrison's point as "undoubtedly correct" (p. 1413). Cox et al., whose experiments were directly attacked by Harrison's criticism, suggest that Harrison had implicitly introduced a cardinal utility function, which was incompatible with subjective expected utility theory. Regardless of the question as to whether this is actually the case (which Harrison 1992, p. 1438 doubts), this objection ignores the actual criticism that payoff functions that are too flat violate the requirement of dominance and also that of salience. Strangely enough, Vernon Smith, who formulated these criteria when he introduced the induced value theory, is one of the authors in Cox et al. (1992) and Cox et al. (1982).

A point of criticism raised by Merlo and Schotter (1992) against Harrison's criticism is worth taking a closer look at here. They argue that the point Harrison makes is undoubtedly correct, but only really plays a role if the subjects actually perceive the payoff function as flat. In their view, two arguments speak against this. On the one hand, there are people among the subjects who approach their tasks analytically. Merlo & Schotter describe such people as "theorists", who calculate the optimal solution on the first run of an experiment and then play it consistently. In contrast to this, there are subjects who try out what might be the best strategy during the experiment (these are the "experimenters"). It is clear that Harrison's criticism can only apply to the second group and only if it becomes clear during this testing that the payoff function is flat. In fact, the information obtained by the subjects during an experiment may not suffice to make the payoff function sufficiently clear.

Merlo & Schotter's objection cannot be dismissed entirely, but it does not change the significance of Harrison's criticism. It would not be good methodological practice to use flat payoff functions and then hope that the subjects will not notice that it is actually hardly worth considering in depth what the rational (payoff-maximizing) strategy is.

In summary, founded on the discussion surrounding Harrison's criticism, our recommendation is to take the safe side when it comes to methodology. A payoff function that is too flat can be criticized and with some degree of probability the criticism is justified. A steep payoff function is safe from such criticism and is suitable for taking into account the criteria of dominance and salience, the importance of which no one in experimental research actually doubts. For this reason, we recommend the use of sufficiently steep payoff functions.

Question

In game theory, the terms "utility function" and "payoff function" refer to almost identical things. However, in experimental research when we refer to payoff functions, we mean something else. What exactly is the difference?

The terms "dominance" and "salience" are used to describe payoff functions. What are the differences in this case? Can a payoff function be dominant but not salient, or salient but not dominant?

2.2.3 Is It Okay to Take Money from Subjects of Experiments?

The title of this section might seem a bit provocative, but it addresses an important methodological problem fairly accurately. In economic contexts, but also in other important situations in society, it is possible that decisions taken by people result in losses. Sometimes it is even the case that people may only be able to exert an influence on how high a loss is and are no longer able to avoid it altogether. An important and interesting question is whether decision-making behavior in the event of losses mirrors that of gains or whether there are systematic differences. The only way forward experimentally is to conduct experiments in which subjects actually face the risk of loss or even have to accept a loss with certainty. In such a case, the experimenter takes money away from the subject. Is he allowed to do this? Should he do this?

It is sometimes claimed that it is unethical to take money from subjects of experiments if they make losses in the laboratory. The reason for this assessment is that the subjects incur costs by participating in an experiment and can therefore expect a "fair" reward. They also come to the laboratory expecting to earn money there and are extremely disappointed if they lose money. However, these ethical concerns can easily be eliminated by pointing out in the invitation that it is possible that in the experiment losses may occur, which have to be borne by the subjects. It is highly likely, however, that it will not be the experimenters' twinges of conscience that prevent them from carrying out experiments in which the subjects have to pay. There is a much more mundane reason behind this. A laboratory that carries out such experiments will very quickly encounter major problems in recruiting subjects. The motivation to be available for an experiment decreases rapidly if the subject expects to be asked to pay at the end.

This poses a dilemma for the experimenters. On the one hand, it is important to find out how people react to possible losses. On the other hand, experiments must be designed in such a way that the subjects end up receiving money and not having to pay anything. A popular method of overcoming this dilemma is to design experiments in such a way that, although there is a possibility that losses may occur in individual parts of the experiment, on average there will be no loss at the end of the experiment. A nice example of this variant is the experiment by Rydval and Ortmann (2005). The question was whether negative payoffs in what is known as the "stag-hunt game" led to an improvement in the players' ability to coordinate.

Starting from a stag-hunt game with positive payoffs, the payoff was transformed by affine transformations into one with negative values. The two ◘ Tables 2.1 and 2.2 provide an example of this:

◘ Table 2.2 is obtained by subtracting 60 from the values of the first table and then multiplying the result by 3. Transformations of this kind (affine transformations) do

Table 2.1 Payoffs of a stag-hunt game

	C	D
A	80, 80	10, 50
B	50, 10	50, 50

Table 2.2 Transformed payoffs of a stag-hunt game

	C	D
A	60, 60	−150, −30
B	−30, −150	−30, −30

not change the strategic structure of the game and should therefore have no impact on the choice of strategy. In both games, (A, C) is the payoff-dominant solution and (B, D) the risk-dominant solution. It is well known that subjects find it difficult to coordinate to find the efficient solution and the question that interested Rydval and Ortmann was whether this would change when using a game in which the risk-dominant strategy would surely lead to a loss. In fact, this was the case. The attempt to avoid losses and the expectation that the other player would do the same led to a more frequent realization of the payoff-dominant solution. This result is not so important for us. The way in which losses were introduced into the experiment is of greater importance. The subjects not only played the game of Table 2.2, but also five different games in total, with only two of them suffering losses. As a result, there were only two subjects who actually realized a loss in the experiment. One of them never appeared to settle the loss and the other was given the option of settling the loss or not. The authors write: "*Individual rationality suggests what happened.*" (Footnote 5 on p. 104).

Incorporating losses without any actual losses being incurred is one way of avoiding the dilemma described above. Cachon and Camerer (1996), for example, followed a similar approach. They offered their subjects the possibility of an "opt-out" which allowed them to avoid losses. However, such approaches are compromises since they only reflect real losses in a limited sense. Not paying for all the losses that result in an experiment can bias decisions. A frequently implemented alternative to this is to pay the subjects a sufficiently high "show-up fee", from which the potential losses can be paid (see ▸ Box 2.2). Kroll et al. (2014) offer another possibility. They investigated the Nash bargaining solution in an experiment in which losses were also negotiated. The trick is that losses were not represented by negative payoffs, but by waiting times that the subjects had to accept before they received their payoffs. The subjects thus experienced a "loss of time", for which there was naturally a monetary equivalent without the unpleasant situation of having to "ask the subjects to pay". On the contrary, at the end of the waiting period a payoff took place, i.e. the experiment ended with a positive experience, which should help to maintain the reputation of the laboratory and not complicate the recruitment of further subjects. However, this procedure also leads to a loss of control

2

on the part of the experimenter over the size of the subjects' losses, which can only be determined exactly if one knows the exact opportunity costs the waiting time incurs for the subjects.

The closest approximation to achieving real losses is probably offered by using the following approach. First of all, an experiment is carried out in which it is certain the subjects earn money. This experiment serves to provide them with an income that can later be used for the actual experiment, in which losses can occur. In order for the whole thing to work, however, a certain amount of time has to elapse between the two experiments, so that the subjects really do regard the money they have earned as their own money and not as the initial endowment provided by the experimenter. However, finding the right time span between the two experiments is not easy. If it is too long, the subjects will have forgotten that they recently earned money in the laboratory and now have to return a part of it. If the time span is too short, they will not really see the money as their own money, which they have to sacrifice to compensate for the loss. The approach has another disadvantage. Experience shows[5] that it takes only 3 days for subjects to regard the money they have received as their own and they may then react very indignantly if the experimenter expects them to pay for participating in an experiment.

Question

In total, there are four "tricks" that allow experiments to be carried out in which losses are incurred without the subjects actually having to pay anything out of their own pockets. Name and describe them.

In general, it must be concluded that it is very difficult to implement losses in laboratory settings. Subjects react to this and there is a not inconsiderable risk of harming one's own experimental pool with such experiments.

Box 2.2 Should a Show-Up Fee Be Paid?

A "show-up fee" is a payment subjects receive for participating in the experiment, regardless of which decisions are made there. Apart from possible income effects, such a fixed amount has no direct effect on the experimental behavior. So why pay it? One possible function of this payment is to create a kind of buffer in experiments in which losses may occur, cushioning the impact of possible losses by ensuring that the subjects receive a positive payoff in any case. More often, however, the show-up fee is used to ensure that even those subjects who end up with a less profitable role in the experiment receive an appropriate reward. Generally, there will be a certain budget available for the payments to the subjects, which then has to be split between the show-up fee and the payoffs, depending on the individual decisions. There is no rule according to which the apportionment should be carried out, but there is a conflict of objectives which one should be aware of. The higher the show-up fee, the lower the incentivizing effect of the "decision-dependent" payoff. Therefore, a prudent balance must be struck between the objectives of not disappointing the subjects, on the one hand, and providing sufficiently strong incentives, on the other. The show-up fee can, however, also be used in a completely different way. Anderson et al. (2008) generate asymmetry between the subjects in a public good experiment due to different amounts of fixed payments. Incidentally, if this is made public, contributions to the public good will fall overall.

5 This experience was gained by talking to an experienced experimenter who had often conducted such experiments.

2.2.4 The House Money Effect

Monetary incentives are usually created in the experiment by, in a sense, pressing money into the hands of the subjects, who then can use it in the experiment. The basic idea here is that the value of money does not depend on where it comes from. Whether you work hard for 10 euros, find it on the street or win it in a lottery, it makes no difference to the quantity of goods you can buy for that money. So why should 10 euros received as a gift be worth less than the 10 euros earned? This view stems from rational choice theory. The notion that money always has the same value cannot be shaken, and under the requirements of the neoclassical rational choice model it would simply not be reasonable to value endowed money any differently from earned money.

As early as 1994, however, Loewenstein and Issacharoff showed in an experiment that the question of where income comes from is important for the valuation of this income (Loewenstein and Issacharoff 1994). It obviously makes a difference whether the money used in an experiment is "your own money" or money provided by the experimenter. This endowed money is something like a windfall profit, i.e. income that simply lands in your pocket without you having to do anything about it. Imagine you are shopping in the city with a friend. During a break, you go to a restaurant to have a cup of coffee, and there is a gambling machine in the restaurant. Your friend persuades you to put one euro into the machine and you win 50 euros straight away. How will your shopping tour continue? As it would have without the windfall profit of 50 euros? Or will you spend more? And will you give some of your winnings to your friend? Compare this situation with one in which you ignore the gaming machine, but earned 50 euros more the day before by doing extra work. In both cases you have the same amount of money at your disposal – do you behave the same way?

It is quite obvious that the unexpected gain changes your behavior – even if this is difficult to reconcile with rational behavior. It can be assumed that after a windfall profit, the propensity to consume increases just as much as the willingness to take risks. If that is the case, then of course this is highly relevant for the design of the payoffs in an experiment. The only question is how to experimentally test the effect of monetary gifts. How do we get people to use their own money in an experiment? Such an experiment is likely to make recruiting subjects quite difficult. For this reason, a different approach is taken.

Instead of the subjects having to spend their own money, the experimenter has them perform a task for the money they receive. The type of task can be freely chosen. We will deal more extensively with the most common methods used in this context in ▶ Sect. 2.7.2. At this point, we will confine ourselves to the example of having subjects answer quiz questions so that they can be paid according to their results in the quiz. The crucial point is that the subjects no longer have the feeling that they have been given the money to use in the experiment. This is not quite the same as using their own, self-earned money, but if it turns out that in this sense money that is not endowed is treated differently from money that is, then it is safe to assume that it is the "house money effect" that describes this.

Clark (2002) investigated the effect of house money in a public good experiment. In such experiments, it is often observed that the subjects cooperate more by making significantly higher contributions to the public good than rational choice predicts. One possible explanation for the strong deviation from the Nash equilibrium could be that people are more willing to think about the group and cooperate when using house money. However, Clark finds no significant difference between the contributions of the groups with and without house money. This result was re-examined 5 years later by Harrison (2007), who was able to show that a house money effect can be demonstrated in Clark's data by looking at the individual data and taking into account that they have the structure of panel data.[6] Cherry et al. (2005), however, do not find a house money effect in a public good game. It is therefore possible to state that the effect is not very pronounced in this connection, if it exists at all. In any case, it does not provide an explanation for the comparatively high contributions in public good experiments.

Kroll et al. (2007) investigate the house money effect in a best-shot experiment with heterogeneous actors. This is a sequential public good game, which, in terms of the experimental results, is known to be very different from simultaneous public good games. While in the latter, as has already been mentioned, there are significant deviations from the Nash equilibrium, this is not the case in the best-shot game, where subjects' behavior is in relatively close agreement with the Nash prediction and where a clear house money effect can be seen. Kroll et al. ascribe the different strengths of the house money effect in the two experiments to the fact that both the Nash equilibrium in the simultaneous public good experiment (all players make no contribution) and the efficient solution (all players invest their entire endowment in the public good) are symmetrical, while the equilibrium in the best-shot game is asymmetrical. It is not clear, however, why symmetry should play such a major role.

Muehlbacher and Kirchler (2009) are also able to detect a house money effect in a public good experiment. They do not compare the contributions with and without house money, but vary the effort that must made to "earn" the experiment money. It turns out that the willingness to cooperate is less when the effort is high, so people have to work hard for their money.

A particularly pronounced house money effect has been observed in dictator game experiments. Cherry et al. (2002) show that if two circumstances coincide, the amount given to the receiver actually drops to almost zero. First, the subjects have to work for the money they can then share between themselves and the receiver. Second, the allocation is completely anonymous, i.e. in a double-blind design.[7] This is an observation that is quite plausible, as it is unclear why people give money to strangers in dictator game experiments but hardly do so in their real lives. The double-blind design could eliminate a possible experimenter demand effect, and avoiding the use of house money could make it easier for subjects to convince themselves that there is no obligation to give anything to the other person. Remember our example of winning on the gaming machine on the shopping tour. Most people would at least invite the friend to join them for dinner afterwards – which they might not have done without the house money, assuming their total income did not change. Cherry and Shogren (2008) are able to confirm the house money effect in the dictator game experiment as well.

Oxoby and Spraggon (2008) extended this finding by adding an interesting facet. In their baseline treatment, the subjects were given the money as usual. In this case, the dictators gave, on average, about 20% of their endowment to the receivers. This is a

---

6 The statistical analysis of panel data differs from that of a time series, where the individual observations are independent. We will discuss this point briefly in ▶ Chap. 4.

7 This means that neither the subjects among themselves nor the experimenter can observe what a subject is doing. How such designs can be achieved and what effects they have will be explained in more detail later in ▶ Sect. 2.5.2.

value that has been similarly demonstrated in many other dictator game experiments. In the second treatment, the dictators had to work for the money they could then hand out. The result was identical to that of Cherry et al., with the dictators keeping almost everything to themselves. In the third treatment, it was the receivers who were obliged to earn the money, which was subsequently made available to the dictators. In this case, the dictators were in fact more generous, giving over 50% to the receivers. Without further evaluation of this behavior, it is again evident that the question of where the money comes from may play an important role. Carlsson et al. (2013) were able to show that in a field experiment a clear difference occurs between windfall money and earned money. In the case of the field experiment, too, dictators who benefited from a windfall profit gave significantly more than those who had to earn the money.

The methodological lesson that can be drawn from these findings is that the way in which subjects obtain money can play a role. If the house money effect is to be avoided, it is necessary to have the subjects work for the money. This is not always absolutely essential, though. If house money is used in all the experimental treatments, the effect could be neutralized to an extent and it might be possible to conduct causal analyses of the remaining differences between the treatments.

Important

In laboratory experiments, the subjects do not have to invest their own money. Rather, they are usually provided with money by the experimenter that can then be used in the experiment. This money is called house money. Basically, it is possible that people behave differently when they handle endowed money than when they use money they have earned themselves. When this happens, it is referred to as a house money effect. It is possible that such an effect will occur, but it is not necessarily the case. While pronounced house money effects have been found in dictator games, this effect seems to be only relatively weak in public good experiments.

If it is desirable to eliminate house money effects, it is advisable to have the subjects carry out a task for which they are then paid. In this way, endowed money becomes earned money and the house money effect is mitigated or, ideally, disappears altogether.

2.3 The Subjects of the Experiment

Ultimately, an experiment revolves around the people participating in the experiment, the subjects. Their behavior determines the outcome of the experiment, and all the interest of the researchers working in the laboratory is focused on their decisions. In view of this importance, no mistake should be made in the selection and treatment of the subjects involved. That sounds easy, but there are some important things to keep an eye on.

2.3.1 Is It Permissible to Lie to Subjects of Experiments?

Experiments in which the subjects can suffer losses are very rare and even rarer are experiments in which they actually have to pay something. Most seldom, however, are experiments in which subjects are lied to – at least in experimental economic research. At first sight, this seems to be self-evident, since lies seem to be at least as unethical as asking for money. In light of this, it should be clear that such a thing is simply not done. On closer inspection, however, it can be seen that honesty in the laboratory is a specialty of economists and that there are other disciplines which are far from being as strict about this as economics. It is therefore worth taking a look at why economists insist on honesty and why, for example, experimental psychologists often fail to do so. So what speaks for and against the manipulation of subjects by the experimenter?

If readers of this book intend to publish an experimental work in an economics journal, they can only be strongly advised against lying to the subjects. Such a "deception" would immediately push the probability of acceptance of the paper to zero for the simple reason that the editors of economics journals are extremely strict on this issue, and not without good cause. There is a very broad consensus within the scientific community of experimental economists that deception cannot be tolerated. The justification for this is essentially an argument based on game theory.

If lying were to occur in experiments, it would not happen in secret, but would be stated in the respective publications. Furthermore, after such an experiment, a debriefing in which the subjects would be informed that they had been lied to would typically take place for ethical reasons. This would mean that the experimenter's dishonesty would always become public. However, according to the argument, this would lead to the experimenters gaining a reputation of not being honest. This, in turn, would have disastrous consequences because, if the subjects were to suspect that they were being lied to in the laboratory, how would it possible to monitor their preferences? If the experimenter did not know which game the subjects thought they were actually playing, he could, strictly speaking, no longer draw any conclusions from their behavior. Such a scenario must be prevented and can only be achieved by the experimenters defending their reputation of being honest.

When subjects enter an economics laboratory, it is safe for them to assume what is written in the instructions for the experiment is what actually happens. There is no way of being sure of that in a psychology laboratory. Does that change the behavior of the subjects? The question is whether the economists' assumption that deception "corrupts" the subjects is justified or not. Bonetti (1998) is one of the few people who have doubts about this, suspecting that subjects' finding out that they are being lied to has hardly any effect. This provoked vehement protests, for example from Hey (1998) and McDaniel and Starmer (1998). It is true there are studies in psychology that do not find any significant effects of deception.[8] However, these experiments have been conducted with psychology students, whose experience it is that they are regularly lied to. If this means that it is not possible to draw any conclusions from their behavior, then of course this also applies to the experiments investigating the impact of deception.

Economists face a certain dilemma when trying to experimentally test the effect of deception. In order to be able to carry out such an experiment, they would also have to experiment with a design in which lying takes place – but this is what is forbidden by the methodological standard they feel obliged to follow. Jamison et al. (2008) found a way around this dilemma. At the University of California at Berkeley there are two pools of experimental subjects. In one of them (used by economists) what has come to

8 See Ortmann and Hertwig (2002) for a survey.

be known as the "no-deception rule" prevails, whereas in the other one it is permissible to use deception. This opens the way for lying without corrupting the pool of subjects. In their experiment, there were two designs, one with deception followed by a debriefing and one without deception. The deception consisted of the subjects playing against a computer and not, as explained in the instructions, against another person. The subjects played a "trust game", in which a player could send money to a partner (who then received three times the amount sent) in the hope that in the second step the partner would send money back and share the total payoff fairly. After 3 weeks, the subjects were invited again by another experimenter and three experimental designs were then used: the prisoner's dilemma game, a dictator game and a lottery choice. The research question was whether the deception in the first experiment had an influence on the willingness to participate and on the behavior in the second experiment. It was found that both were the case and that the "deception effects" were particularly pronounced in women. Deceived women were less willing to take part in the second experiment, especially if they were unlucky enough in the first experiment not to have had their trust in their partner rewarded. The subjects in the second experiment were more inconsistent in their lottery choices when they had been lied to in the first experiment and were less generous in the dictator games. In addition, women who had been deceived were less willing to cooperate in the prisoner's dilemma.

This seems to demonstrate that deception can actually lead to behavioral changes that make it difficult to properly interpret the results of an experiment. How should the behavior in dictator games be interpreted, when selfish behavior can be a result of the experience of having played against a computer in the past rather than a human being as promised? Barrera and Simpson (2012) replicated the experiment with a few modifications. For example, the subjects had to participate in both experiments (otherwise there were no credit points). This was done to eliminate the self-selection observed by Jamison et al. Unlike Jamison et al., Barrera and Simpson do not find any effect of deception.

Alberti and Güth (2013) also attempted to circumvent the dilemma mentioned above and to investigate the effect of deception in an experiment. Their trick was to use a subject as an experimenter who can either be honest or deceptive. In this way, the actual experimenters are taken out of the line of fire, so to speak, because it cannot be inferred from the experiment that experimenters might not be telling the truth. Alberti & Güth showed that deception did not have any effect in this case. This suggests that deception which is part of the behavior of a subject is perceived differently from deception for which the experimenters or "the laboratory" are responsible.

An interesting opinion on whether or not deception is permissible is provided by Cooper (2014). On the one hand, he certainly argues along the lines of the argument of game theory presented earlier, but, on the other hand, he also discusses cases in which deception may not be so bad. For example, in those cases in which it would be very difficult to carry out the study without lying to the subjects, or when there is a large amount of knowledge to be gained from the experiment, but the detrimental effect of the deception is comparatively small. The question is, however, how (and by whom) should a decision be made as to when results are important enough or when an honest experiment is difficult enough to justify deception? Much the same is true of Cooper's statement that, as an editor of the "Journal of Experimental Economics", he would strictly reject papers in which deception occurs but would be much more generous as an editor of the "Journal of Wine Economics". He conjectures: "*The reputation of experimental economics is probably little affected by papers published in JWE, and evidence for an indirect effect is weak in any case.*"(Cooper 2014, p. 113). Here, too, the question arises as to when a journal is sufficiently distant from the experimental scene to avoid undesirable reputational effects. Perhaps it is a good idea to avoid all these difficult considerations and judgments by simply following the general rule that deception is not allowed at all. Then it is not necessary to worry about the rather ambiguous experimental findings. By being honest, you are always on the safe side (almost like in real life).

An important question does remain to be resolved in this connection, however. When does dishonesty begin? Is it already deception if the subjects are not informed about everything? For example, in some experiments it is important that the subjects do not know in advance how long the experiment will take or how often it will be carried out. In the case of public good experiments, for instance, a study was conducted into what happens when the subjects, after they have run through the experiment once (over ten rounds, for example), surprisingly run through the same experiment again. What is of interest here is whether, after such a restart, the subjects continue where they left off in the last round of the first run-through (that is, with generally very low levels of cooperation), or whether they start again and behave similarly to the start of the first experiment.[9] What is decisive is that the subjects do not know that there will be a second run-through when they start the first one. There are many variations of subjects' receiving incomplete information. The convention is not to regard this as deception. The rule could be formulated as follows. Everything that is said to the subjects must be true. However, the whole truth does not always have to be told all at once. Cooper (2014) puts it this way: "*Deception is generally considered a sin of commission rather than omission.*" (p. 112).

Question

In each of the following cases decide whether or not the subjects have been inadmissibly manipulated (i.e. deceived).

1. In an experiment involving a group of subjects (for example, a public good experiment), the subjects are told that the group consists of 60 people. Actually, it consists of only six people.
2. At the beginning of the experiment, the subjects are told that the game will be played exactly once. However, once this has been done, the game is repeated.
3. The game is explained to the subjects without any reference to how often it will be played. It is then played three times and the subjects do not know whether they will be expected to play one more time.

2.3.2 Are Students the Right Subjects?

We are not aware of any statistics on this, but we are fairly certain that well over 90% of all laboratory experiments are carried out with student subjects. And this not only applies to laboratory experiments in economics. Is this the right choice? This is a

9 In fact, this is exactly what they do (cf. Andreoni 1988).

question which is constantly being raised with concern. Is it really possible to learn something about the behavior of people in general from the behavior of students? Or are students actually too "special", i.e. not sufficiently representative? Some critical voices fear precisely that.[10]

Let us start by explaining why students are so popular with experimenters. They simply have many advantages (Feltovich 2011). First of all, they are readily available, being represented in large numbers at universities and blessed with a relatively large amount of freely available time. That is why they can take part in an experiment, for example, at 2 in the afternoon or at 10 in the morning. Another advantage is that it can be assumed that students generally understand relatively easily and quickly what is expected of them in the experiment. A well-executed experiment ensures that the subjects have understood the instructions and know the rules of the game. It is quite certain that students are able to do this. From the point of view of the experimenters, it is also an advantage that students are often short of money and therefore gladly take the opportunity to earn something by participating in an experiment. The relatively low opportunity costs in terms of time mean that the monetary incentives set in the experiment do indeed carry a high weighting.

These advantages are to a degree mirror images of the disadvantages of conducting experiments with non-student subjects. Recruiting the latter is much more difficult and time-consuming. If they are working people, only laboratory hours after work can be considered. In other words, it is necessary to get people to spend their scarce free time in the evening in the laboratory instead of at home with their families. In addition, it is difficult to establish initial contact. Students can be recruited relatively easily in lectures and are therefore usually represented in large numbers in a database of experimental subjects. Recruitment can be done at the push of a button or with a few clicks. This is not the case for non-students. The normal procedure is to make a random selection from the phone book and then either call or write a letter. Both are costly and the success rate is not very high, which is why a large number of contacts are needed to gain the required number of subjects.

We have said elsewhere that the opportunity costs of the subjects of the experiment are the lower limit for the payoffs that should be made in the experiment. These costs are significantly higher for employed people than for students, with experiments with non-students, therefore, always being more expensive than those with students. The more difficult recruitment and higher costs increase the effort required for an experiment, but they do not make experiments with non-students impossible. Some other difficulties are harder to resolve. First of all, there is no control whatsoever over the background of the subjects, their experiences, their level of educational attainment, etc. Although a lot of questions can be asked afterwards, at the recruitment stage it is not clear who it is that is being invited. This especially applies to the attitude of the subjects towards an experiment and the experimenters. Two effects are possible. In particular, non-academically trained individuals could be intimidated by the atmosphere in a laboratory and the fact that they are at a university. This could cause them to be subject to a particularly pronounced experimenter demand effect because they try hard to do exactly what they think is expected of them in this situation.[11]

---

10 One example is Harrison and List (2004).

11 We will come back to this point in ▶ Sect. 2.5.1.

The opposite effect is also conceivable. One of the authors of this book once experienced that in an experiment with non-students, a small number of subjects left the laboratory immediately after studying the instructions and made comments such as "waste of taxpayers' money!", "Such nonsense!" and similar things – in one case underlined by loudly slamming the doors. Students are by and large familiar with the sense and purpose of experiments, but for non-students it is quite possible that this method can occasionally cause them to shake their heads.

In view of the advantages that students have as experimental subjects and the disadvantages that non-students have, it is not surprising that experiments with non-students are so rare. So far, however, we have only listed the advantages of student subjects. We have up to now concealed their biggest disadvantage. Inviting students into your laboratory may create a double selection bias.[12] First, students differ systematically from the average population. On the one hand, they are younger and better educated; on the other hand, they do not have the experiences of an average adult. For instance, they have no professional experience, do not know what it is like to pay income tax or to negotiate for their salary. These systematic differences make it difficult to transfer the decisions observed by students to the average population.

The second selection bias comes into play when students participate voluntarily in experiments. It cannot be ruled out that only certain types of students participate in experiments. What is particularly worrying is that this self-selection process affects the preferences of the subjects, both their *risk preferences* and their *social, or other regarding, preferences*.

Definition

The term "risk preference" refers to an individual's attitude towards risk. The person can be risk-seeking (risk-loving), risk-neutral or risk-averse. The precise definition of these terms will be described in ▶ Sect. 2.4.1.

"Social preference" refers to a person's tendency to deviate from purely self-interested behavior in certain situations. This may be because the welfare of others also plays a role (altruism), for example, or because people do not like the fact that benefits are very unequally distributed (inequity aversion). Reciprocity, i.e. the desire to "reward good with good", can also be a reason for showing "pro-social behavior".

Harrison et al. (2009) are mainly concerned with possible selection bias with respect to risk attitudes. Since the subjects of experiments are randomly assigned to the individual treatments, they can be lucky or unlucky in how they are classified. For example, in a dictator game experiment one can either play the role of a dictator or the receiver. The

12 There is now a large body of literature on the selection biases that can be associated with student subjects. For example, Harrison et al. (2009), Feltovich (2011), Anderson et al. (2013), Cleave et al. (2013), Exadaktylos et al. (2013), Falk et al. (2013), Abeler and Nosenzo (2015), Belot et al. (2015) as well as Cappelen et al. (2015).

former is certainly more advantageous than the latter. This risk aspect could discourage risk-averse students from participating in an experiment or entice more risk-averse students to take part in the experiment. The result would be a selection of subjects who are prepared to take above-average risks. Harrison et al. point out that the effect can, however, also go in the opposite direction if generous show-up fees are regularly paid in a laboratory, i.e. appearance fees that a person can be sure of as soon as he participates in the experiment (see ► Box 2.2). High, guaranteed payoffs could attract higher-than-average risk-averse individuals to participate.

It is of course possible to experimentally test the various hypotheses on potential selection biases. Harrison et al. have done this and they conclude that there are in fact such effects. In the case of randomized experiments (with uncertain payoffs), the willingness to take risks generally increases within the group of subjects, although there is no change in the proportions of relative risk aversions. The risk-averse individuals become just as willing to take risks as those who are already risk seeking. Harrison et al. have also demonstrated the opposite effect, i.e. using show-up fees results in groups of subjects who are less risk seeking than those who do not receive such fixed payments.

The finding of Harrison et al. is put into perspective by a study by Cleave et al. (2013). In their experiment, 1173 students initially took part in a lecture hall experiment, which included among other things a lottery choice. Afterwards, all the students who had participated in the lecture hall experiment were invited to a laboratory experiment. In this way, the group was divided into two subgroups: those who accepted the invitation to the experiment and those who declined the invitation. In the last step, the risk attitudes in both groups were determined with the data from the first experiment and it was found that there was no significant difference. This finding suggests that there is no systematic selection that leads to predominantly risk-seeking people participating in an experiment.

It is not only the risk attitude of the subjects that is suspected of playing a role in the selection process; social preferences could also have an impact. It is conceivable that people with strong social preferences are less likely to take part in experiments involving money and income motives. However, the opposite notion would not be absurd either, since participation in an experiment also serves science (and thus the general public) well. Falk et al. (2013) have investigated whether this leads to a selection of students who tend to have a stronger prosocial attitude. To this end, they first carried out a field experiment that involved some degree of social behavior on the part of the subjects. In the second step, the subjects were invited to a laboratory experiment. It turned out that the question of how prosocial one was in the first experiment did not have a significant impact on the decision to participate in the second experiment. The authors conclude from this that it can be assumed that there is no selection bias associated with voluntary participation in an experiment.

Anderson et al. (2013) come to a similar conclusion with regard to selection bias in relation to non-students, i.e. (as the authors write) "adults". The experiment, in which the subjects revealed more or less pronounced "other-regarding preferences", was carried out with three groups: college students, adults who voluntarily responded to a call for participants (and who could therefore have been influenced by selection bias due to self-selection) and a group of truck drivers who were in a training program. In the last group, 91% of all those solicited took part in the experiment, so it is unlikely that a selection bias occurred. The low opportunity costs of participation are likely to have played an important role here. As with Falk et al. (2013), this experiment did not show any significant difference between the voluntary adults and the truckers, for whom no selection bias had taken place. In fact, there is some evidence that as far as social preferences are concerned, no selection bias is to be feared when recruiting subjects.

So far, we have only talked about the second selection effect resulting from students volunteering for an experiment. What about the first one? Are students different from non-students? Exadaktylos et al. (2013) conclude that there is no significant difference in social behavior between student volunteers and non-student volunteers. These findings admittedly contrast with a whole series of observations showing significant differences between students and non-students. For example, Falk et al. (2013) found that in a trust game, non-students paid back significantly higher amounts to the first movers than students did. In the experiment already mentioned by Anderson et al. (2013), it can also be seen that students behaved much more selfishly than the adult volunteers and the truck drivers. Cappelen et al. (2015) examine social behavior in a dictator game experiment and in a trust game experiment. They also find significantly more pronounced social preferences in a group of subjects consisting of representative persons (of Norwegian society) compared to a group of students. Belot et al. (2015) come to the same conclusion. They also find that students are more likely to be able to think strategically than "normal citizens". Belot et al. conclude that:

> *Experiments using students are likely to overestimate the extent of selfish and rational behavior in the general population.* (p. 26).

The question arises as to whether this conclusion can be drawn quite so easily. We have already pointed out that non-student subjects may be exposed to a strong experimenter demand effect. This could play a decisive role when it comes to the question of whether a person should act more selfishly or socially. Imagine being in the position of a person who has never studied and who, as a non-student, is invited to an experiment in a scientific laboratory. Not familiar with the experimental methods in economics and social sciences, such a person will probably be very uncertain about what exactly is expected of him. Moreover, he will not know how far the experimenters can be trusted, how closely he is being observed and whether non-social behavior is penalized after all. In view of this uncertainty, the question "What do they want from me?" inevitably arises. In such a situation, it is perhaps very difficult to act selfishly, even though that is exactly what the person would do if they were not sitting under observation in a laboratory. Students, especially those who regularly take part in experiments or who are informed about experiments in lectures, are likely to find it easier at this point because they are familiar with the situation and have learned from experience that their behavior in the laboratory only involves monetary consequences. Such an asymmetric experimenter demand effect cannot be ruled out, but it is, of course, not certain that it exists. To clarify this issue, it would be necessary to conduct experiments with an experienced pool of non-students to check whether experience leads to an increase in the willingness to behave selfishly.

Bortolotti et al. (2015) observed an interesting difference between students and non-students. From the experimental literature, it is known that possibilities to punish in public good games lead to a strong increase in the willingness to cooperate (Gächter and

Fehr 2000). Bortolotti et al. (2015) have also observed this effect when using students as subjects. With non-students, however, this effect failed to materialize and threats of punishment remained without effect. As a result, the sequence of the ability to cooperate was reversed: without punishment, non-students cooperated more closely than the students, and with punishment it was vice versa.

So far, we have been able to establish that there are differences between the behavior of students and that of non-students, although it is not entirely clear how marked these differences are. Thus, the question of the transferability of observations made with students to the general population has not yet been settled conclusively. The point remains that students lack a whole series of experiences that the general population has already had in their lives. In addition, sometimes it is questioned whether conclusions about the behavior of experts can be drawn from student behavior. This is a truly important question because the phenomena that are often of interest in experimental research are those where experts make decisions in the real world. This is the case, for example, in negotiations, at the stock exchange, in medical practices or in many of the decisions made in companies. Should experiments that reproduce specific situations therefore be carried out with experts who are very well versed in such situations?

The results are mixed. For example, it has been shown that experts perform better than students when the experiment requires exactly the kind of decisions they have practiced in their everyday lives. Feltovich (2011), however, points out that the use of experts in experiments can also be counterproductive. If the decision situation is not completely identical to that of everyday life, experts tend to use their trained routines even though they do not exactly fit the decision problem. This means that it takes them a while to adapt to the new situation as compared to everyday life and therefore they do not perform better than students in such experiments.

All in all, students are not the worst possible choice. The differences to the rest of the population tend to be moderate, while the use of experts in experiments is not unproblematic. Therefore, using students as subjects may well represent a good alternative in the vast majority of cases. This does not rule out the possibility that there may be specific questions in which it seems advisable to conduct experiments with a more representative population. However, it is important to be aware that this can massively increase the effort involved in recruiting. But even if we remain within the group of students, there are a few things to consider when it comes to recruiting, because students are not always the same.

2.3.3 What Role Does the Student's Subject of Study Play?

Since the famous article by Marwell and Ames (1981), it has been known that economics students behave differently from non-economics students. Above all, students of economics seem to be less cooperative than students of other disciplines. Frank et al. (1993) confirm this observation. It is not really important for the methodology of experimental economics research to assess this observation. However, it is essential to know that selection bias occurs in experiments if attention is not paid to the students' subject of study when choosing participants for experiments. For example, if the number of economics students in a public good experiment is significantly higher in one treatment than in another, this selection of participants could lead to effects that do not occur if both treatments are carried out with non-economists or only with economists.

We do not want to leave it at this comment alone, however, because it is worth taking a closer look at the effect of economists. Rubinstein (2006) further fuelled the discussion on this effect and gave it a new direction using an experiment that did not involve strong monetary incentives, but was essentially based on a survey. Economics students and students of other faculties were faced with a fictitious corporate decision, having to put themselves in the position of a manager who has to decide how many workers are to be employed in the future. The decision was based on a table or formula that indicated the relationship between the employment level and the net profit of the enterprise. In order to realize the maximum profit, a large number of employees would have to be dismissed due to the company's difficult economic situation. Rubinstein examined the question of whether the number of layoffs depended on which subject the participants studied and on whether they were provided with a table or a mathematical formula as a basis for their decision. The result was clear: economists dismiss more people than non-economists and the formula leads to more redundancies than the table.

The conclusions Rubinstein draws from his findings are very far-reaching. For example, he calls for a critical examination of mathematical education in economics because it tends to result in students not developing a "social conscience" or "social responsibility". There seems to be a fundamental lack of these two things among economics students, which leads Rubinstein to take a very critical view of economics courses.

The question is, however, whether the Rubinstein experiment really does allow such conclusions, since there are some methodological problems that this experiment raises. For example, it could be that the economics students in particular have been exposed to a strong experimenter demand effect. A degree in economics is often completed with the aim of working in a company at some point in the future and making decisions there. A well-trained manager should first and foremost have an eye on the interests of the company. Anything else would be rightly criticized by his employer as mishandling the situation. The students of economics are therefore familiar with the perspective of management and the experiment is therefore an ideal opportunity to see whether they have learned something during their studies. In addition, students of economics learn that a dismissal is not necessarily a fateful blow if the labor market can be assumed to be functioning properly. The fact that Rubinstein attributes the difference between economists and non-economists to their education and fails to consider selection bias is also a point of criticism. Perhaps those who decide to study economics have different attitudes to others per se.

Brosig et al. (2010) repeated the Rubinstein experiment and modified it in such a way that it was possible to investigate whether the problems mentioned above actually came into play. An important modification was that the experiment was carried out with beginners in the first semester as well as with advanced students of higher semesters. In addition, the subjects were asked to put themselves in the position of a manager who is on the verge of retirement and only has one more decision to make and that is how many people should be dismissed. Through these two modifications, it was possible to test whether economic education is responsible for people giving little attention to the fate of workers and whether the specific role in which decisions had to be made had an impact. The findings revealed that there was no significant difference in the behavior of the first-year students and the advanced students (neither economists nor non-economists). If at all, the older students laid off somewhat fewer people than the freshmen did. This would suggest selection bias occurred and contradicts Rubinstein's thesis that studying changes people. The role that the subjects were expected to adopt also had an impact. When taking on the role of a person who is about to retire and whose career no longer depends on doing what the employer expects of the person, economic students lay off significantly fewer employees than they do in the role of active managers.

Nevertheless, the overall results of Brosig et al. (2010) show that economics students are more willing to lay people off than non-economics students. Moreover, presenting the decision task with the help of a mathematical formula leads to more people being fired and, more specifically, the maximum number of layoffs being chosen more often. Yet, there could also be an experimenter demand effect behind this. When presented with a profit function, it is self-evident for someone who is mathematically trained that, first of all, it has to be maximized. Few people would understand this kind of presentation as an instruction to derive a table from the function. And it is also obvious that the subjects would develop the idea that the experiment is all about correctly determining the employment level that results in maximum profit. This demand effect (see ▶ Sect. 2.5.1) could explain why even non-economists often achieved the maximum profit when the formula was presented to them.

Ockenfels and Weimann (1999) have a slightly different view on the question of how economics students behave. Their experiment is concerned with the question of whether students from eastern and western Germany behave differently. To this end, they conducted experiments in two cities in the west and one in the east of Germany, in each of which students from different faculties participated. A public good experiment and a solidarity experiment[13] were played. Both the students' subjects of study and the gender of the students were controlled for. The results showed that an interaction effect between the subject of study (economics or not economics) and gender existed in all three cities. In the solidarity experiment, there was no difference between female economics students and non-economists, but certainly between male economics students and non-economists. This suggests that the "economics effect" is limited to male students. This rather incidental finding could explain why studies investigating gender effects have not always yielded clear results. If the subject of study is not controlled for, it may happen that a more or less high proportion of male economics students leads to gender effects, but sometimes it may not.

The main question in the experiment of Ockenfels and Weimann (1999), however, was whether the different cultural backgrounds of students who grew up in a socialist or a capitalist system played a role. This is the question that we will address in the following section.

---

13 This will be explained in more detail in the next Section.

2

Important

The vast majority of experiments are carried out with students as subjects. Generally speaking, there is a risk of selection bias because students may behave differently from the general population of people. There is also clear evidence of this, but the differences do not seem to be very marked. Students have significant advantages over non-student subjects because it is relatively easy to recruit them, they have a high level of understanding and it is comparatively easy to create monetary incentives that are noticeable.

In the process of deciding how to recruit the subjects, it must be borne in mind that even when drawing on students, selection processes can play a role. There is less need to worry about risk attitudes and social preferences, but it is important to control the students' subject of study because economics students might behave differently from non-economics students.

2.3.4 Cultural Differences

The experimental method has long been used worldwide. The papers that result are published internationally and are recognized by the international scientific community, for whom national borders are not really important. But all this does not change the fact that the subjects of the experiments are almost always recruited within the region in which the laboratory is located. This raises the question of whether experimental results obtained in different countries can be easily compared. In other words, does it matter which country the subjects come from and what cultural background they have? In the meantime, there are a large number of studies on this issue and it is not possible to report on all of them here. The tenor that emerges from all the literature is that cultural background plays an important role. This is best explained using a few examples.

Let us begin with a study that subsequently acquired a certain degree of political topicality, not to say controversy. Csukás et al. (2008) investigated a trust game with student subjects in each of four countries: Brazil, Greece, Hungary and Russia. First, the study examined the level of trust of the first movers (trustors), which was reflected in the amount of money given to the second movers (trustees). The higher the amount sent, the more likely the sender is to trust that the trustee will compensate him or her at the second stage of the game by sending something back. Second, it examined the extent to which the trustees behaved reciprocally and justified the trust placed in them by sending back a significant proportion of the money they received to the trustor.

Csukás et al. found that there are differences between the countries. One finding was that reciprocal behavior was very strong in Russia, much stronger than in the other three countries. However, the most important and meanwhile somewhat contentious finding is that Greek students differed significantly from non-Greek students. In Athens, much less was sent to the receiver than in other countries. The Greek trustors did not trust that they would get a lot back from their compatriots. As it turned out, this mistrust was justified, since the trustees sent back significantly smaller shares of the money received than the trustees in other countries. The trust game enables the subjects to consideraby increase the total payoff by trusting each other and proving to be trustworthy. The extent to which trust and reciprocity are expressed in the form of behavior makes it possible to realize this increase of the total payoff. Obviously, the capacity to do this was significantly weaker in Greece than in the other countries.

Studies in which cultural differences are investigated usually have to contend with methodological problems resulting from the need to ensure that experiments in different countries differ only in terms of the cultural background of the subjects involved. This is not trivial because two experiments taking place in two countries can also differ in other dimensions, for example frequently in the language spoken and possibly also in the currency in which the money is paid out. The language difference may necessitate the use of different experimenters who speak the national language, thus leading to a further difference. All these problems, which can render intercultural comparisons more difficult, did not occur during the cultural comparison experiment of Ockenfels and Weimann (1999) because it compared the behavior of eastern and western German students. Both groups spoke the same language and had the same currency. The experimenters were also the same. The only difference was the cultural background of the subjects. The eastern German students who took part in the experiments all grew up and were socialized in a socialist system (the experiments took place in 1995/96). The western Germans, on the other hand, grew up under capitalist conditions.

As already mentioned, two experiments were used for the intercultural comparison: a public good experiment played in groups each consisting of five subjects and a solidarity game experiment that ran as follows. Three subjects formed a group, but remained anonymous (double-blind design). By rolling a die, each subject had a 2/3 probability of winning an amount X. Before rolling the die, however, the subjects had to decide how much of their winnings they would give to the potential losers (if they won). They therefore had to state how much they would give to a loser and how much to two losers. Only then did the die decide whether a subject was one of the winners in the group or one of the losers.

The results of the East-West comparison were unequivocal and quite surprising. It turned out that the eastern German subjects were significantly less cooperative (in the public good experiment) and also gave much less to the losers in the solidarity game than the western German subjects. Obviously, the formative influence they had experienced in a socialist system did not have the effect one would expect. On the other hand, the experiment showed very clearly that market systems do not morally deform people. In fact, the opposite seems to be the case. In order for decentralized systems to work, a certain degree of willingness to cooperate and solidarity may be required.

Brosig-Koch et al. (2011) carried out an identical replication of the solidarity game experiment of Ockenfels and Weimann (1999). In contrast to the first experiment, it could no longer be said that the eastern German subjects of the second experiment had grown up under the conditions of a socialist system, because they were, on average, just over 2 years old at the time of the German reunification. Surprisingly, however, the finding was the same. The data were almost identical to those collected in 1995/96. This suggests that the cultural background of parents plays an important and enduring role in the development of social norms.

An important observation in the experiments of Ockenfels and Weimann (1999) and Brosig-Koch et al. (2011) was that the subjects in the solidarity experiment made very precise predictions about the average amount of money that the other subjects would give. The social norms governing gift behavior were evidently well known to all those involved and differed between eastern and western Germany. This finding was also obtained in a similar way in a completely different location. Goerg et al. (2016) carried out an experiment comparing cultures in a particularly volatile place. They conducted the trust game experiment with Palestinians and Israelis. The most important result was that both groups held very clear positions of trust and reciprocity and that these positions were very different. While the Palestinians were exceedingly trusting and gave large amounts to the trustee, the Israelis were considerably less trusting. The situation with regard to handing money back was similar. While the Palestinians returned large amounts, the Israelis were clearly more restrained. All these findings applied both to mixed pairings (Israeli plays Palestinian) and to the experiments in which the two groups were not mixed. Cultural differences of this kind can lead to considerable conflicts. For example, if an Israeli is in the role of trustor and a Palestinian is in the role of trustee, the Israeli will give little according to his "level of trust". The Palestinian may not interpret this as the "normal behavior of an Israeli", but as discrimination.

In the meantime, intercultural studies have made it clear that social norms can be assumed to develop very differently on a local level. It should be noted at this juncture that this is an important insight in two respects. First, it once again demonstrates that human decisions are also massively influenced by social norms – and not just by a rational, materialistic process of evaluation. Second, it again renders invalid the rational choice theory's claim to provide an explanation for all behavior. It appears that certain behavioral differences can often only be explained by resorting to local social norms. It is an important research question of the future for which type of decisions this is valid and for which the universal claim of the rational choice model can still be made. However, given all the cultural differences revealed by the experimental research conducted so far, it should be emphasized that these are mostly limited to differences in degree and only very rarely include qualitative differences. In other words, the *patterns* of human behavior demonstrated in laboratory experiments seem to be very similar across cultural borders.

Question
How do you judge the external validity of the experiments on cultural differences presented in this section?

2.4 Preferences, Payoffs and Beliefs

2.4.1 Risk Behavior in the Laboratory

In the context of the economic model of rational choice, a person's decision is basically understood as an act of choosing from a well-defined set of alternatives, taking into account the respective restrictions. The prerequisite for this is the existence of a preference, which transforms the alternatives available into an ordering. This preference ordering is represented by a utility function that assigns values to the elements of the set of alternatives according to their position in the preference ordering. In this way, three different types of choices are presented: first, the choice between alternative bundles of goods; second, the decision on when to consume (consumption today or in the future);

and thirdly, a choice between lotteries, i.e. between a number of different alternatives involving risk.

The first decision is taken on the basis of a preference ordering for bundles of goods, while the second decision presupposes the existence of a *rate of time preference*. This rate is used to place a valuation on current and future consumption. As a rule, the shift of consumption into the future means that it is necessary to temporarily forego consumption, which leads to a utility loss. Depending on the extent of this loss, the rate of time preference indicates how much higher future consumption must be in order to correspond to current consumption.

The third choice presupposes that the decision-maker has an idea of how he evaluates the *risk* associated with different lotteries. This is referred to as a risk preference. Based on expected utility theory, in economics three classes of risk preference are frequently distinguished. Risk neutrality occurs when a decision-maker is indifferent between choosing a lottery and a guaranteed payoff that corresponds exactly to the expected value of the lottery. In this sense, the decision-maker does not pay any attention to the risks associated with the lottery. Risk-averse decision-makers prefer the guaranteed payoff to the lottery with identical expected value because by doing so they can eliminate the risk. Risk-seekers, on the other hand, prefer the lottery to the guaranteed payoff because they value the chances of winning offered by the lottery.[14]

What does all this have to do with experimental research? In the laboratory, all three types of preferences (goods preference, time preference and risk preference) can in principle play a role. Using the induced value method, we have already explained how preference for goods is modeled. Time preferences rarely play a role in experiments because decisions in the laboratory usually have consequences immediately and not first at some time in the distant future. Therefore, knowledge of the relevant time preference is not usually so important for conducting an experimental study – unless it is specifically the focus of the investigation. The picture is entirely different when it comes to risk preference. In many situations, it is essential for the experimenter to know the risk attitude of the subjects. This is quite obvious when models on the risk attitudes of the actors are being tested. For example, if an auction model requires that bidders behave in a risk-neutral manner, and if the Nash equilibrium is based on this assumption, the model can only be tested in the laboratory with subjects who are actually risk-neutral. Testing the model with risk-averse subjects may mean that the Nash equilibrium is not observed, although the model might have been confirmed with risk-neutral subjects.

The question is, of course, whether risk preferences cannot be induced in a similar way to preferences for bundles of goods. Roth and Malouf (1979) describe a method offering a relatively easy way to do this. Let us assume that the payoff in an experiment does not consist of money but of lottery tickets for a binary lottery, i.e. a lottery in which there are only two possible payoffs (frequently a non-zero payoff with a probability of x

---

14 Strictly speaking, the second derivative of the utility function is decisive for the distinction between risk neutrality (second derivative equals zero), risk aversion (second derivative is negative) and risk seeking (second derivative is positive). There are also characterizations of risk preferences that are based on higher derivatives of the utility function, such as prudence (positive third derivative of the utility function, see Kimball 1990) and temperance (negative fourth derivative of the utility function, see Kimball 1992).

and a zero payoff with a probability of $1 - x$). Assuming strict self-interest and complete rationality, the subjects in this experiment should maximize the number of *expected* lottery tickets – which is nothing more than risk-neutral behavior. By simply restructuring the payoffs it does appear to be possible to induce risk neutrality. This would of course solve the problem that has just been discussed, because if risk neutrality can be induced, it is no longer necessary to determine the risk preference of the subjects.

Unfortunately, Selten et al. (1999) quite badly undermined the belief that this method works. They conducted experiments in which they varied the payoff mechanism by providing either money or tickets for a binary lottery. The reference point was the behavior that would have been expected in the event of risk neutrality. It was found, in fact, that risk-neutral behavior could not be induced with monetary payoffs. The lottery tickets, however, were not any better in this regard. On the contrary, the deviations from the risk-neutral decision were even greater than in the treatment with money. Harrison et al. (2013) made another attempt to save the cause. In their experiment, they were able to show that when very simple non-repeated games were considered, the use of lottery tickets actually led significantly more frequently to risk-neutral behavior. However, even though the rise in the rate was significant, it was only a very slight increase of just 14%.

It is highly debatable whether a similar result would emerge for repeated games, since Harrison et al. reported a finding in another paper (Harrison et al. 2015) that is a cause for concern in this respect. In order for risk neutrality to be induced, it is necessary to fulfill an axiom that says something about how individuals maximizing expected utility behave when dealing with a compound lottery, i.e. a lottery whose winnings are lottery tickets. The Reduction of Compound Lotteries Axiom (ROCL) says that little attention is paid to the fact that there are several lotteries in play. Ultimately, only the final payoffs weighted with the probabilities of the occurrence of the different lotteries are considered, i.e. the compound lottery is equivalent to the simple lottery obtained by combining the probabilities of the choice of a lottery and the corresponding payoffs.

Let us take an example. In a game, the respective probability of being able to win lotteries L^1 or L^2 is $\frac{1}{2}$. There is a $\frac{1}{3}$ probability in L^1 of winning 10 euros and a $\frac{2}{3}$ probability of 0 euros. L^2 pays 5 euros or one euro, each with a probability of ½. This results in the compound lottery

$$L^G = \left\{L^1, \tfrac{1}{2}; L^2, \tfrac{1}{2}\right\}$$

The ROCL states that a decision-maker is indifferent between this compound lottery and the lottery

$$L = \left\{0, \tfrac{1}{3}; 1, \tfrac{1}{4}; 5, \tfrac{1}{4}; 10, \tfrac{1}{6}\right\}.$$

L is obtained by multiplying the probabilities for the outcomes $\{0,1,5,10\}$ in both lotteries by the probabilities for the occurrence of the lotteries. The sum of the resulting probabilities is 1. The experiment by Harrison et al. (2015) shows that the ROCL does seem to hold in simple decision-making situations. However, if several decisions are to be made and at the end a random determination is carried out as to which of the decisions is relevant for payment, the ROCL is violated. This means that in experiments with

Table 2.3 Choices in the Holt-Laury method

Lottery A		Lottery B				
p($2.00)	*p($1.60)*	*p($3.85)*	*p($0.10)*	Expected value A	Expected value B	Difference
0.1	0.9	0.1	0.9	1.64	0.48	1.17
0.2	0.8	0.2	0.8	1.68	0.85	0.83
0.3	0.7	0.3	0.7	1.72	1.23	0.49
0.4	0.6	0.4	0.6	1.76	1.60	0.16
0.5	0.5	0.5	0.5	1.80	1.98	−0.17
0.6	0.4	0.6	0.4	1.84	2.23	−0.51
0.7	0.3	0.7	0.3	1.88	2.73	−0.84
0.8	0.2	0.8	0.2	1.92	3.10	−1.18
0.9	0.1	0.9	0.1	1.96	3.48	−1.52
1.0	0.0	1.0	0.0	2.00	3.85	−1.85

Based on Harrison and Rutström 2008, p. 46

more complex decision-making situations the theoretical justification for the assertion that binary lotteries induce risk neutrality no longer exists.

Given these findings, it is unlikely that we can get around revealing the risk preferences of the subjects. A whole series of methods for doing this have been intensively discussed in the literature. It should be noted that this literature is now so extensive that one could easily write a textbook on the question of risk preferences.[15] All we can do here is concentrate on the most important methods and attempt to sketch out their advantages and the problems associated with them.

The most widespread and best-known method is the multiple price list (MPL), which was used especially by Holt and Laury (2002) and is therefore also known as the Holt-Laury method. In this procedure, the subjects have to make a series of choices between two binary lotteries. Lottery A has payoffs that are relatively close to each other, for example $2.00 and $1.60.[16] Lottery B has more divergent payoffs, such as $3.85 and $0.10. The ten choices between the two lotteries differ in terms of the probabilities of the payoffs. Table 2.3 provides an example:

Up to and including the fourth choice, lottery **A** has a higher expected payoff than lottery **B**, i.e. a risk-neutral decision-maker should choose lottery **A** for the first four decisions. A risk-averse decision-maker will not switch over to lottery **B** at the fifth decision

15 For example, the contribution of Harrison and Rutström (2008) in volume 12 of "Research in Experimental Economics" contains about 150 pages on this topic alone.

16 The numbers and the Table 2.3 are taken from Harrison and Rutström (2008).

and possibly not even at decision six to a maximum of decision nine. After all, risk aversion means that a person is willing to accept a lower expected payoff if it reduces the risk. The line as of which the decision-maker's risk aversion crosses over from **A** to **B** provides information on the extent of the decision-maker's risk aversion (Harrison and Rutström 2008, p. 47). The Holt-Laury method is often used in conjunction with the random-lottery incentive system, which means that not all lines of ◘ Table 2.3 are played and paid off, but only a randomly selected one. This payoff mode does not change the incentive compatibility of the method. Given that a random move determines which line is played, the best answer is to make a choice that corresponds to the risk preference that actually exists, thus ruling out (in the case of rational behavior of the subjects) the possibility of strategic behavior, in which a different lottery is chosen instead of the one actually preferred. Nevertheless, the random-lottery incentive system has its pitfalls and has been the subject of intense methodological debate. This is due to the fact that it is not only used in connection with the Holt-Laury method, but is also generally considered to be a method for avoiding income effects, since only one of the decisions taken by the subjects actually leads to a payoff. This is a fine thing if it is guaranteed that this type of payment cannot lead to a change in the behavior of the subjects. But this is precisely the point of contention, and is why we will look at the method in more detail in the next section.

The Holt-Laury method has several advantages that explain why it is used relatively frequently. For one thing, it is easy to understand and easy to use. In addition, there are to a certain extent built-in checks that can be used to determine whether the subjects have understood the procedure. For example, a subject should not choose **A** in the last line – unless he prefers a safe \$2.00 to a safe \$3.60. Moreover, the decisions should be consistent. Having switched from **A** to **B**, those who have understood the procedure and who behave in line with expected utility theory should not change back again. Another advantage is that the method is incentive-compatible.

In addition to the Holt-Laury method, there are others that can always be used to identify risk preferences. We would like to present one of them in more detail, seeing as it is also very common and is not only used to uncover risk preferences. In the Becker-DeGroot-Marschak (BDM) method, which has been in existence since Becker et al. (1964), the subjects are required to participate in a lottery and state their "selling price", i.e. indicate the minimum price for which they are willing to sell the lottery ticket. The subjects are informed that a "purchase price" will be randomly selected from a relevant interval, for example, between the minimum and maximum payoff of the lottery. If the selling price is higher than the purchase price, the lottery is played; if it is lower, the lottery ticket is sold to the experimenter at the purchase price. As a result of the random draw, the method is incentive compatible. Given that the purchase price is independent of which selling price is chosen, the weakly dominant strategy in this game is to state the true valuation of the lottery as the selling price. These prices can then be used to draw conclusions about risk preferences. Thus, risk neutrality implies that the selling prices correspond to the expected payoffs, while risk aversion, that they are lower and risk seeking, that they are higher.

Question

Is it necessary to control risk preference even in experiments that do not involve the testing of a model?

Is it possible to answer this question in general terms, or is it necessary to examine it on a case-by-case basis?

The BDM method can be used quite generally to determine the willingness to pay for goods in an incentive-compatible manner. However, this presupposes that the subjects have understood that the weakly dominant strategy is to specify the true valuation as the price. Although meeting this requirement cannot be taken for granted, it is relatively easy to explain it using examples. If the instructions are formulated with due care, there should be no difficulties in understanding the BDM method, at least. Both the Holt-Laury and the BDM methods are simple to use and reveal the risk preferences of the subjects of experiments with comparatively high reliability. Many other methods discussed in the literature are variants of these two methods. Harrison and Rutström (2008) also mention three other methods which differ conceptually from the ones described above, but which are only briefly mentioned here because they have either weaknesses or no noticeable advantage over the Holt-Laury and the BDM methods.

The random-lottery incentive mechanism stems from Hey and Orme (1994) (Harrison and Rutström 2008, p. 50 ff). In this method, subjects are required to select one lottery from two lotteries on a recurring basis. The lotteries have fixed payoffs[17] and vary in their probabilities. The data thus obtained can then be used to estimate an expected utility function whose functional form provides information on the risk preference. In contrast to the methods discussed hitherto, it is not possible to draw a direct conclusion on the risk preferences – despite the comparatively high effort involved in the method. Binswanger (1980, 1981) proposed a method in which the subjects are presented with lotteries that realize a higher and a lower payoff, each with a probability of ½. They are arranged in such a way that not only the expected value increases, but also its variance. The subjects are required to select one of the lotteries, which is then carried out. This method is similar to that of Holt and Laury, but uses constant probabilities. Finally, Harrison and Rutström (2008) discuss the tradeoff method of Wakker and Deneffe (1996), which has the considerable disadvantage of not being incentive compatible.

Important

There is some evidence that risk preferences cannot be induced in the laboratory. However, since it is often necessary to control risk preferences, they need to be determined. A whole series of methods are available for this purpose, with the most commonly used being the Holt-Laury and BDM methods, both of which have the advantage of being easy to use and incentive compatible.

2.4.2 Selecting the Payoff Mechanism

As a rule, economic experiments use monetary payoffs to create incentives in the laboratory, which are assumed either to be effective in the model (to be tested) or to play a role in real decisions. This raises not only the question of how large the incentives should be, but also how they should be paid. In an experiment examining a person's one-off decision without interacting with other subjects, it is clear that the amount paid

17 In Hey and Orme, this is 0, 10, 20 and 30 pounds sterling with 100 decisions having to be made.

off is the result of the subject's decision. Even if the experiment involves a game in which n people are involved, but who only make a decision once, the amounts resulting from the n decisions are usually paid out.

2

Even in these simple cases, another mechanism called the "between-subject random-lottery incentive" can be used. Suppose a group of n people is involved in an experiment that involves a single decision and there is no strategic interaction between the people. The payoff rule could then stipulate that only $m < n$ of the n subjects will be paid off. The reason for such a mechanism could be the desire to conserve the limited experimental funds or to obtain as many observations as possible with the available funds. This is not possible, however, if the (average) size of the expected payoff to the subjects is still based on their opportunity costs (see ► Sect. 2.2.2). Moreover, the question arises as to whether the use of such a mechanism influences the result of the experiment or not.

This question becomes much more important when the subjects make several decisions. In the last section, we dealt with experiments that deal with this very issue. In order that information about the risk preference of the subjects can be obtained, they are usually required to perform several lottery comparisons. However, repeated decisions or several similar decisions are not an exclusive feature of experiments to reveal risk preferences. On the contrary, they can be found in many contexts.

At first glance, one might think that in such cases it is the gold standard to pay off all the decisions of all the subjects. Whether this standard is achieved solely depends on the funds available. But this point of view is wrong because the "pay-each-task" payoff method is only acceptable if it is ensured that the subjects of this method treat each individual decision as if they *only had to make that one decision*, thus making it necessary to examine each decision in isolation. However, there are good reasons for believing that in many cases this just cannot be guaranteed. Two effects can prevent this isolation hypothesis from being fulfilled.

First, *income effects* can lead to decisions later in the experiment taking place under conditions that differ from those prevailing at the time of earlier decisions. If every decision is paid off individually, a subject can calculate how much he or she has already earned. The following example shows that this could have a major impact on decision-making behavior. Suppose a subject is aiming to earn at the very least the opportunity cost of participating in the experiment. Suppose further that this is an experiment in which social preferences could play a role. Then it is quite plausible that the subject may behave differently as soon as the opportunity costs are covered. This means, however, that the isolation hypothesis is not fulfilled because how the subject acts when making a decision depends on what happened in the previous decisions.

The second effect that is capable of violating isolation is the *portfolio effect*. This means that in the case of decision under risk, the combined effect of decisions can lead to different results than if all individual decisions are taken separately. Take as an example the two-stage choice between two lotteries **A** and **B**, with the former being less risky than the latter. A risk-averse decision-maker would choose (**A**, **A**) for *isolated* decisions while a risk-seeker would choose (**B**, **B**). However, if the decision-maker can form a portfolio of both lotteries, it is possible that (**A**, **B**) has a higher expected utility than (**A**, **A**) and the risk-averse decision-maker therefore prefers (**A**, **B**) (Cox et al. 2015).

Wealth effects and portfolio effects can occur in repeated decisions in many cases and should therefore be eliminated by appropriately choosing the payoff mechanism. But what is an appropriate choice? Cox et al. (2015) list a number of mechanisms that could be applied here: "pay all decisions sequentially"; "'pay all decisions correlated' at the end of the experiment"; "'pay one decision randomly' at the end of the experiment"; "'pay 1/n (of the amount) of all decisions correlated' at the end of the experiment, where n is the number of tasks"; "'pay all decisions independently' at the end of the experiment"; and a "'one task' design in which each subject makes only one decision (and is paid for the outcome)".

The random-lottery incentive mechanism (RLM) is extremely popular and can be used either between-subject (not all subjects are paid off) or within-subject (all subjects are paid off, but not all decisions). Originally, this payoff mechanism was used solely to prevent wealth effects and portfolio effects. The fact that it is better value for money due to it being possible to obtain more observations for the amount of money indicated by each choice was rather a fortunate by-product (Harrison and Rutström 2008 p. 116). To begin with, however, it is an unresolved issue as to what effects RLM has. Is the isolation assumption still valid? And what impact does it have if each choice possibly has only a very small probability of being relevant in terms of payoff? The between-subject variant could be seen as particularly critical, since it implies that a whole series of subjects end up having to go home without any payoffs. Are the incentives really still sufficiently strong? Moreover, what effect does such a payoff system have on the profitability of the subjects? One has to bear in mind that there is a conflict of objectives when it comes to designing this payoff method. For example, if only one subject out of 30 is paid off and this person only receives payment for one decision out of 100, the probability that a concrete decision will actually be paid off is only 1/3000 = 0.00033. Of course, the payoff per decision can be set very high. Let us suppose that the duration of the experiment is 1 hour and the opportunity cost per subject is €10. Then the average payoff should be €10, i.e. the lucky player who is selected will, or can potentially, receive €300 per decision without the budget of the experiment being exceeded. The prospect of having a 1/30 = 0.033 chance of earning an impressive 300 euros in 1 hour is perhaps not bad and can certainly motivate people to participate in the experiment. However, it is likely that this will be susceptible to selection bias since such a risky payoff mechanism is particularly appealing for risk-seeking students. Risk-averse people will tend to avoid the laboratory under the conditions of such a RLM.

In comparison, the within-subject variant of the RLM is relatively unproblematic. A possible disadvantage could be that the significance of an individual decision decreases, because the probability of an individual decision having an impact on the payoff is only $1/N$ if one in N decisions is to be paid off. This effect can be counteracted by paying off a randomly selected round N times. In this way, wealth or portfolio effects can be prevented without reducing the incentive for the subject to make an effort in each round. However, these advantages come at a cost, as this procedure is just as expensive as paying off each round. Of course, cost-cutting measures are possible. For example, the randomly selected round could be paid off not N times, but only M times, where $1 \leq M \leq N$. Depending on the size of M, there is either a stronger "saving effect" or a greater "incentive effect" (also see the comments in ▶ Sect. 2.2.2).

2.4.3 Eliciting Beliefs

The great advantage of the experimental method is that it allows decision processes to be observed under controlled conditions. By systematically changing individual parameters in the experimental treatments, we obtain behavioral data that provide information on how the conditions under which choices are made influence the behavior of the subjects in the experiment. There is admittedly one constraint we need to accept. The behavior we observe is the result of individual calculations (perfectly rational or boundedly rational) in which two factors that we cannot directly observe play an important role: the *preferences* and the *beliefs* of the subjects. It may not be possible to deduce, from the behavioral data, what contribution these two things made to the decision. A simple example (Manski 2002 and Schotter and Trevino 2014) may help to illustrate this. In an ultimatum game, the first mover (proposer) must decide how much to offer to the second mover (responder). Let us suppose that we observe that a proposal is made to divide the "pie" into equal parts. An obvious conclusion might be that the proposer has strong "social" preferences and therefore proposes a fair split. However, it could just as well be that the proposer *believes* the responder will react very aggressively and reject an unfair offer. If the proposer is sufficiently risk averse, the *belief* that the responder will behave in such a way can prompt him to propose 50:50. He can be very confident that this proposal will be accepted.[18] In this case, it may well be that the proposer does not attach the slightest importance to fairness, but his belief nevertheless leads him to submit a "fair" offer.

This example shows that there may well be situations in which it would be advantageous to know what the subjects' beliefs are. Of course, skillfully varying the design also enables us to gain an idea of the role these beliefs play. For instance, the ultimatum game can be converted to a dictator game by depriving the responder of the possibility of rejecting the proposer's offer. This means the latter does not need to form a belief about what the former will do. The difference between the behavior in the dictator game and in the ultimatum game then allows us to estimate the extent to which the belief that the responder will behave aggressively influences the action of the proposer (see e.g. Forsythe et al. 1994).

This method is not perfect, however. It presupposes that the subjects perceive the two games more or less equally. In particular, the respective design must not lead to the subjects developing different conceptions of which behavior might be "appropriate" or "socially desirable". Therefore, even if variations in design can be used to reveal the role of beliefs, eliciting these beliefs may still be interesting.

When considering the possibility of eliciting beliefs, two important questions arise: first, how best to do this, and second, does eliciting have any effect on the actions of subjects? If the answer to the second question is affirmative, then there is a problem. If the act of eliciting alters subjects' behavior, not even having successfully elicited the beliefs will assist in knowing what beliefs would have formed the basis of a decision had there not been any elicitation of the beliefs. At first sight, there is a straightforward solution to this problem, and that is by eliciting the beliefs *after* the subjects have made their decision. This does have some drawbacks, however. For example, there is uncertainty as to whether the beliefs will not then be adjusted retrospectively. It could well be the case

18 Indeed, such proposals are almost always accepted. See e.g. Güth and Kocher (2014).

that it is not the beliefs that are then the basis of the decision, but rather that the decision that has already been taken determines what beliefs are reported when subsequently elicited. This risk is, of course, diminished if there are monetary incentives to state the true beliefs. This brings us back to the question of *how* to elicit beliefs.

In principle, it is possible to elicit subjects' beliefs in the presence or absence of monetary incentives. If incentives are employed, a so-called scoring rule is frequently used. This is a payoff mechanism that depends, on the one hand, on the beliefs a subject reports and, on the other hand, on their true beliefs. Following Schotter and Trevino (2014), we consider by way of illustration a binary random variable, which can assume either the value A or the complement A^c. Let p be the probability with which subject i expects A to occur, and let r be the probability the subject reports. The scoring rule then consists of a lottery whose payments S_A and S_{Ac} are realized from the point of view of the subject with the probabilities p and $(1-p)$, respectively, and whose value depends on the reported probability r:

$$L_{A,\,Ac} = pS_A(r) + (1-p)S_{Ac}(r)$$

Scoring rules which are suitable are those in which the value of the lottery is at a maximum exactly when the subject chooses $p = r$, i.e. when the subject reports the probability that he actually assumes to be the true one. Scoring rules with this property are referred to as proper. Perhaps the best-known proper scoring rule is the quadratic scoring rule, abbreviated QSR, the payoffs of which are defined as follows:

$$S_i(r) = \alpha - \beta \sum_{k=1}^{n} (I_k - r_k)^2,$$

where n is the number of possible events that *can* occur (but only one of which actually occurs) and I_k is an indicator function that takes the value of 1 if event k occurs and 0 otherwise. Each mistake made by the subject is punished with a "penalty" of

$$-\beta(I_k - r_k)^2$$

The quadratic ensures that the penalty is always negative. Someone who knows for certain that event i will occur would choose $r_i = 1$ and $r_j = 0$ (for all $i \neq j$). The QSR possesses the property of being proper as long as the subjects are risk neutral. If the subjects are risk averse, the scoring rule must be adjusted. There are several ways of doing this. Offerman et al. (2009) present a comparatively elaborate and complicated approach. First of all, the subjects have to undergo a process of testing to determine their risk preferences more precisely and then specific individual adjustments are made to the respective scoring rule. It is easier to use a stochastic scoring rule in which the payoffs $S(r)$ are not amounts of money but lotteries. How this works is explained using, as an example, the Becker-DeGroot-Marschak method, with which not only the willingness to pay can be determined, but also probability estimates.[19]

19 Schotter and Trevino (2014), p. 107, also see Holt and Smith (2009) and Karni (2009).

Holt and Smith (2009) use the following design. Two urns, A and B, are filled with red and black marbles. One-third of A's marbles are red and two-thirds are black, while B's ratio is the reverse. Marbles are drawn from one of the urns (with replacement) and the subjects are to state what they believe to be the probability of the draw coming from urn A. For this purpose, a number R between 0 and 100 must be stated. Then a number t is randomly drawn from the same interval. If $R \geq t$ the subject receives the amount V, provided it was actually urn A, and 0 otherwise. In the case of $R < t$, a lottery is played in which the probability of winning V is $t/100$ and the probability of winning 0 is $(1- t/100)$. Suppose that the subject assumes with a probability of p^* that it is urn A. It is easy to see that the best strategy is to choose $R = p^*$. Selecting a smaller value could result in $R < t < p^*$ being drawn. However, this would mean that the lottery is being played with a smaller probability of winning the prize V (from the point of view of the subject) than it would have if $R = p^*$ were chosen. It is also not an advantage to report an $R > p^*$. As a result, this procedure is indeed proper, irrespective of the risk attitude of the subjects.

It is therefore possible to elicit the beliefs of subjects in an incentive-compatible manner – irrespective of their risk preference. But is this the reason why this should be done? After all, such an elicitation requires some effort. Do we even need a procedure that provides incentives? Would it not suffice to simply ascertain the beliefs? And if the beliefs are elicited, what can be expected of the answers in terms of quality? In addition, there is still the question of whether eliciting beliefs does not in fact alter behavior.

With regard to the last point, there is some evidence indicating that eliciting beliefs beforehand does not lead to a change in behavior.[20] However, Schotter and Trevino (2014) find that eliciting beliefs can result in the subjects learning the game more quickly. If an experiment is repeated, in the presence of elicitation of beliefs it may be possible to observe patterns of behavior occurring at an early stage that otherwise would only occur later in the course of the experiment.

The question remains whether the effort needed to use sophisticated methods for eliciting beliefs justifies their use. Trautmann and van de Kuilen (2015) addressed this question and subjected a number of methods to a test. They used a variant of the ultimatum game in which €20 could be split by the proposer, who then had to choose one of only six different divisions of this amount: [(20, 0); (16, 4); (12, 8); (8, 12); (4, 16); (0, 20)]. The proposer's assessment of the probability of acceptance of the various offers was determined. The responders were asked to estimate the probability of the proposer choosing the individual splits.

The study tested how well incentive-compatible methods compare to simple elicitation, which the authors call "introspection" because the purpose is merely to put oneself in the other person's position and not to earn money with as accurate an estimate as possible. A total of six different methods were tested with regard to their internal and external validity. The external validity described how well the subjects were able to estimate the true probabilities (the empirical frequencies). The internal validity was determined using two measures. On the one hand, the consistency of the individual

20 Nyarko and Schotter (2002), Costa-Gomes and Weizäcker (2008), Ivanov (2011) and Schotter and Trevino (2014).

behavior was checked against the elicited beliefs; on the other hand, the additivity of the elicited probabilities was tested. For example, when the responders indicate how likely they think each of the six possible proposals will be chosen, these probabilities should add up to 1. In addition, both the probability of the proposer choosing (12, 8) and the probability of the proposer not choosing (12, 8) were elicited. Proposers were also asked about their beliefs concerning the probabilities of acceptance and rejection of this division. Here too, the answers should add up to 1.

The methods up against the introspection method were the quadratic scoring rule (QSR), a QSR corrected for risk aversion, a probability matching method similar to the Becker-DeGroot-Marschak method, and a method called outcome matching. The purpose of the last effort was to elicit the certainty equivalent (CE) for a lottery that pays a certain amount when a certain event occurs and zero otherwise. For example, this event could consist of a responder accepting the proposal (12, 8). The lottery pays, for example, 15 euros if this happens and 0 otherwise. Provided that the decision-maker is risk neutral and expects that there is a probability p of the proposal being accepted, the certainty equivalent of the resulting lottery is identical to its expected value: $15p + (1-p)0 = 15p = CE$ and thus $p = CE/15$. The probability p assumed by the decision-maker can thus be calculated directly by eliciting the certainty equivalent. This was done by comparing the lottery with ascending amounts that are paid out instead of the lottery. The amount as of which the sure amount is preferred is the certainty equivalent.

The last method to enter the contest was one in which outcome matching is corrected in such a way that it is also applicable for risk-averse decision-makers (i.e. the property of properness is retained).

The actual experiment consisted of three stages. In the first stage, an ultimatum game was played using the full strategy method[21]; in the second stage, the expectations of the subjects were elicited; and, in the last stage, the risk attitudes were measured with a simple lottery choice. The subjects did not receive any feedback after the first two stages, and payoffs were made after the third stage by randomly selecting one of the three stages and *one* of the decisions made there. This procedure was used to prevent something that could arise when subjects are required to make decisions about their own behavior and also to state their expectations. If both are rewarded, the subjects may use this fact to obtain a sure payoff by means of hedging. Blanco et al. (2010) illustrate this strategy in a simple 2×2 coordination game: two subjects have the choice between *A and B*. If both subjects choose the same, there is a payoff of x, otherwise 0. At the same time they should state their expectation about the other subject's choice and, if they are correct, they also receive x. If a subject chooses A and says he expects B, he can secure the payoff of x. However, he is then no longer reporting his true beliefs. This form of hedging can always occur when the subjects are paid for indicating their beliefs. Therefore, if monetary incentives are used, the payoff mechanism should ensure that hedging is avoided.

---

21 The full strategy method will be discussed in ► Sect. 2.7.1. In this method, subjects specify complete strategies instead of reacting to a specific move made by the other subject. In the ultimatum game, for example, responders must specify as of which proposal of the proposer they are willing to accept the proposal.

The findings of Trautmann and van de Kuilen are a little sobering. Let us start with internal validity. If the proposers are asked to specify both the probabilities of acceptance and rejection for the proposal (12, 8), the two probabilities almost always add up to values greater than 100%. The same applies to responders when asked about the probability of this proposal and the probability against it. Regardless of the method used to elicit beliefs, the sums are in the region of 105%. In the case of the proposers, all the mean deviations of 100% are significant at the 5% level. Given that the two probabilities were directly elicited together, this observation is astonishing. The sum of the probabilities is, however, at least close to 100%. The situation is entirely different for the sum of the six probabilities that the proposers state for the six possible splits of the proposal, where the average sum is more than 200%! And no significant differences can be found between the six methods tested, i.e. the incentivized elicitation methods do not perform better than introspection. On the contrary, the difference between the sum of the six probabilities and 100% is smallest when a simple survey is made without financial incentives.

The second measure of internal validity is the consistency of the choices with regard to expectations. In this respect, introspection performs significantly worse than incentive-based methods. However, there are no significant differences between the methods that use incentives.

The last measure to be applied is external validity. This is determined by comparing the reported beliefs of the subjects with the relative choice frequencies actually measured. The sum of the average squared deviations (Brier Score) is used as a measure. The data show that there are no significant differences in the predictive accuracy between the different elicitation methods. Simple elicitation, which does without any monetary incentive, performs just as well as the complex methods, such as the QSR or the corrected QSR.

Against this background, it is a moot point whether the financial and logistical efforts associated with the use of elicitation methods with financial incentives are worthwhile. All in all, there are comparatively few advantages to be gained. Whether it really is worth the effort can only be decided in individual cases. In general, however, the accuracy with which beliefs are determined could depend on what the beliefs formed pertain to. When it comes to predicting how other subjects will act and if their actions do not have much influence on one's own decision, incentives can be helpful in encouraging the subjects to make an effort to form beliefs. If, on the other hand, the actions of others are important for one's own decision, then there may be less need for external motivation.

The results of Trautmann & van de Kuilen differ in this regard from earlier findings in which it had been shown that if the formation of beliefs only concerned an individual decision, no significant distortions due to simple elicitation were detected, i.e. the introspection method worked well in these cases (Offerman and Sonnemans 2001). However, this could not be confirmed in situations where the subjects were involved in strategic interactions with others, i.e. monetary incentives helped to improve the formation of beliefs (Vieider 2011). Trautmann & van de Kuilen disagree, because in their ultimatum game the predictive accuracy of the introspection method is just as good as that of the incentivized methods. Obviously, the absence of monetary incentives in experiments with strategic interactions *can* be a problem, but it *need* not be. All this can be summarized in the following recommendation:

Important
If eliciting beliefs is an important element of the experiment and if the formation of beliefs takes place via a complex process, then it is advisable – to be on the safe side – to choose an incentivized method of eliciting beliefs. Nevertheless, in many cases simple elicitation without monetary incentives should be sufficient. Risk aversion of the subjects is in principle a problem, but the experimental findings indicate that this does not play a particular role in quantitative terms when eliciting beliefs.

2.5 The Influence of the Experimenter

People interact in an experiment, with two forms of interaction being distinguishable: vertical interaction between the experimenter and the subject(s) and horizontal interaction between the subjects. The first interaction is unavoidable, while the second may or may not exist, depending on the design. This section will focus on the interaction between the experimenter and the subjects. The effects and ramifications that need to be considered in horizontal interaction will be discussed in more detail in ► Sect. 2.6.

The experimenter influences what happens in an experiment through different channels. Some are obvious, such as the instructions given to the subjects by the experimenter, or the exercises used to test whether the subjects have understood the experiment. Others are less obvious, but just as important. Thus, the experimenter can consciously or unconsciously exert social pressure or certain expectations can be generated in the subjects as to the purpose of the experiment and what behavior is now expected of them. The frame of the experiment also plays a role, i.e. the question of which "story" to use to present the task the subjects are to face.

Regardless of which channel the experimenter is currently using, it is important that it is done consciously and that the influence on the subjects is such that it is in line with the objective of the experiment. Therefore, in the following the channels are considered individually in order to enable them to be used in a deliberate and purposeful manner.

2.5.1 The Experimenter Demand Effect

In laboratory experiments the interaction between the experimenter and the subject is inevitable (even if it is through the design developed by the experimenter). It cannot therefore be a question of avoiding any kind of interaction, but rather of designing it in such a way that it does not lead to any distorting influence on the behavior of the subjects (experimenter demand effect), thereby curtailing the interpretability of the data obtained.

Zizzo published an article in 2010 in which he investigated the sources and impacts of experimental demand effects in a very structured and thoughtful manner. Much of what follows is based on that paper, although we do not share every conclusion that Zizzo drew.

What is the typical experimental situation in which experimenter demand effects can take place? As a rule, it is a laboratory experiment carried out by a scientist with students, thus creating a hierarchical situation. This is not only because the experimenter has a higher standing on the rankings of scientific qualifications, but above all because the person conducting the experiment has the status of an expert in relation to the experiment, while the subjects are, so to speak, lay persons. Zizzo points out that the subjects come to the laboratory with a desire to understand what is involved and what they should do. It is plausible to assume that subjects think in this way, although there may be subjects who have an other or no specific expectation at all. The experimenter is the expert from whom the subjects learn what is being played, and, consequently, what their roles and tasks in the experiment are. As a result, any cue emanating from the experimenter can be interpreted as information about what is at stake. The *explicit* cues in the instructions play just as much a role as the *implicit* cues that the experimenter issues unconsciously or "by mistake".

Zizzo distinguishes between *cognitive experimenter demand effects* and experimenter demand effects caused by *social pressure*. The former occur because the experimenter has to explain the experiment to the subjects. Understanding this explanation is a cognitive process and it may well happen that *how* it is explained leads to it being understood in a particular way, for example what is *appropriate* behavior in the experimental situation. Experimenters should be aware of the fact that subjects may take every word seriously and, therefore, that every word used by the experimenter should be carefully considered. For instance, if examples are used to illustrate the payoff function or other elements of the experiment, anchoring effects may occur. The numbers used in the examples must therefore be carefully chosen. Moreover, cognitive experimenter demand effects also lurk in places where they are not immediately suspected. For instance, in some experiments great importance is attached to assuring the subjects that they will decide completely anonymously. In double-blind designs, it is furthermore ensured that even the experimenter cannot observe what the individual subject is doing. It is obvious that the subjects ask themselves why so much importance is attached to anonymity. The answer could be that it is presumably because it facilitates selfish behavior. So it is this behavior that the experimenter probably wants to create in his experiment?![22]

A very suitable candidate for demonstrating potential experimenter demand effects is the dictator game experiment. We have already dealt with this experiment elsewhere and have seen that the results in dictator game experiments can depend very much on individual elements of design. There is some evidence to suggest that this high variability of the results is connected with fact that experimenter demand effects have a particularly strong influence in this experiment. This may already apply to the basic design of the game, in which a subject (the dictator) receives a sum of money and is informed that there is another subject next door to whom any share of the money just received can, but does not have to, be given. The potential experimenter demand effect in this situation is obvious. With the design and the instructions, the experimenter draws the subjects' attention to the fact that this is an experiment in which the willingness to give something is tested. Giving is the thing to do, because it seems to be what the experimenter wants to see. This would explain the generally high amounts that are given in

22 See Zizzo (2010), p. 83, footnote 11.

such experiments and stands in strong contrast to the observation that people in the real world almost never hand out gifts to strangers for no reason whatsoever.

The fact that the amounts allocated in the dictator game experiment can possibly be attributed to an experimenter demand effect, at least to a certain extent, sheds a new light on experimental setups in which these allocations decrease. Particularly remarkable are the already mentioned results of Cherry and Shogren (2008), which show that the allocations drop very sharply when the dictators have to work for the money they can share. The experimenter demand effect becomes particularly evident in the variation of the experimental design undertaken by List (2007) and Bardsley (2008), who allowed the dictators in their experiments not only to give money to but also to take it away from the other subject. As a result, there were practically no more allocations to the receiver. Quite the opposite, in fact, the dictators took money from the receivers. The experimenter demand effect interpretation of this finding goes as follows. By providing the dictators with the opportunity to take something away from the second subject, the experimenters changed the interpretation of the experiment completely. It was no longer about "giving" and the question of how generous one acts in this situation, but about "taking" and the question of how much restraint one shows.

Zizzo identifies another important source of potential cognitive experimenter demand effects. This is referred to as the strategy method. As announced earlier, we will discuss this method in more detail in ▶ Sect. 2.7.1. It should nevertheless be noted that using this method could certainly have an influence on how the game is perceived by the subjects. For example, the strategy method forces the subject to look at practically the entire strategy space of the other subjects, because for every possible choice from this space it is necessary to give some indication of what the response will be. This can lead to the other subjects' actions becoming the focus of attention much more than is the case without the strategy method.[23]

In addition to cognitive experimenter demand effects, in which the understanding of the experiment is influenced by the way in which the experiment is explained, undesirable manipulation of the subjects may also result from social pressure, which can arise both between the subjects and vertically from the experimenter. There are many reasons why people succumb to social pressure. For example, a role may be played by the desire for *conformity*, or by *social acceptance*, which is experienced when acting in accordance with a social norm.[24] It is quite possible that there are also subjects who attach great importance to being *nonconformist* and therefore oppose any social pressure. While it may not be too bold a hypothesis to suggest that nonconformists are rare, the widespread desire to conform to social norms is well known.

Experimenters can create different forms of social pressure and, as is the case with cognitive experimenter effects, this can happen consciously or unconsciously. A very direct form of social pressure arises when the experiment is conducted by an

---

23 In Brosig et al. (2004), this aspect of the strategy method is used, among other things, to explain the differences between "hot" (without strategy method) and "cold" (with strategy method) experiments. See ▶ Sect. 2.7.1.

24 See, for example, the work of Krupka and Weber (2013), which shows that allocations in the dictator game can be interpreted as the price to pay for being able to be in accordance with a social norm.

experimenter who is perceived outside the laboratory solely in the role of the teacher, and who may also be a successful researcher. Imagine a student who attends a professor's lectures and takes her exams. Perhaps this student has also read one or two of the professor's research papers published in international journals. She has come to know and appreciate her academic teacher as a competent person to be respected and now meets her again in the laboratory, where her professor is now explaining the experiment to her and all the other subjects and in the process behaves completely neutrally, i.e. she does not reveal through any facial expressions or verbal cues that she has any particular expectation of the subjects. Nevertheless, our student subject might feel under some pressure. She may not want to disappoint the professor, and even if a double-blind design ensures that the professor cannot observe what the subjects are doing, the desire could encourage her to make the experiment successful for her professor. The fact that social pressure actually arises in this situation becomes clear when one compares this situation with that in which the experiment is carried out by a young employee whom our student has never seen before. There is hardly any reason for our student to do this person a favor or to seek to gain recognition.

Brañas-Garza (2007) used an experiment to study the effects of a professor or an associate conducting an experiment. He used a dictator game that was carried out in two variants, one with a neutral description of the game and the other with the same description, but supplemented by the sentence "*Note that your recipient relies on you*", which was written in capital letters under the instructions. Brañas-Graza describes the two instructions as different frames for one and the same game and does not explicitly mention the experimenter demand effect associated with these frames in his paper. Zizzo (2010) suspects – and we agree with this assumption – that a strong experimenter demand effect is associated with the final finding. In both treatments, the experiment was carried out once in the classroom by Brañas-Graza himself and once in the laboratory by an employee. In both cases, the supplementary information resulted in the dictators becoming significantly more generous. The effect was, however, much stronger in the case of the professor than in the case of the employee. Although a classroom experiment cannot be directly compared to a laboratory experiment,[25] the effect is in the direction assumed and at least does not contradict the hypothesis that the authority of an experimenter or his position in the academic hierarchy is likely to be a significant factor in the strength of a possible experimenter demand effect. We will return to the Brañas-Graza experiment Brañas-Garza (2007) when we talk about the effect of frames (▶ Sect. 2.5.3).

Zizzo (2010) also makes use of some examples to show that it is sometimes difficult to separate the impact of an experimenter demand effect from other effects. Suppose an experiment is testing a relatively complicated game-theoretical model, for example a public good experiment with a non-linear production function and an interior solution. The experimenter assumes that not all the subjects will able to determine the game equilibrium ad hoc. Apart from other things, the ability to solve the game depends on the subjects' previous knowledge of game theory, and this knowledge varies considerably.

---

25 This is also pointed out by Zizzo (2010).

In this case, it is clear that the experiment not only tests the game-theoretical model, but also the subjects' ability to calculate the Nash equilibrium. An obvious solution to this problem could be to explain to the subjects before the experiment what the equilibrium looks like, that this equilibrium is inefficient and what the efficient solution would look like. With this information, the subjects should then be capable of making a choice between the individually rational solution and the collectively rational solution in the experiment.

At first sight, the experimenter has merely ensured that the subjects have a better understanding of the game-theoretical model to be tested. It cannot be ruled out, though, that this may also have triggered an experimenter demand effect. By explicitly pointing out that the game equilibrium is not efficient, it could give the subjects the impression that it is *socially desirable* to deviate from the equilibrium. What is more, this impression is created in a hierarchical relationship. The experimenter approaches the subjects as an expert and "explains" the game. It is therefore not clear whether the subjects' behavior is due to their increased understanding of the game or to the fact that they have followed the presumed instructions of the experimenter.

This example shows an interesting conflict of goals. On the one hand, the fact that the subjects are likely to be overwhelmed with the understanding of the experiment is a real problem. After all, a successful experiment requires that the participants in the experiment know exactly what the consequences of their decisions are. On the other hand, the attempt to solve this problem leads to a new problem, a potential experimenter demand effect. We will come back to this point and the example later.

The types of experimenter demand effects based on social pressure that we have discussed so far are rather subtle. There are, of course, also some that are much more direct. The instructions that the subjects receive at the beginning of an experiment are ideally suited to creating massive experimenter demand effects. The language used, for example, is suspected of doing this. It is possible to describe things in an emphatically neutral way or to "load" them with valuations to a greater or lesser extent. However, closer inspection reveals that it is difficult to distinguish between the different effects. Two examples may serve to clarify this point.

Liberman et al. (2004) report on two public good experiments, which were identical except for the names of the games provided to the subjects. One was a "Community Game" and the other was a "Wall Street Game". The names actually had a huge influence on the results, with much more cooperation in the Community Game than in the Wall Street Game. In the experiment by Burnham et al. (2000), too, altering only one word triggered substantial effects. In their experiment, two players could significantly increase their payoffs compared to the equilibrium payoff if player 1 trusted player 2 and player 2 acted reciprocally, thus vindicating the trust. In the first treatment, the other player was called the "partner", while in the second treatment the word "opponent" was used. The word "partner" led to significantly more trust and trustworthiness at the beginning of the experiment. Admittedly, both declined in later rounds.[26]

---

26 See also Abbink and Hennig-Schmidt (2006).

The decisive question in both cases is what effect is actually present. Is it a particular value judgment associated with the respective terms, or is it an experimenter demand effect? In the latter case, when a game is called "Wall Street Game", the subjects might have the feeling that the experimenter wants to test how well they can assert themselves. If the game is called "Community Game", the experimenter might want to know how well the subjects perform as social beings. If the other player is called a partner, the experimenter apparently wants to test the ability to cooperate. If, on the other hand, the other player is designated an "opponent", then competition is evidently at issue and it is a matter of asserting oneself.

Abbink and Hennig-Schmidt (2006) tested the effect of non-neutral language in the instructions by conducting an experiment dealing with corruption. In the non-neutral variant, there was a *company* that could provide a *state official* with a private payment. The official could then either *issue a permit*, or *not issue a permit*. In the neutral version of the instructions, however, there were only *player 1* and *player 2*, a *transfer* and the option to choose *Y* or *X*. So while the non-neutral description plainly made it clear that bribery was the issue, the neutral version was much more abstract. It turned out that the different descriptions of the situation had no influence on the result. A possible explanation for this could be that the experimenter demand effect in this experiment is relatively small. In both variants of the experiment, it may not be entirely clear to the subjects what the experimenter's expectations are or what kind of behavior the experimenters want to see. This would suggest that the impact is so strong in the two earlier examples because it is very obvious there what the experimenter expects. It follows that when formulating instructions, it is less a matter of influencing behavior by means of a certain frame than of avoiding massive experimental effects that are generated by the selection of a certain frame.

Social pressure can also be established very explicitly in the instructions. Binmore et al. (1985) provided a prime example of this. They wrote in their instructions:

> *How do we want you to play? YOU WILL BE DOING US A FAVOUR IF YOU SIMPLY SET OUT TO MAXIMIZE YOUR WINNINGS.*

In order to appreciate how problematic this instruction is, it is necessary to shed some light on the background of the paper. It dealt with the fundamental question of whether non-cooperative game theory is suitable for predicting human behavior or not. Güth et al. (1982) had triggered this debate with their paper on the ultimatum game, in which they show that the subjects on average did *not* play the non-cooperative equilibrium. Binmore et al. sought to provide evidence to the contrary and to prove that subjects do indeed behave as game theory predicts. This is exactly what they ask their subjects to do. It should be clear that this is an experimenter demand effect that distorts the interpretability of the experimental results.

Important

A basic distinction is made between *cognitive experimenter demand effects* and those caused by *social pressure*. Both types have several channels through which they can exert their influence. Cognitive experimenter demand effects arise mainly because subjects try to determine what the experiment is about and what the appropriate behavior is from the information they receive from the team conducting the

experiment. Not only the information conveyed by the instructions plays a role here, but also other cues that may be sent unconsciously to the subjects.

Social pressure can arise because there is a natural divide between experimental subjects and experimenters that stems from the fact that experimenters are experts in the experiment and the experimental subjects are laypeople. In addition, the leader of the experiment is often at a higher level on the ladder of academic qualifications.

The question arises whether the existence of an experimenter demand effect invariably results in a restricted interpretation of the experimental findings. Zizzo (2010) shows very clearly and plausibly that this does not always have to be the case. It depends on the direction in which the experimenter demand effect acts and in which direction the effect that is to be experimentally investigated goes (hereinafter referred to as the experimental effect). Zizzo identifies three cases. The experimenter demand effect can be orthogonal to, exactly opposite to or in the same direction as the experimental effect.

The aforementioned example of Binmore et al. (1985) is a fine example of the experimenter demand effect and the experimental effect acting in the same direction. The experiment was intended to show that subjects behave rationally in the game-theoretical sense and this is the very thing the experimenter demand effect requires of them. This is the classic case in which the experimenter demand effect can be highly problematic. The reason is simple. If it turns out that the players are actually maximizing their payoff (which was actually the case), it is not clear whether they are doing so because it corresponds to their very own wishes and preferences or because they have been vehemently urged to do so by the experimenter.

Hoffman et al. (1996) point to another example in which the experimenter demand effect acts in the same direction as the experimental effect – but without actually referring to an experimenter demand effect. They investigate the impact on giving behavior in dictator games of the more or less great "social distance" between the subjects and the experimenter. We will return to this work in the next section when dealing with double-blind designs, since Hoffman et al. (1996) were the first to look at them in economic experiments. In their work, however, they not only tested double-blind arrangements, but also repeated a famous dictator game experiment of Forsythe et al. (1994). They were able to replicate the relatively high offers observed in that experiment. Then they repeated the experiment of Forsythe et al. with a comparatively slight variation. The instructions of the Forsythe et al. experiment stated that the dictator "*has been provisionally allocated*" $10 and that the task was to decide "*how to divide*" this amount. These two formulations were deleted and replaced by a simple description of the game situation. The result was that in the experiment with the new instructions, the proportion of dictators who decided not to give anything increased significantly. Even if they do not call it that, Hoffman et al. (1996) thus demonstrated a distinct experimenter demand effect that acted in the same direction as the effect that was to be demonstrated in the experiment. Forsythe et al. wanted to show that non-trivial offers are made in dictator game experiments. But this is exactly what the two formulations implicitly called for by stating that the dictator's endowment was only "provisional" and that it had to be decided how the amount should be "divided up".

It is not clear, however, whether an experimenter demand effect must always be detrimental. It is quite conceivable that the experiment shows that the experimental effect occurs, although the experimenter demand effect has acted in the opposite direction. An example of this also exists and is described in the following.

Sturm and Weimann (2007) conducted an experiment similar to the already mentioned public good experiment with an interior solution. The experiment dealt with the behavior of countries in climate negotiations and the question of what abatement activities they commit themselves to if there is a leader among them. In a separate session, the subjects were fully informed about the theoretical basis of the experiment. They then knew the Nash equilibrium and the efficient solution. There was a leader in the experiment who was the first to decide on the abatement. The other subjects then learned of his decision and made their decisions simultaneously. The experiment ran over ten rounds. In terms of game theory, the leader, being in the role of a Stackelberg leader, makes a *smaller* contribution to the public good than when the entire group decides simultaneously. The predicted experimental effect is therefore a reduction in the provision of the public good in the case with a leader as compared to the situation in which everyone decides simultaneously.

It is highly probable, though, that instructing the subjects combined with the design of the experiment triggered an experimenter demand effect. Since the difference between the equilibrium and the efficient solution had been explicitly pointed out, it may also be clear in which direction this potential experimenter demand effect was working. The subjects were given the impression that it was important to find out whether the existence of a leader would *help* to achieve the efficient solution. The background for this experimental design was the fact that in the climate policy debate the position is repeatedly held (especially in Europe) that a leader is needed if countries are to be persuaded to become active in climate protection themselves.

In order to model this political perspective, the potential experimenter demand effect was contrary to the game-theoretical prediction that leadership *reduces* the chances of success of international climate negotiations.[27] The result of the experiment, on the one hand, point to the impact of the experimenter demand effect, but also, on the other hand, indicate that the experimental effect prevailed despite the experimenter demand effect. Initially, the leaders tried to boost cooperation by making higher contributions than in the equilibrium. However, their success was only modest, and as the experiment progressed, their willingness to cooperate faltered and their behavior approached equilibrium. The *observed* effect thus went in the opposite direction to the predicted experimenter demand effect and confirmed the experimental effect. This, however, makes the finding even more powerful. Although the experimenter demand effect should have led to a positive influence of leadership, it did not. This reinforces the result that the mere existence of a leader in this public good situation does not lead to cooperation.

An experimenter demand effect that counteracts the experimental effect need not be at all harmful. However, the experimental effect will be smaller than without the existence of the experimenter demand effect and it is also possible that the experimenter

27 The experiment was actually also presented as a climate negotiation, i.e. the subjects were asked to imagine representing their country at an international climate negotiation. For the effect of such frames, see ▶ Sect. 2.5.3.

demand effect manages just to neutralize or even overcompensates the experimental effect. As a consequence, the real experimental effect resulting from the experimental design cannot be observed, although one would have been present without the experimenter demand effect. It is therefore necessary to carefully consider all possible experimenter demand effects associated with an experimental design.

The least problematic are experimenter demand effects that are neither positively nor negatively correlated with the experimental effect. Zizzo (2010) speaks of *orthogonal* experimenter demand effects in this context. This is the case when the subjects cannot guess what the real purpose of the experiment is. Even then, there may still be an experimenter demand effect, but it does not shift the behavior of the subjects in a particular direction, rather it has a neutral impact so to speak. What can be done to create this kind of experimenter demand effect?

First of all, it cannot be avoided that experimenters come into contact with subjects. Avoiding experimenter demand effects is not always possible, either. Sometimes it even has to be consciously accepted that such effects occur. This is especially true when certain frames of an experiment play an important role – for whatever reason. If we are aware of the fact that a certain frame may also lead to a certain experimenter demand effect, then it may become necessary to consider which is the lesser of two evils. Avoiding an experimenter demand effect by always avoiding any frame is certainly not a good research strategy. In some experiments it makes sense to embed the decision in a certain context. It is much more important to be aware of the potential experimenter demand effects and to handle them with great care, similar to other elements of the experimental design. Nevertheless, we should know how experimenter demand effects can effectively be reduced (Zizzo 2010 p. 88 ff).

Since contact between experimenters and subjects can lead to social pressure, it is advisable to try to minimize this contact. It is definitely advisable to minimize the vertical distance on the ladder of academic qualifications. It is not an indication of laziness when professors leave the actual laboratory work to their assistants or support staff. In order to avoid cognitive experimenter demand effects, experiments should be designed in such a way that the subjects are not immediately aware of what is to be investigated with the experiment. If the research question is to be answered by comparing different treatments, it is advantageous to have subjects participate in only one treatment at a time. Such "between-subject" designs make it easier to conceal the purpose of the experiment from the subjects. Subjecting the participants of the experiment to several treatments may result in the variation of the treatments revealing what the objectives of the experiment are. If it is necessary for the experimenter to provide information that focuses on a certain type of behavior in the experiment, this information may need to be conveyed in a way that does not highlight this behavior. Putting up a smokescreen is also sometimes called for. An example may help to illustrate this.

Brosig-Koch et al. (2011) aimed to compare the behavior of students born in western and eastern Germany. At the eastern German university where the experiment was conducted, however, the proportion of subjects coming from western Germany was almost 50%.[28] Since

28 In contrast, the proportion of eastern German subjects at the western German university was well below 2%.

a homogeneous group of eastern Germans was required for the experiment, the origin of the subjects had to be checked during recruitment, which was carried out using a database containing information on potential subjects. Unfortunately, the individual data of the potential participants did not contain any information about their place of birth. If the invitation had indicated that only eastern German subjects were allowed to participate, this could have triggered a strong experimenter demand effect. For this reason, the following procedure was used. Brosig-Koch et al. sent a very extensive questionnaire with many questions of all kinds to a large number of students. One of the questions concerned the place where the person in question graduated from high school. The answer to this question made it possible to distinguish between eastern and western German students and to invite a homogeneous group of eastern German students without them being aware that their background played a role in the recruitment process.

Important

Experimenter demand effects can act in different directions. The reference point is the experimental effect expected in the experiment. The experimenter demand effect may be in the same direction, opposite or orthogonal to the experimental effect.

The most problematic is the experimenter demand effect that acts in the same direction as the expected experimental effect. In such a case, it is difficult to decide whether what is observed is due to the experimenter demand effect or to the experimental conditions. If the experimenter demand effect runs in the opposite direction, it can just offset the experimental effect and no clear effects can be detected. The least problematic are experimenter demand effects that are orthogonal to the experiment effect. They may not influence the behavior of the subjects in a way that hinders the interpretation of the results of the experiment.

Zizzo (2010) goes on to recommend avoiding possible frames, i.e. not telling stories that are more or less realistic with the experiments. However, this is advice that we do not believe should be followed unconditionally. Frames can have meaningful functions. This thought leads us to the topic of the section after next, which deals with the framing of experiments. Prior to this, we will discuss the double-blind designs we announced earlier.

2.5.2 Double-Blind Design

In the previous section, we already referred to a study by Hoffman et al. (1996) in which the effect of a double-blind design was investigated for the first time. The background for this was the finding that relatively high allocations could be observed in dictator games. Hoffman et al. suspected that the "social distance" that prevails in an experiment plays an important role. They see this as the degree of reciprocity that people adopt in a given social interaction and vary this distance by using a double-blind procedure for a dictator game experiment. This is an experimental design that ensures that the experimenters cannot observe how the individual subject acts and that also maintains anonymity between the subjects. This is generally achieved by having the subjects drawing identification numbers randomly and in a concealed manner. As a result, the experimenters know how, for example, subject number 17 behaved, but not who number 17 is. In ▶ Sect. 3.3.2 we will discuss how to ensure that this anonymity can be maintained even when the subjects are paid off.[29] A single-blind design means that the subjects cannot observe each other, but the experimenter sees what the individual person is doing.[30]

Hoffman et al. found that the variation of social distance actually led to changes in giving behavior in dictator games. How is this finding to be interpreted? And does it mean that double-blind designs are necessary to obtain reliable results? The work of Hoffman et al. (1996), to which there is a precursor (1994), which unfortunately suffered from methodological weaknesses, provided the inspiration for a whole series of further works which examined whether a double-blind effect can also be observed in other games and whether what Hoffman et al. observed could actually be attributed to reciprocity between the experimenter and the experimental subject.

It is essential to see double-blind designs in close conjunction with the topic of the last section – the experimenter demand effect. This is necessary because it cannot be ruled out that the use of a double-blind design itself will trigger an experimenter demand effect. If experimenters explicitly draw the attention of their subjects to the fact that they are acting anonymously and cannot be observed by the experimenter, then it is obvious that the subjects will think about why it is so important to the experimenter that they can act without being observed. In a dictator game, for example, this could indicate that the objective of the experiment is to prove that less is given to the receiver under conditions of anonymity – and this could promote this very behavior.

This is precisely the line Barmettler et al. (2012) pursue. In three standard experiments (dictator, ultimatum and trust games), they compare a single-blind and a double-blind design, but without pointing out in the instructions that it is not possible in the double-blind design to observe what the subjects are doing. That this is actually the case is a direct consequence of the design of the experiment and, above all, of the modalities of the payoff method. The experimental design was relatively intricate, so the reader is referred to the paper of Barmettler et al. (2012) for details. The result of the experiment confirms that no difference can be found between a double-blind design and a single-blind design when the double-blind design does not explicitly state that the anonymity of the subjects is ensured. This applies to the dictator game experiments as well as to the ultimatum game experiments and the trust game experiments. This suggests that an experimenter demand effect may be linked to specific double-blind designs.

Another observation, however, suggests that double-blind designs are particularly effective where a strong experimenter demand effect is expected. This is especially likely to be the case in dictator game experiments (see ▶ Sect. 2.5.1). The trust game also possesses elements that are similar to the dictator game. In the second stage of the game, the

---

29 This is not exactly a trivial problem, because a receipt from the subject is usually required for the settlement of the money from the experiment.

30 It should be noted that the term double blind used here is not identical to that used in medicine, for example. In drug studies, this means that neither the physician administering the drug nor the patient participating in the study knows whether the drug to be tested or a placebo is being administered.

second mover can, like a dictator, decide how much of his entire endowment he wants to return to the first mover. Cox and Deck (2005) show that this second stage is affected by a double-blind design in which explicit reference is made to the anonymity of the players. Under double-blind conditions, the second movers ("dictators") return less. In contrast, the double-blind design does not change the behavior of first movers. The same applies to public good experiments. Laury et al. (1995) show that there is no significant difference between cooperation rates under double-blind conditions and single-blind conditions. It is quite possible that in strategic games the interaction between the subjects is much more important than that between the experimenter and the subject.

An experiment by Fischbacher and Föllmi-Heusi (2013) dealing with the willingness to lie also shows that the double-blind design does not lead to changes in behavior. The subjects were able to roll a die without being observed and then reported the result to the experimenter. The payoff increased directly in relation to rolling the numbers 1 to 5, while a roll of 6 resulted in a payoff of zero. It was found that the numbers reported did not follow the statistically expected uniform distribution over the six values, but that the numbers with the highest payoffs were approximately twice as frequent as those with the low payoffs. The double-blind design consisted of the subjects being unobserved not only when rolling the die, but also while reporting "their" number. A possible explanation for the fact that this double-blind design had no effect could be that a sufficiently high degree of anonymity is already guaranteed by the concealed roll of the die. A subject reporting a "5", for example, may actually have rolled that number. The experimenter cannot possibly identify this person as a liar.

This shows that complete anonymity of the experimental subjects is not absolutely necessary to ensure that they are not influenced by the experimenter. It is not only the fact that subjects can be observed that generates experimenter demand effects, but also the *design of the experiment* in all its details. Under certain circumstances, a double-blind setup will tend to cause an experimenter demand effect rather than avoid one. Therefore, one should be careful with double-blind designs and be aware of their possible consequences.

The question remains whether it really is, as Hoffman et al. (1996) claimed, reciprocity in the relationship between the subject and the experimenter that determines behavior in certain experiments. To put it crudely, this assertion amounts to claiming that giving takes place in a dictator game experiment primarily because the subjects think they are in a reciprocal relationship with the experimenter. They give some of the pie away and expect goodwill and recognition from the experimenter in return. From this perspective, there is not much room for real altruism towards the receiver. Eckel and Grossman (1996) did not believe this and attempted to refute it with a double-blind dictator game experiment, in which there were two treatments. In the first, the receiver was another student subject, while in the second it was the American Red Cross, a non-profit charity. It was found that in the first case only 10.6% of the endowment was given away. This value is approximately the same as that observed by Hoffman et al. (1996) in their double-blind design. In the second treatment, however, this value rose to 30.1%! Eckel and Grossman concluded from this that real altruism could be observed even under double-blind conditions. This is not a compelling conclusion, however. One could ask why it is that students donate 30% of their income to the Red Cross *in experiments* but are unlikely to do so in real life. The reason could again be a strong experimenter demand effect. The first treatment investigates how much money the subject will give under anonymous conditions to a completely unknown person who is not known to be "needy" in any way. The second treatment examines the share of endowed money that a subject passes on to a charity to help people in need. The second treatment seems to be intended to test something different from the first. Consequently, subjects might judge the two situations differently and draw different conclusions concerning the question of what is expected of them in the experiment.

The question of whether reciprocity is behind the double-blind effect has also been investigated in other experiments, for example by Charness and Gneezy (2008) or Bohnet and Frey (1999). However, these experiments did not focus so much on the interaction between the experimenter and the subject as on the interaction between the subjects. Therefore, we will deal with these quite interesting experiments in more detail later in ► Sect. 2.6.1.

Important

Double-blind designs should ensure that the decisions of the subjects cannot be observed by those conducting the experiment. The background to this is the assumption that if the behavior of the individual subjects can be observed, this leads to an experimenter demand effect conveyed through social pressure.

However, the findings suggest that an experimenter demand effect may also be linked to a double-blind design in which the anonymity of subjects is particularly highlighted. This occurs because the explicit emphasis on the fact that the subjects are not under observation leads the subjects to reflect on why establishing such a high degree of anonymity is so important to the experimenters. Therefore, when using a double-blind design, it is not advisable to explicitly point out that this is intended to achieve anonymity.

2.5.3 The Frame of the Experiment

The frame of an experiment is the way in which a specific decision problem is presented to the subjects. Framing effects are the changes in the subjects' behavior that occur solely because the presentation of the decision problem is varied without changing the problem itself and its solution. Despite what appears to be a clear definition at first glance, the term "framing effect" has some of the character of an umbrella term owing to many changes in the design of the experiment and the resulting outcomes being subsumed under it. The realization that framing effects exist is very old. As Pruitt (1967) pointed out, different descriptions of the prisoner's dilemma lead to different levels of cooperation.[31] The phenomenon of preference reversal in the valuation of lotteries is also famous, for example in Lichtenstein and Slovic (1971). In their experiment, the subjects were faced with the choice of playing either a so-called P-bet (a high probability

31 The game can be presented either in the usual normal form or in the "decomposed game" version, in each case indicating how the choice of one's own strategy affects one's own payoff and the payoff of the other player.

of winning a modest amount or zero) or a $-bet (a modest probability of winning a large amount or zero). Generally, the subjects chose the P-bet. However, when the subjects were asked to state the lowest price they would accept to sell the bets, the $-bet got the higher price. These classic examples demonstrate the principle of the framing effect. The same game or lottery is involved, but the behavior in the game and the valuation of the lottery depend on the way it is presented.

In the recent literature, two types of framing effects play a special role. The first occurs when only the name of a game is changed (label frame). We have already referred to the following example in the previous section. Whether you call a public good experiment "Community Game" or "Wall Street Game" makes a major difference. It is no coincidence that we also took this example when we discussed the experimenter demand effect, because, as we will see, there is a close relationship between the experimenter demand effect and the framing effect. Unfortunately, this does not always receive the attention we think it deserves in the literature.[32] We will come back to this point later.

The second framing effect that has attracted much attention is what is named the valence frame. This means that certain terms are loaded with respect to the values or preconceptions associated with them. The standard example again concerns the public good game, which can be played in a "Give" or a "Take" treatment (Dufwenberg et al. 2011). In the Give frame, the individual members of a group each receive an initial endowment (z_i), which they can either keep or pass on to any part of a joint project (the public good). In the Take frame, the entire initial endowment (i.e. the sum of the z_i) is in the joint project and the subjects can withdraw money up to the amount of z_i. Obviously the same decision problem is involved in both cases, but the experimental findings show that significantly more is invested in the public project under the Give frame than under the Take frame.

The observation that the results of experiments can be strongly influenced by the respective frame has led to the emergence of *neutral* frames as a standard – at least when it comes to testing general models. This means that names that could be given to an interaction or the persons involved are consciously avoided and that the description of the experiment is designed as value-free and neutral as possible. Abbink and Hennig-Schmidt (2006) point out that this means that very abstract experiments are carried out which are supposed to answer very concrete and practical questions under certain circumstances.[33] Critics have some doubts whether this can be achieved and refer to the very limited external validity of experimental studies (Eckel and Grossman 1996). Thus, the question of whether framing effects actually occur with the frequency and intensity generally assumed is of some importance.

The experimental evidence on framing effects in relation to important standard experiments is, unfortunately, not entirely clear. We will first present some examples

32 An exception, which has already been mentioned, is Zizzo (2010).

33 They refer, for example, to Irlenbusch and Sutter (2006), who use an abstract experiment to describe the behavior of EU states, or to Erhard and Keser (1999), who in the same way attempt to model joining a trade union.

of observations that at first sight may seem contradictory. Then we will discuss why a framing effect actually occurs and which channels play a role. This consideration will resolve the previously mentioned contradictions and put us in a position to give a recommendation for handling frames and to put the critical significance of framing effects into perspective somewhat.

Two standard economic experiments are of particular interest in connection with framing effects: public good experiments and dictator game experiments. Many (but not all) experimental economists suspect that the dictator game is particularly susceptible to framing effects. For example, experiments have shown that an important role is played by social distance (Hoffman et al. 1996), the design of the strategy space (List 2007; Bardsley 2008) or the origin of the money to be distributed (see ▶ Sect. 2.2.4). Fehr and Schmidt (2006) therefore conclude, "*... the dictator game seems to be a rather fragile situation in which minor factors can have large effects*" (p. 659).

A framing effect in the dictator game experiment is explicitly investigated by Brañas-Garza (2007). Brañas-Garza finds that when the sentence "*RECUERDA el esta en tus manos*"[34] is added to the instructions, the amounts given to the receiver are higher. He interprets his findings in such a way that this sentence creates a frame that reminds the dictator that he is in an advantageous but unfair situation. Dreber et al. (2013) provide evidence of the opposite. In their experiment, both the label frame and the valence frame of dictator game experiments are systematically varied. The names used are "Giving Game" and "Keeping Game", with the actions being called either "transfers" or, depending on the circumstances, "give" or "keep". In addition, the starting point is varied, i.e. sometimes the receiver had the initial endowment and the dictator could take the amount he wished to have for himself, while at other times the dictator had the money and could give some of it to the receiver. Finally, the information that the receiver had was also varied. Sometimes the receivers knew that they were in a dictator game, sometimes they had no idea where the money came from (if they had received some). The finding obtained by Dreber et al. (2013) is clear: no framing effect can be discerned. The transfers in the different frames hardly differ from each other. In their conclusion, Dreber et al. refer directly to Fehr and Schmidt's (2006) assessment quoted above: "*...our current view is the polar opposite of Fehr and Schmidt's (2006) hypothesis,...*".

We have already pointed out the effect of label and valence frames in public good experiments earlier in this section. In order to vary the game names, Dufwenberg et al. (2011) chose a neutral description in one case (Neutral) and the public good was called a Community project in the other. Together with the Take and Give frames, they had a 2×2 design. The contributions to the public good in the four treatments of the 2×2 design did not differ very much. Only one in six comparisons between the treatments produced a significant difference: the Give-Neutral treatment had significantly higher contributions than the Take-Community treatment.

---

34 Literally translated: Remember that he is in your hands. What is meant here is the receiver of the dictator's transfer.

At this point we should also remember the experiment by Abbink and Hennig-Schmidt (2006), in which it was shown that a realistic frame for a corruption experiment yielded results that were no different to those with a neutral frame. This finding also shows that the question of how often and to what extent framing effects work is not easy to answer, since it is possible to find both experiments in which the findings are very robust to changes in the frame and those in which the results are very sensitive to altering the frame. In order to deal with this enigma, it makes sense to consider why the frame of an experiment can have any effect at all.

Let us assume that when subjects enter a laboratory and receive instructions for an experiment, they first try to understand what the experiment is about and what behavior is expected from them. The frames of the experiment then serve as an orientation aid for the subjects. What is the name of the experiment? What is the name of the activity I need to perform? What conclusions can be drawn from the type of task I am faced with here? Questions of this kind will occupy the subjects. It should be borne in mind that the subjects assume that the frame – i.e. the answers to their questions – was set by the experimenter. The person who wrote the instructions and designed the experiment thus provides the information that the subjects use to make sense of the experiment. This means that each frame – no matter how it is designed – is always associated with a potential experimenter demand effect. If one accepts this consideration, the question of whether a change of the frame impacts on the behavior can also depend on whether this alters the potential experimenter demand effect and whether this in turn has any impact. Of course, the behavior of subjects is not only determined by experimenter demand effects. Ideally, their influence is rather small and the effect of monetary incentives dominates the decision. Nevertheless, when designing an experiment, one should at least be aware of the potential connection between frames and experimenter demand effects.

Subjects' behavior can also depend on their beliefs concerning other subjects' behavior. Dufwenberg et al. (2011) point out that the frame of an experiment may also play an important role with regard to the beliefs of the subjects. If the theoretical framework is extended and psychological game theory (see e.g. Geanakoplos et al. 1989) is taken into account, it is quite possible that a subject's behavior will also depend on second-order beliefs. Dufwenberg et al. use the following example to demonstrate this. Suppose a dictator in a dictator game experiment is guilt averse, i.e. he wants to give the receiver as much as the receiver thinks he will get. If the dictator tries not to disappoint the receiver in this respect, then he must form a belief concerning the Receiver's beliefs and structure his allocation accordingly. The frame of a dictator game can indeed influence this second-order belief. Dufwenberg et al. propose two names they expect would lead to a guilt-averse dictator giving very different amounts: in a "*let's-split-a-grant game*" presumably half would be given, whereas in a "*German tipping game*" only very little.[35]

Using four different frames of a public good game, Dufwenberg et al. observe not only the contributions of the subjects, but also their first- and second-order beliefs, and they can show that the behavior in the four treatments is compatible both with the

35 We suspect that the authors use the name "German tipping game" because tips in Germany are smaller on average (because waiters are better paid) than in the US.

assumption of guilt-averse subjects and with the assumption of reciprocal behavior. Both indicate that first- and second-order beliefs had an impact on the subjects' actions.

We have identified two channels through which a frame could act. On the one hand, it might generate a belief about what the experimenter wants from the subjects and in this way might convey an experimenter demand effect. On the other hand, it allows beliefs to be formed about what other subjects will do and what they believe. Beliefs regarding both the experimenter's goals and other subjects' behavior can have an impact on a subject's own behavior. A third channel through which the frame of an experiment can influence behavior is the activation of norms of behavior. The frame creates a certain context in which the subjects are to make decisions, and this context may be connected to social norms that are important for the subjects. If such norms are only effective in certain contexts, then a kind of "norm-setting" effect can arise directly from the frame of an experiment. In this way, too, variations of the frame can produce changes in behavior.

Let us take another look at the framing effects in the dictator game and the public good experiment. Brañas-Garza (2007) was able to create a strong framing effect in a dictator game, while Dreber et al. (2013) found that different frames had no effect on dictators' behavior. One way to explain the difference between the two findings is to use the strength of the experimenter demand effect and the activation of norms as an explanation. Brañas-Garza (2007) appeals immediately and directly to the conscience of the dictators and makes it very clear that the experimenter would not consider it appropriate for the dictator to exploit his position and not give anything to the recipient. At the same time, he creates a context in which altruistic behavior seems appropriate since a corresponding social norm exists (would the experimenter otherwise be so explicit?). The frames used by Dreber et al. have a considerably less appellative character for the subjects. From the dictators' point of view, the basic design of the dictator game may already be associated with a strong experimenter demand effect, for it is quite obviously a test of the willingness to "share" or to be "altruistic". Whether it is a matter of deciding whether something is to be given (Give frame) or kept (Keep frame) does not greatly change the basic design of the experiment and the basic message of the experimenter. A more pronounced framing effect might have appeared in Dreber et al. had the game been called "Take Game" (in contrast to Give Game), because in this name there is a request that has a different thrust to "give" or "keep" (see the experiments of List 2007 and Bardsley 2008 already described earlier).

In public good experiments, in particular, an experimenter demand effect and an activation of norms could play an important role in the creation of a framing effect. It should be clear that the designation of a game as "Wall Street Game" generates a completely different expectation of the experimenter's intention than the designation "Community Game", thus radically changing the context of the decision (including the norms applicable there). In Dufwenberg et al. (2011), who compare the contributions made to a neutrally framed public good to that made to a public good framed as Community project, this framing effect does not work, however. The authors argue that the term "community" in Germany (where the experiment took place) has undergone a change in meaning resulting in a more negative connotation. This explanation is a little speculative, but if it is correct, it could explain the declining contributions, since the experimenter is then asking the subjects to contribute to a project that has a negative perception.

The potential experimenter demand effect is also of importance with regard to the effect of the frame on subjects' beliefs regarding other subjects' behavior since it directly establishes common knowledge. All the subjects know that they all have received the same instructions and possess the same information. If a frame has an effect, then it is a rational expectation that this is not only present in one person but also in the other subjects. By addressing all the subjects in the same manner, the frame creates a basis for the formation of beliefs, which reinforces its underlying appellative effect. If a public good experiment is presented as a "Give Game", every individual subject knows that not only she is being called on by this frame and the associated message of the experimenter to give something to the project, but that the same request also is being made to all the other subjects. Conditionally cooperative subjects are only willing to contribute to a public good if they can be sure that others will do the same. This belief can be generated by the frame and the potential experimenter demand effect of a Give Game. This also proved to be the case for Dufwenberg et al., where the Give frame also had a significantly positive effect on contributions to the public good.

What then are the implications for handling experimental frames? The first is that anyone who performs the experiment should be aware that there is no experimental design that does not contain a frame and none where experimenter demand effects can be completely eliminated. If an experiment is designed to be as neutral as possible and the contact between the experimenter and the subjects is reduced to the absolute minimum, the subjects could also draw conclusions about the use and purpose of the experiment and the kind of behavior that might be appropriate. This also makes it clear that striving for a frame that is as neutral as possible does not always make sense. It may be justified in many cases, but if researchers want to use an experiment to draw conclusions about behavior in a specific real-world decision context, it becomes questionable. If a frame works in the real world, for example because it activates certain norms, but this frame does not occur in the experiment, what can we learn from the experiment about the real world? To put it another way, attempting to minimize potential experimenter demand effects entails the risk of preventing the influence of norms that may play an important role in the real world.

In any case, when designing an experiment, one should ask oneself what kind of experimenter demand effect or norm activation could be associated with the selected frame. Of course, this applies even more if non-neutral frames are used. At the same time, an answer should be sought to the question of how the frame (and thus the norm or the experimenter demand effect) may influence the subjects' beliefs regarding other subjects' behavior. All these considerations should then be taken into account when interpreting the results of the experiment.

One last thought strikes us as being very important: the frame and potential experimenter demand effects should be regarded as normal design elements of an experiment. This includes the fact that these may be elements whose effects can be tested experimentally. Just like other design variables, these two closely linked elements of design are available when it comes to creating meaningful experimental designs. This would be entirely in the sense of Loomes (1999), whom Abbink and Hennig-Schmidt (2006, p. 104) quote: "*It may be rather more useful to try to study the impact of context than to pursue the impossible goal of eliminating it*". Attempting to comply with Loome's demand admittedly poses a not inconsiderable problem. Although it is possible to experimentally study the effect of different frames, in the end behavior in one experiment is always being compared with behavior in another experiment. What is lacking is a comparison with behavior in the "natural" or "real-world" environment, i.e. that behavior occurring under frames that is not determined by an experimenter, but rather by the world in which we live.

Important

The frame of an experiment provides information the subjects might use to gain an idea of what the experiment is about and what behavior is appropriate. Since the frame is consciously designed by the experimenter, a potential experimenter demand effect is therefore associated with *every* frame. This is the first way in which a framing effect has an impact.

The second way the frame has an impact is that it can also influence the beliefs of the subjects about other subjects' behavior. This is all the more the case because the frame directly creates common knowledge.

A third way it impacts arises because a frame can be accompanied by the activation of social norms. It is important to note that such norms can also have an influence in the real world. If a real phenomenon is to be simulated in the laboratory, a corresponding frame should therefore be included.

2.5.4 Instructions and Comprehension Tests

All the elements of an experimental design must be communicated to the subjects of the experiment. This is done in the instructions, which are either provided verbally or distributed in writing to the subjects. Two important questions are of interest here. First, how can the instructions be conveyed in such a way that it is certain that all the subjects have actually taken note of and understood them, and second, how can potentially distracting effects be eliminated? Both questions are relevant not only with regard to the actual instructions, i.e. the description of the experiment, but also to the control questions which can (and should) check whether all the subjects have really understood how the experiment works.

Ideally, instructions should be in writing and distributed as a document to the subjects. An important reason for this is that it is then certain that the subjects can look at the instructions again during the ongoing experiment if anything is unclear to them. This also rules out variations in the presentation of the instructions from session to session that can undoubtedly take place if the instructions are communicated verbally (even if arises simply through a variation in the emphasis of some words). However, by providing the instructions verbally, it is possible to ensure that they are common knowledge for all the subjects. In other words, the subjects know that everyone knows that everyone knows… that everyone knows what is in the instructions. It is therefore not at all unusual for the instructions to be distributed in writing, and also to be read out. As far as the content of the instructions is concerned, there are three points to bear in mind:

1. The description of the experiment should be as short and concise as possible. The reason for this is very simple: people – including students – often shy away from reading long texts. In the age of Twitter and Facebook, this trend has become more

2

pronounced. Therefore, the instructions should not be too long. Of course, the attempt to keep the instructions short should not be at the expense of comprehensibility – which brings us to the second point.

2. The description of the experiment must be as simple and understandable as possible. An important question in this context is whether examples should be used. These have the advantage that they make things transparent, but they have the disadvantage that an experimenter demand effect or an anchoring effect may be associated with them. Therefore, when using examples, it is important to consider very carefully what values to use. In any case, it is advisable to use several examples that use different, ideally randomly drawn values to avoid the effects described above.
3. Instructions are the point where experimenter demand effects could be generated or norms might be triggered. This is something to be aware of, i.e. when writing the instructions it is important to remember that signals are being sent to the subjects who could possibly use them to interpret what they should do.

Handing out written instructions and leaving it at that leaves it up to the subjects whether they read the instructions carefully or not. The subjects are then in a sense left to themselves to decide how intensively they will deal with the description of the experiment. There are likely to be major differences between the subjects in this respect. While some subjects may be quite content to just skim the instructions, others will read them carefully and thoroughly and really attempt to understand the experiment. If a little more certainty that everyone has read the instructions completely is desired, it is advisable to read them out loud to the subjects. Experience has shown that when the experimenter reads the instructions to the group, almost all the subjects look at the sheet and read along. This technique is, therefore, relatively effective in ensuring that everyone has read the instructions.

How should we deal with questions that the subjects still have after they have received the instructions? We recommend that questions not be asked publicly. For this reason, reading the instructions out loud should not be concluded by asking the group if anyone has a question but rather by pointing out that questions can only be asked in the strictest confidence and then answered one-on-one between the subject and the experimenter. Only if it transpires that a certain point was either not understood or misunderstood by several subjects is it advisable to provide "public" clarification to all. As a rule, this indicates that the instructions are not yet optimally designed and should be improved accordingly. Ideally, such things occur in the pilot experiment and can then be corrected before the actual experiment.

Why is it not advisable to have questions asked publicly? The problem is that there is no control over what is asked. As a consequence, questions might be asked that are not about understanding the experiment, but rather about giving an indication of individual expectations or behavior or how one should behave. An example should serve to clarify this point. Suppose a public good experiment is being carried out. A rather disastrous question, for example, would be: "*Am I right in saying that we should all invest all our initial endowment in the public project?*" Questions of this kind can have a very strong impact on the conduct of all the subjects in the experiment, because they act as a kind of coordination device. Since it is not possible to know what kind of questions will be asked, it is better not to allow any public questions at all. This eliminates the risk of a question such as the one mentioned above rendering an entire session worthless. Furthermore, questions should always be answered in the same way so that there are no differences between the treatments. It has proved helpful to prepare a catalog of possible questions and the corresponding answers in advance, supplementing this list if necessary during the course of the experiment. This catalog can then be distributed to all the experimenters involved because sometimes, due to time constraints, it is not one experimenter alone who answers all the questions in the experiment.

The saying that trust is good but control is better also applies to experimenters. It is therefore a good idea to check whether the subjects really have understood the experiment. Control questions are therefore important, but they also entail the risks already mentioned. They can trigger experimenter demand effects, activate norms or lead to anchoring effects. The last mentioned, in particular, seems very obvious. Let us again consider a public good experiment. An obvious control question would be to specify to the subjects the contributions of the "others" and their own contribution and then have them calculate how high the payoffs will be. This makes it relatively easy to check effectively whether someone has understood the game or not. But which numbers should be chosen? The risk of setting an anchor that the subjects will use as a guide when making their decisions later on cannot be discounted.

Roux and Thöni (2015) investigated the issue of possible anchoring effects and experimenter demand effects arising from control questions. They used an experiment on the Cournot oligopoly due to the fact that the calculation of the equilibrium in this game is not straightforward and control questions are therefore particularly important. Experiments were conducted with 2, 4, 6 and 8 players. In the control questions, the subjects were given the average production quantity of the other players and their own production quantity. The numbers used were randomly determined in all the runs of the experiment. But only in half of the cases did the subjects know that the numbers were the result of a random draw. In the other half, by contrast, the subjects were led to believe that they were made up by the experimenter. In this way, possible anchoring effects as well as a potential experimenter demand effect could be investigated.

The findings of this experiment were surprisingly clear. Neither anchoring effects nor an experimenter demand effect could be detected. The authors left no stone unturned in their efforts to detect these. However, neither for the first round nor for later rounds was it possible to observe any effects. This result is somewhat reassuring because it shows that experimental results are by no means so fragile and sensitive that all it requires is to ask one particular control question to influence the result of the experiment. It should be noted, however, that experimenter demand effects in experiments in which the subjects are in competition with each other tend to be weak. It might therefore have made more sense to investigate the effect of control questions for example in a coordination game experiment.

Despite this reassurance, it is worth playing it safe when asking control questions. Either the values should be selected randomly and the subjects informed of this or different values containing all kinds of indications should be used. Alternatively, the researchers can also have the subjects themselves generate numbers prior to the control questions, which are then used in the control questions. Perhaps such measures are not necessary at all, but implementing them does not cost much and one can rest assured that the control questions have not triggered an unwanted effect. There is one point, however, that must be taken into account: all the subjects should be given the *same* control questions. This means that if values are determined randomly, then this should be done once for all the subjects and not for each one individually. This ensures that the group of subjects is homogeneous in terms of subjects' previous experience.[36]

Question

For computerized experiments, it would of course also be conceivable to provide the subjects with the instructions "online" and to dispense with printed instructions. How do you rate this method?

2.6 Interactions Between the Subjects

In addition to the interaction with the experimenter, there may of course also be interactions between the subjects in experiments. This is obviously evident in experiments that involve strategic interactions. There are, however, different types of exchange between subjects that go beyond purely strategic interaction. Well-known examples of these are the reputation effects that can accompany the identification of individuals and the effects of communication. Whether such effects are possible or not depends on the experimental design. In any case, it is important to be able to assess their impact when deciding whether or not non-anonymous interactions between subjects should be possible.

What are in fact the arguments for designing experiments in such a way that the subjects remain anonymous to each other? The most important reason is specifically to prevent reputation and communication effects. Anonymity is frequently sought due to fears of losing control over the experiment if they are permitted. As a result, however, the experimental context sometimes differs markedly from the context in which real interactions take place.

A good example of this is experiments that deal with coordination problems. One of the workhorses in this field is what is named the *minimum effort coordination game.* This involves a group of players who are required to complete a task together. To this end, each individual can make a smaller or larger effort, which generates costs. The payoff to all the members of the group depends on how large the minimum effort made by an individual member of the group is. In other words, the weakest link in the chain decides. In this game there is a payoff-dominant equilibrium, which consists of all players making the maximum possible effort. Note that there is no free-rider option in this game, i.e. the best response is to maximize one's effort when everyone else does the same. However, there is a risk that the effort will be wasted because one or another group member has not made the maximum effort. Therefore, a risk-dominant equilibrium is for all the players to make the least possible effort. The question is what kind of equilibrium does coordination ultimately lead to when the game is repeated. It has been well known since the work of van Huyck et al. (1990) that groups of more than 4 members are generally not capable of coordinating on the payoff-dominant equilibrium.

36 Of course, the values can also be drawn at random for each subject. In order to control for these heterogeneous previous experiences, however, a correspondingly large sample is required (see chapter four).

This insight has led to a large number of subsequent experiments addressing the question of the conditions under which coordination performance improves. This is an interesting question, but is it also a question that arises in any real situation? The vast majority of the experiments dealing with the coordination problem were conducted anonymously, i.e. the group members did not know each other and had no opportunity to communicate with each other. Is it conceivable that in the real world there are situations in which 6 or 7 people in a group have to complete a common task, the weakest group member decides on the remuneration for everyone and this happens *in complete anonymity*? It would be difficult to find an example of this. Sometimes reference is made to situations in which it is not desirable to meet personally, for instance in the case of implicit price agreements. However, this is not a good example either, seeing that such agreements are primarily about solving a cooperation problem rather than a coordination problem.

It is indeed a limitation to use experimental designs that are known not to exist in such a way in real situations. Instead of categorically excluding reputation and communication effects, it may therefore be a sensible strategy to consciously allow them so as not to ignore what may be an important aspect of real decision-making environments. Of course, in this case it is important to know what effects communication and reputation can have.

2.6.1 Reputation Effects and Social Distance

Does reputation really play a big role? And if so, how are reputation effects triggered? How much social interaction is necessary for subjects to start thinking about their reputation? And how does the reputation effect differ from what triggers a reduction in social distance? Alvin Roth suspects that face-to-face interaction is needed to initiate socially trained behaviors: "*Face to face interactions call into play all of social training we are endowed with*" (Roth 1995, p. 295). But how strong must this interaction be?

Brosig et al. (2003) investigated how communication between subjects and reduced anonymity affect behavior in a public good experiment. Such a game was played over 10 rounds in groups of four players. Before the actual experiment, there was a "communication phase" that varied over a total of seven different treatments. One variation was that the four group members could see each other on a screen for about 10 seconds, with each member being able to visually identify the other three group members. This should actually suffice to trigger a reputation effect since every subject can expect to be recognized by the other three players on campus later after such identification. For this reason, what happened in the experiment could still be the subject of conversations, discussions or other forms of interaction later on. In principle, it cannot be ruled out that this will influence behavior in the experiment. However, it turned out that an exclusively visual identification of the other players had no impact at all. The contributions of the groups in which the subjects saw each other did not differ statistically from those of the groups acting in complete anonymity. Even the smaller social distance associated with knowing what the others look like had no influence on behavior in the experiment.

Bohnet and Frey (1999) made a very similar finding in a dictator game experiment. This experiment was double blind, i.e. the experimenter could not observe dictators' giving behavior. The interaction between dictator and receiver, however, was varied. All the experiments took place in a classroom. In the first treatment, as a baseline treatment, the experiment was conducted anonymously. The receivers did not know who the dictator was, and the dictators were not informed about which fellow student was the receiver. The second treatment involved one-way identification. This was achieved by having the receiver rise from his or her seat and thus be identified by the respective dictator. As in Brosig et al. (2003), this simple identification had no effect. Admittedly, in view of the fact that the receivers remain completely inactive in a dictator game experiment, there is also no reputation effect. In light of the above, what happened in the third treatment is quite astonishing. The one-way identification described above was repeated there, but this time the receiver said their name and mentioned their favorite hobby. Although *no* reputation effect could occur in this arrangement either, the allocations increased significantly. It does apparently matter how familiar the other person is. Social distance is important, at least in dictator game experiments.

That reputation effects also play a role, however, became clear in the fourth treatment used by Bohnet and Frey (1999). This involved mutual identification, with both the dictator and the receiver rising from their seats and being able to identify each other. As a result, 71% of dictators then decided to split the amount of 13 francs equally. In the one-way identification it was only 39% (without naming) or 16% (with naming) who did this. It is surprising, however, that in this experiment in the two treatments with the highest contributions there was a relatively high number of "super-fair offers". This is understood to mean giving more than half the endowment to the receiver. Such behavior is very rare in both dictator and ultimatum game experiments. The fact that Bohnet and Frey have many such observations may indicate that the classroom situation played a role.

Charness and Gneezy (2008) sought to separate social distance from possible reputation effects. To this end, they paired subjects from two different universities (Tilburg and Amsterdam) in a dictator game experiment and an ultimatum game experiment. Even though the partners in the respective pairs became acquainted with each other, there was no reason to expect that they would ever meet again after the experiment. As a consequence, reputation effects could be largely ruled out. Two treatments were compared and both experiments were played anonymously as baseline treatments. In the second treatment, the subjects were given the surname of the other player. First names were not given to avoid possible gender effects. It was found that this action had a strong impact on the dictator game experiment. If the names were known, considerably more was given than in the anonymous situation. In contrast, the ultimatum game experiment showed no effect. The offers of the proposers remained unaffected by the smaller social distance.

He et al. (2016) investigated the effect of social distance in the prisoner's dilemma game. The authors compared two treatments: one with a 10-second visual identification of the two players *before* they interact and one in which the players can only visually identify themselves *after* interaction. On the basis of the finding that neither the beliefs of the players nor their behavior differ between the two treatments, they conclude that social distance plays no role in this decision-making environment. However, other studies on the trust game find positive effects of social distance on behavior (see, for example, Buchan et al. 2006 or Charness et al. 2007). In an decision-making environment that has some similarities with the trust game, Brosig-Koch and Heinrich (2018) demonstrate that the decrease of social distance affects behavior particularly in those cases in which subjects are not allowed to make specific promises.

The fact that social distance can influence laboratory behavior suggests that even outside the laboratory it is be important how anonymously people act or how close they get to other people (see also Brosig-Koch and Heinrich 2018 whose study is based on both, laboratory and field data). This should be taken into account when deciding which interactions are to be permitted in the laboratory. Strict anonymity makes the experimenter's life easier because it ensures that the conditions of interaction can be well controlled. A reduced social distance is always associated with a potential loss of control. It is important to be aware, however, that anonymity can lead to certain types of behavior that do not occur with lower social distance. If the real phenomenon to be studied experimentally is not characterized by strict anonymity, experiments conducted anonymously are subject to a considerable loss of external validity.

2.6.2 Communication Effects

Communication may play a part in many of the topics addressed in the last sections, whether it is communication between the subject and the experimenter, or communication between the subjects. It is now time to turn to a discussion of this topic.

Controlling communication

No matter how communication between the subjects is to be designed, it is important that the experimenter retains control over how the subjects interact. This involves not only the experiment itself, but also what happens before and after the experiment. In ▸ Chap. 3, we will explain in more detail how this control can be established when it comes to the practical implementation of experiments. However, it should already be pointed out here that it may be advisable to ensure that uncontrolled communication can be ruled out as far as possible when recruiting subjects. The same applies to the way the subjects enter the laboratory and the way they leave the laboratory after the experiment. A complete control of communication requires that all these steps are included.

The Conflicting Objectives of Control and External Validity

The basic problem that arises in connection with communication among subjects can be described as a conflict of objectives. If experiments are played in complete anonymity, greater control over the interaction is achieved, as effects triggered by communication can then be avoided. This facilitates the interpretation of the results and eliminates the need to isolate and identify the effects of communication. Unfortunately, completely removing communication between the people involved means that we are far removed from many real contexts in which people are active. Complete anonymity and complete lack of spoken communication are rarely found in the "real world" at the same time. Even if we try to justify these conditions with the fact that people in larger groups act more or less anonymously, this is not really convincing, because even then people usually exchange information with other people. It is important to realize at this point that a very artificial handling of anonymity and possibilities to communicate can lead to an experimenter demand effect. Subjects in the experiment could ponder why it is so important that they remain anonymous and not be allowed to talk to anyone. It cannot be ruled out that they may then conclude that this is to facilitate or promote certain behaviors – which then appear to be desirable.

On the other hand, this does not of course mean that there are never situations that are best reproduced in the laboratory using treatments that are anonymous and without communication. For example, it can be argued that the actors in (online-) markets often make decisions alone, without interaction with other people. Nevertheless, the conflict of objectives tends to remain. The more one strives for control over the interaction of the subjects, the greater the distance from real contexts. Basically, however, it should be noted that this conflict of objectives is mainly due to the fact that economic experiments are generally intended to test a theory. This requires control and experimental economists accept the artificiality of the decision-making environment.

The reason most experiments do not allow communication is that there is often the concern that communication can have many very different effects and that, if it is allowed, the ability to interpret the results of the experiment in a meaningful way is lost. On the other hand, fear of the lack of control over communication effects has led to the study of economic phenomena in speechless anonymity. Even with the best will in the world, one cannot imagine that in reality such phenomena take place even remotely under such conditions. For example, there will likely be very few negotiations in which those who negotiate never exchange words and who moreover do not know each other. Against this background, the question arises whether concerns about giving up too much control when allowing communication are really justified. It must be kept in mind that there are different forms of communication and that different techniques can be used which differ greatly in terms of control over the effects of communication.

Forms of Communication

Communication can be used for different purposes. It can be used to transmit information that the communication partners possess. But it can also be used to gain a visual (gender, appearance, facial expression) or acoustic (dialect, emphasis) impression of the communication partners. Communication can be uni-, bi- or multidirectional. It can be face-to-face or without eye contact and messages can be spoken, written or conveyed with gestures. Even within these forms of communication there are still many possible variations. For example, face-to-face can mean that the subjects sit at the same table and talk to each other, but face-to-face can also be achieved by means of a video conference. Written messages can be communicated through a chat program or with handwritten messages.

Further distinctions are possible. For example, the permitted communication content can be limited or unlimited. In the first case, only discussions relating to the task set in the experiment might be permitted, or the subjects may be allowed to talk about everything except the experiment. If, for example, the written form is chosen precisely because communication is permissible but reputation effects are to be excluded, it should be strictly forbidden to send messages that allow conclusions to be drawn about the sender's identity.

Finally, the experimenter has to decide in what form and to what extent the communication should be recorded and evaluated. If, for instance, a video conference is recorded, it is possible to evaluate not only the contents of the communication, but also the gestures and facial expressions of the subjects. With the aid of suitable software, such an evaluation is now also possible by computer. Eye tracking makes it possible to determine the way in which people perceive information. This makes even unstructured face-to-face communication considerably easier to monitor.

Box 2.3 Emoscan
Advanced facial recognition techniques can be used to automatically capture emotional states. With the help of such "emoscans", for example, emotional reactions to certain information content or, more generally, emotional states during communication can be monitored.

Communication Effects

The analysis of communication effects should take place against the background of the economic evaluation of communication. The focus here is on the game-theoretical concept of "cheap talk". In general, this means communication that does not affect players' payoffs. This form of communication can have behavioral effects if the interests of the players are sufficiently similar. However, if players have conflicting interests – such as in the prisoner's dilemma – this form of communication should not influence their actions.[37] Kartik (2009) defines the term cheap talk even more clearly. For him, cheap talk is a strategic interaction when it is not possible for players to check the truth of the information they receive from other players and when it is possible to lie without incurring costs. It remains to be seen whether the latter condition is important and, in a certain sense, critical.

From a game-theoretical point of view, experiments in which players have conflicting interests and communication between each other are completely harmless – at least if this communication is merely cheap talk. Since cheap talk is not supposed to change behavior here, it can be ignored. If we follow the definition of Kartik (2009), however, then it is no longer so clear when cheap talk can still be assumed in such situations and when not. This will depend on whether or not the liar incurs costs. Since psychological causes for such costs – which cannot be directly observed – are also possible, it is therefore conceivable that communication is not "cheap" at all, although at first sight it appears to be so. Thus, also from a theoretical point of view, it cannot be ruled out that communication may have an effect in a great many contexts and games.

What do the experimental findings look like? Does communication between the subjects have an effect and do the different forms of communication work in the same way? In order to answer this question, we will refer to the literature and present some examples in which communication effects have been studied using different games. As always in this book, the aim is not to give a survey of the literature, but to illustrate important methodological aspects by means of examples.

At which points is it relatively easy to imagine that communication between the subjects has an effect? The first thing that comes to mind is the experiments that deal with the

---

37 For a detailed discussion of the behavioral effects of cheap talk predicted by game theory, see Crawford (1998).

coordination problem, such as the *minimum effort coordination game*, which we talked about briefly at the beginning of this section. There we argued that it really does not make any sense to carry out such experiments anonymously and without the possibility of communication, because such situations are hardly likely to be found in the real world. The reason behind this was of course the expectation that the coordination problem would be more or less resolved if those who were faced with it could communicate with each other. In fact, according to the game-theoretical prediction, communication in the minimum effort coordination game can also have behavioral effects due to the common interest of the players to achieve the payoff dominant equilibrium. For example, if players mutually promise to put in their best effort, there is no incentive for players to falsely state the level of effort they intend to play. Lying is not a rational strategy in this game. This lends a high degree of credibility to the pronouncements, which in turn enables the players to use communication to make the payoff-dominant solution a kind of focal point that everyone is guided by.

Riechmann and Weimann (2008) extended a classic minimum effort coordination game with a communication phase in which the subjects had the opportunity to talk to each other about the game using face-to-face communication. The effect was unequivocal. After such a round of talks, all the groups were able to coordinate on the payoff-dominant equilibrium. Groups that were previously unable to talk to each other, however, failed in this task. In a similar experiment, Blume and Ortmann (2000) also come to the conclusion that the efficient solution is realized much more often with communication than without. In fact, the coordination problem seems to disappear (at least if all the players have similar interests) when the players are given the opportunity to consult with each other. This once again reinforces the point we made at the beginning of the section. If there is no longer a coordination problem, if communication is permitted and if the latter is precisely what happens in real-world situations, why conduct anonymous experiments with the minimum effort coordination game?

Question

What exactly is the difference between a coordination problem and a cooperation problem?

It is not only in pure coordination games that it is advantageous to be able to mutually coordinate behavior. Even when it comes to striking cartel agreements, there is reason to believe that being able to consult with each other could have an impact on the formation and stability of such agreements. In this case, however, - assuming strict selfishness - we are dealing specifically with a game with a dilemma, which means that the interests of the players are not similar. Although everyone has an interest in the others abiding by the agreement, the individual would prefer to deviate from it. Fonseca and Normann (2008) examined whether this is actually the case. They conducted an experiment in which 2, 4, 6 or 8 players were in Bertrand competition. Each treatment was played in two variations, one without communication and one with the possibility to consult with each other via a chat program for 1 minute. Although this is not exactly an excessive form of communication and although theoretically it should not trigger any behavioral effects, it did have a clear impact. With chat, the prices that companies set were higher than those without chat and corporate profits increased when communication was possible. However, communication did not affect all the player numbers equally. The impact was comparatively weak with 2 and 8 companies and strong with 4 and 6 companies. This is because, even without communication, the prices of two companies were already significantly higher than in the treatments with more than two players, with the result that the communication effect was limited from the outset. As the number of companies increased, it became increasingly difficult to coordinate on high prices, yet it was still effective even for 8 companies, with a price there of 55.2 as opposed to 1.1 in the treatment without communication.

Box 2.4 Bertrand-Oligopoly
The Bertrand model describes the competition between oligopolists who offer a completely homogeneous product, i.e. their products are perfect substitutes. Price is the sole strategic variable. The only Nash equilibrium in the pure Bertrand competition between two symmetric suppliers is that both suppliers choose a marginal cost price. This means that even with two suppliers, unrestricted price competition leads to competitors competing down to the marginal cost price. The model provides a good understanding of why companies often make great efforts to avoid pure price competition.

In the examples given so far, the subjects are in a strategic interaction in which coordinated behavior can have a very positive effect on the whole group of subjects. What is the effect, however, of communication in the dictator game, where strategic considerations play no role whatsoever? In principle, communication can lead to a change in strategic conditions, for example because reputation effects play a role, or it can cause subjects to take a different view of the decision-making situation. Greiner et al. (2012) try to isolate the last point in their experiment. This is achieved with a three-person dictator game experiment in which the dictator remains completely anonymous, but the two potential receivers can send him messages. In addition to a baseline treatment without communication, a second treatment shows the dictator a video image of the two receivers. In the third treatment, one of the receivers is permitted say something to the dictator. Communication is therefore strictly unidirectional and completely eliminates reputation effects. Having made their decision, the dictators were asked to evaluate the receivers according to six different criteria.[38] Communication effects can only occur in this experiment if "viewing" the receivers or the video message change the perception of the decision-making situation. This is, in fact, the case, but in a quite differentiated form.

The mere identifiability of the receivers did not result in an overall increase in donations. The dictators did, however, differentiate more between the two receivers, with those who were better rated receiving more generous amounts. If one of the receivers had the opportunity to send a verbal message to the dictator, more was given to that receiver. This was not at the expense of the other receiver, however, but at the expense of the dictator. The authors' interpretation of this result is that the social evaluation of the receivers plays an important part in the allocation of the money and that this evaluation is influenced by the direct verbal contact that was possible in the third treatment.

38 This was done with the help of bipolar scales that questioned the following pairs of terms: "active – passive", "lively – dull", "attractive – unattractive", "pleasant – unpleasant", "strong – weak", "influential – not influential".

The minimum effort coordination game considered above is characterized by the fact that there is no conflict between the interests of the subjects, whereas the dictator game is characterized by the fact that there is no strategic interaction between the subjects. In the Bertrand competition, we saw that communication can positively affect the prices despite conflicting interests of the players, which means that the subjects are able to improve their own overall position. Can this observation also be applied to other games in which there is a pronounced conflict between the players? Studies on the ultimatum game yield a mixed answer to this question. In his experiment, Roth (1995) examines how social contacts influence behavior in the ultimatum game. He let the players talk to each other before the experiment. In one treatment, people were allowed to talk about everything, but in another it was forbidden to talk about the experiment. The effect was the same in both cases, i.e. there was a significant increase in equal allocations. Communication thus led to the "fair" solution being chosen more frequently.

Rankin (2003) substantially restricted communication in the ultimatum game, thus obtaining a completely different outcome. In his experiment, the responders were able to make a request to the proposer. This was done while maintaining the anonymity of the subjects. The responders proved to be tough negotiators in this experiment, with the average of all requests being over 50%. At first sight, they did not do themselves a favor, because on average the amounts offered by the proposers were lower in the treatment with the responder making requests than in the treatment without these requests. As a result of these low offers, the number of rejections was also greater with communication than without. On closer inspection, however, it becomes apparent that it is nevertheless a sensible strategy to make high requests, given that a statistical analysis shows that higher requests also lead to higher offers. When the responders have the opportunity to make requests, their position deteriorates rather than improves; it nevertheless makes sense to make a high request under these conditions. Despite 92% of the offers remaining below the responders' requests, the requests still acted as anchors. When given the opportunity to make requests, the responders evidently tried to bluff and at least achieved a higher offer than if they had made a modest request in the first place. Obviously, the effect of unidirectional communication in this experiment is completely different from that of face-to-face communication in Roth's (1995) experiment.

Of particular interest is the question of how communication generally affects dilemmas. Can the findings from the Bertrand experiment also be applied to other dilemmas? To what extent is it possible to increase the ability to cooperate through communication? In answering this question, we would like to start with a study that clearly identifies two central results of research on this question.

Brosig et al. (2003) examined the effects of different forms of communication in a public good experiment. This experiment was played in groups of four over ten rounds, i.e. the subjects had to decide ten times in succession how much of their initial endowment (which was the same in each round) they would give to provide a public good and how much they would keep for themselves. In the baseline treatment, the game was played without any interaction between the subjects. The group members sat in soundproof booths during the experiment and had no opportunity to come into contact with each other before or after the experiment. In the further treatments, a communication phase, involving a total of six different forms of communication, preceded the actual public good game. Three of these consisted of active communication in which players could talk to each other. The other three were of passive nature, i.e. the players could only receive messages.

The first passive form of communication was that players could see each other for 10 seconds on a monitor divided into four areas. They thus received information about the appearance of the other group members. The purpose of this identification was to determine whether reputation effects potentially triggered by showing a person's face were sufficient to change cooperation behavior. The second passive form of communication consisted of the group being played a video of one of the experimenters explaining the game and highlighting the dilemma in which the subjects found themselves. The video also explained that cooperative behavior could lead them out of this dilemma. The background for this treatment was the question of whether unidirectional communication, such as that carried out by the mass media, might influence contribution behavior. The third passive form of communication was chosen for a similar reason. Here the group was presented with a video showing the videoconference of another group that also took part in the experiment. In this conference, the subjects agreed to invest all their endowment in the public good in all the rounds.

The first active form of communication was a telephone conference, i.e. the group members could talk to each other but could not see each other. The content of the discussions was not restricted except for the identities of the subjects. The second stage of active communication consisted of a videoconference where all four group members could see each other on the four quadrants of their monitor. The last active form of communication was a conversation in a separate room where the group members sat together at a table. The content of the discussions was recorded electronically and evaluated later.

The experiment provides two significant results. First, communication has a positive effect on the average contribution to the public good and, second, the effect of communication is strongly dependent on the form of communication. The observation that communication before a public good game leads to an increased willingness to cooperate was made early on. Dawes et al. (1977) and Isaac et al. (1984) as well as Isaac and Walker (1988) are among the early works showing this. This effect also occurred in Brosig et al. (2003), but by no means in all the communication treatments. The mere identification of the subjects had no effect at all on their behavior. Although the other two passive, unidirectional forms of communication positively affected contributions, the effect was restricted to the first few rounds. The situation was similar in purely verbal bidirectional communication. The conference call was only able to increase the willingness to cooperate for a short time in the first rounds as well. However, the picture changed completely in the two treatments in which verbal communication was accompanied by eye contact. Both the videoconference and the discussion at the table resulted in the groups being able to achieve cooperation almost 100% of the time. The analysis of the contents of the communication showed that there was practically no difference between the three active communication treatments. In almost all the conversations, the subjects agreed to 100% cooperation and they made promises to each other to stop cooperating if one member of the group deviated from this solution. Achieving this solution was only possible, however, if the group members had eye contact during the conversation.

Face-to-face communication therefore has a massive effect even if it is cheap talk and (selfish) players have conflicting interests. An important insight is that as long as there is no eye contact, communication is actually not very effective in the sense that it does not lead to a lasting change in behavior. But why does eye contact change this so dramatically? One possibility is that it means that face-to-face communication is no longer cheap, for example, because lying is more difficult or involves costs after such communication. We will return to this point later.

Also in other games, communication can have an effect if it takes place verbally and with eye contact. Ben-Ner et al. (2011) investigated the effect of communication in a simple trust game. The first mover (trustor) and the second mover (trustee) were each endowed with $10, and the amounts given to the trustee by the trustor were tripled. In addition to a baseline treatment without any communication, there were two further treatments. In one of these, the two players were able to make mutually non-binding agreements before making their decision. To do so, they clicked on a row and a column in a matrix containing all possible trustor contributions (presented in 11 rows) and all possible trustee returns, in % (presented in 7 columns). In the second communication treatment, before a making a proposal, the two players could exchange messages for 1 minute using a chat program while remaining anonymous. The proposal was then made using the same matrix that was already used in the first communication system.

It was found that the exchange of numerical suggestions alone had practically no effect on the behavior of the two players. Only when the chat was added did the average trustor contribution increase from $7.66 to $9.21 and the average return rate from 45% to 56%. The increase is mainly due to more efficient and equitable solutions (release of $10, return of $20). However, there were still a number of trustees who decided not to give anything back. The authors explain their results by the fact that the proposals that are made are a kind of self-obligation, which, however, only becomes really valid when it is more or less confirmed by being formulated in sentences. Charness and Dufwenberg (2006) also examined the communication effect in trust games and came to very similar results to Ben-Ner et al. (2011), but came up with a somewhat different explanation. We will come back to this when we consider the possible causes of the effects of communication.

In the context of communication effects in dilemma situations, an experiment by Sutter and Strassmair (2009) is particularly interesting because it draws on the literature on public good experiments, while at the same time establishing a link to the works dealing with collusive behavior (e.g. Fonseca and Normann 2012). Sutter and Strassmair investigate a tournament played between two groups of three players. The groups must decide on the level of an "effort", which is the sum of the efforts of the members and a random shock. The group with the higher effort wins the competition. Since effort is costly for each individual group member, a public good problem arises *within* the group. There is the possibility of a cost-reducing agreement (collusion) *between* the groups. The effect of communication within and between groups is examined.

Communication within the groups leads to increased efforts. The reason is that communication reduces the free-rider problem that groups face. An important role is played here by the fact that low levels of effort can be observed and are "punished" by the other group members with expressions of verbal disapproval. Harbring (2006) showed that communication in a competition between individuals leads to collusion. This does not seem to happen with the groups of Sutter and Strassmair (2009). Here communication between groups does not significantly affect efforts. Combining communication between the groups with communication within the groups leads to an increase in efforts as compared to the arrangement without communication.

Communication, as the explanations so far have shown, has a very strong effect in many experiments. It can facilitate cooperation and makes collusion more likely. It can lead to fair negotiated solutions and can increase trust and trustworthiness in the trust game. The question is, why is it that it leads to these effects? When deciding whether and in what form communication should be allowed in an experiment, it is helpful to know the channels through which communication can influence behavior. It is not clear whether all channels really are known and whether we already have a comprehensive understanding of the effect of communication, but some statements can be made which can claim some plausibility and for which experimental evidence is available.

Important

Communication between people is undoubtedly a very important element of their coexistence and it has a decisive influence on their behavior. Nevertheless, most laboratory experiments take place in the absence of communication. The reason is that communication effects are very difficult to control and this can make the interpretation of experimental results very difficult. Admittedly, this leads to an unfortunate conflict of objectives. By dispensing with communication, treatment effects can be better interpreted, but a very artificial anonymity might arise. This is rarely found in the real world and therefore severely restricts the external validity of the experiments. There are, however, also decision-making situations for which an anonymous situation without communication is quite appropriate. An example of this is when actors make decisions in highly competitive markets.

It is nonetheless an important task to investigate the effect of communication and to develop techniques that measure the effect of communication and isolate it from other effects. The experimental findings on the effect of communication show that it can facilitate cooperation and make collusion more likely. It can lead to equitable negotiated solutions and can increase trust and trustworthiness in the trust game.

2.6.3 Possible Causes of Communication Effects

How and why communication works depends to a large extent on the context in which it takes place and on the form of communication. The question of whether or not eye contact is associated with verbal communication has turned out to be critical for the sustainable effect of communication. The combination of language and visual identification is obviously important. The simple identification of the other person does not in itself make much difference, but the face-to-face exchange of information results in marked changes in behavior. It should come as no surprise that this form of communication plays a special role. For a very long time, face-to-face communication was the only form of communication. Evolutionarily, therefore, it may have played an important role. But it is also of paramount importance in the individual socialization process of each person. Long before learning to use other communication channels, people meet their closest caregivers almost exclusively face to face. These are of course pure plausibility considerations, but they are consistent with the experimental evidence we have reported on.

It is a good idea to take a separate look at the different contexts in which communication effects can be demonstrated. A not inconsiderable number of situations in which players need to coordinate can be characterized by the fact that the interests of the players involved are relatively similar, but that there are unfortunately different equilibria possessing different qualities. The task is then to agree on a particularly preferable equilibrium. In such a situation, communication can be used to create a self-commitment and to disseminate information about this self-commitment. The crucial point here is that such a self-commitment is *credible* because it refers to a Nash equilibrium (Farrell and Rabin 1996). If communication manages to create a kind of focal point and this focal point is a Nash equilibrium, then it is relatively clear why communication has an effect. In such a situation it is sufficient to introduce relatively weak forms of communication. Even unidirectional communication in the case of complete anonymity can be sufficient to achieve coordination. The credibility of messages means that no other ingredients are needed to achieve a communication effect.

This changes when we move to situations in which there is a conflict between the players. This is the case, for example, in public good games. Even in the ultimatum game there are conflicting interests of the players. The self-commitment, which in the public good game, for example, consists of committing oneself to contributing efficiently, is not credible because it does not involve a Nash equilibrium. Nevertheless, we observe in experiment that it works, especially when the communication is face to face (Brosig et al. 2003). If one assumes that at least some of those involved believe that not all the players behave purely selfishly, but act, for instance, in a conditionally cooperative manner – an assumption for which there is certainly experimental evidence (see Fischbacher et al. 2001) – it is possible to derive several equilibria that possess different qualities. The dilemma game then takes on the character of a coordination game. In this case, the task is again to agree on a particularly preferable equilibrium.

A reliable explanation of communication effects can be linked to two points. Either reputation effects that are caused by communication change the strategic situation – and thus the equilibrium – or the personal encounter with the other players changes the attitude towards them or provides additional information that leads to a different perception of the decision situation. The experimental evidence suggests that reputation effects alone may play a rather minor role. The experiments have shown that it is not enough for subjects to be able to identify one another visually in order to trigger behavioral changes. This speaks in favor of the second point that the perception of the decision-making situation changes when communication takes place. A possible explanation for how this could happen can be found on the basis of "psychological game theory", which goes back to Geanakoplos et al. (1989). Charness and Dufwenberg (2006) show that the effect of communication in dilemma situations can be explained by guilt aversion. The theory can be outlined as follows.

Two players, **A** and **B**, are in a situation where player **B** can receive something from player **A** (a transfer payment, a contribution as part of cooperation, or similar).

A assumes that **B** has a certain belief about **A** concerning this. On top of this, **A** himself forms a belief, i.e. **A** can either fulfill or disappoint the expectation that he believes **B** to have of him. In the latter case, the anticipated disappointment of **B** resulting from this may lead to a feeling of guilt in **A**. If people do not like feeling guilty, fulfilling the beliefs serves to ward off this feeling. The decisive factor here is that communication, and in particular face-to-face communication, can change **A**'s beliefs regarding **B**'s beliefs. This is usually done by way of promises made during communication. If, after a conversation with **B, A** believes **B** believes he will receive more from **A** than **A** had assumed before the conversation, the pressure on **A** to increase the amount he gives to **B** in order to avoid feeling guilty increases.

What is decisive in this chain of argumentation is the second-order beliefs that **A** forms. Charness and Dufwenberg (2006) use a variant of the trust game to experimentally test whether this belief changes in accordance with the theoretical prediction of guilt-aversion theory (◘ Fig. 2.1):

In the first move, **A** decides whether to choose an option that pays 5 to both players, or leaves the decision to **B** to either choose an option that pays **B** 14 and **A** 0, or to have the final payoff for **A** randomly determined. **A** has a probability of 1/6 of being left empty-handed with this random move and a probability of 5/6 of getting 12, while **B** has a secure payoff of 10 if he chooses the random move. In this case, both players would receive an expected outcome of 10 and thus twice the payoff they would have received had the game ended if **A** had chosen the option to pay 5 to both players.

A has a belief concerning the probability of **B** being nice and choosing the random move if **A** gives him the opportunity to do so. **B** forms a belief τ_B about **A**'s belief. γ_B measures the **B**'s sensitivity to the guilt he experiences if he does not fulfill **A**'s belief. If **B** does not choose the random move, **A** misses out on a payoff of 10 in the expected outcome. Taking into account the guilt aversion, the payoff of **B** in this case is $14 - \gamma_B\ 10\ \tau_B$. The stronger **B**'s belief that **A** believes **B** will choose the random move and the more sensitive **B** is to guilt, the lower his payoff will be if he ends the game immediately.

In Charness and Dufwenberg's (2006) experiment, after making their choice, the **A**-players were asked to guess what proportion of the **B**-players chose the random move. The **B**-players were subsequently asked what they expected the **A**-players to guess. The experiment was played with and without communication, with the communication consisting of the **B**-players being able to send a message to the **A**-players. The main finding of the experiment was that communication changed the beliefs of the **B**-players regarding the beliefs of the **A**-players and that the outside option was chosen less frequently by the **B**-players. This result is in line with the thesis that it is guilt aversion that

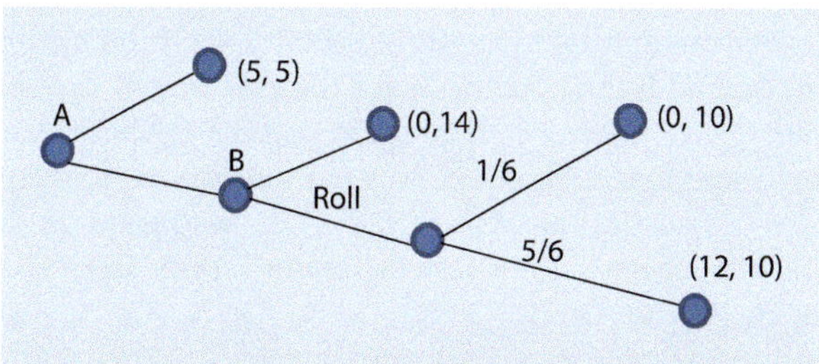

◘ **Fig. 2.1** Game tree in the experiment by Charness and Dufwenberg (2006, p. 1581)

forms the basis for the communication effect. Furthermore, guilt aversion can explain why subjects' promises during communication have a big impact: it is very difficult not to keep these promises in the presence of guilt aversion.

Of course, this does not rule out the possibility that there are other psychological reasons why communication changes attitudes towards communication partners. For example, empathy could play an important role and this in turn does not have to be the sole basis for aversion to guilt, but can also provide grounds for sympathy, for instance. In a somewhat generalized form, Kartik (2009) provides the game-theoretical foundation for the fact that such psychological motives lead to different equilibria, even in the case of rational behavior. The focus here is on the costs associated with lies or failure to keep promises. Where these costs come from remains open. It could be guilt aversion, a general preference for keeping promises (Vanberg 2008) or striving for consistency (Ellingsen and Johannesson 2004). Of course, the social distance between the subjects could also be a factor (to what extend promises as well as arguments reducing social distance between subjects affect behavior is investigated by Brosig-Koch and Heinrich 2018). Perhaps the most important insight of the model is that as the cost of lying increases, the equilibria contain less and less dishonesty. If the costs are low, however, the result is what Kartik calls "language inflation": everyone lies at equilibrium and those who are lied to know this. It is nevertheless necessary to continue to lie, because those who are lied to correct for the lies and so the truth is not rewarded but subjected to the same correction.

In conclusion, it can be said that the effect of communication depends very much on the way in which the communication takes place. When it comes to creating the experimental communication design, there is always a trade-off between internal and external validity – and this trade-off should always be made against the background of the research question that one wants to answer with the experiment.

2.7 Decisions Made by the Subjects

Laboratory experiments are about presenting subjects with questions and observing their decisions under controlled conditions. In a sense, experimenters direct questions to the subjects, who answer them in making their decisions. But how should these questions be formulated? And in which form should the answers be collected? There is no one definitive answer to these two methodological questions, as there are different approaches to take and methods to use and all have their advantages and disadvantages. Therefore, the experimenters first have to make a decision before the subjects do: Which experimental design is the best for our specific experiment? An answer can only be found if the research question on which the experiment is based is known and if the hypotheses for the experiment have been established. Both, the formulation of the research question and the establishment of hypotheses are, therefore, important first steps on the way to a suitable experimental design. The question of the correct statistical analysis of the results also plays an important role. We will discuss this in ▶ Chap. 4. The following sections provide an overview of the options available for eliciting decisions.

2.7.1 Strategy Method Versus Direct Response

It is generally easy to determine the elicitation method to use in experiments involving the decision-making behavior of individual subjects without the occurrence of any strategic interaction. The subjects are presented with a specific decision problem, i.e. they have to make a choice, and it is this choice that is observed. The matter can become much more complex if strategic interactions arise in the experiment. It is, in the first instance, irrelevant whether the game played by the subjects takes place simultaneously or sequentially. For better understanding, however, it is simpler to assume a sequential game.

The normal case is that the players make their moves in the order specified, with the second mover responding to the move made by the first mover, the third mover reacting to that of the second mover, etc. The players thus provide a direct response to the action of the mover before. This method of eliciting the responses is simple and easy to understand. From the point of view of the experimenter, however, it can have a considerable drawback. Let us take the simplest sequential game imaginable. Two players each choose between two possible alternatives. In this case, there are four possible outcomes of the game. Each individual decision that is observed, however, only provides information about one of the four possible paths on which the game tree can be traversed. Suppose the first mover has a choice between alternatives *a* and *b*. If the first mover (for whatever reason) has a preference for *a*, and chooses this strategy in nine out of ten cases, it becomes quite difficult and expensive to collect enough observations in the subgame following *b*.

The strategy method, which essentially goes back to an article by Selten (1967), offers an elegant solution to this problem. Instead of the second mover being presented with the decision of the first mover, he is required to specify a complete strategy. In our simple example, he has to indicate what he will do at the two decision-nodes he can reach. In other words, he must indicate how he will respond in both cases, i.e. if first mover plays *a* and if he plays *b*. The result of the game is obtained by combining the move chosen by the first mover with the corresponding response from the strategy of the second mover. In this way, the experimenter elicits information about behavior throughout the game. In a laboratory experiment, every decision costs money and time. The strategy method promises to be very economical with both. The application of this method does not have any influence on the decision, because why should the second mover in his strategy specify a different response to move *a* (or move *b*) than the one he would choose if he had to provide a direct response to the move of the first mover? This question has been the subject of a whole series of studies and for a long time (until 2003, to be precise) there was much to suggest that it really does not make any difference which elicitation method is used.

In a survey article, Brandts and Charness (2011) compiled all the experiments that deal with whether the strategy method leads to the same results as gathering direct responses. The majority of the studies listed there find no difference, but there are four studies that do find clear and significant effects.[39] For the sake of illustration, we will present one of them in more detail.

---

39 A whole series of studies provide different findings, i.e. the strategy method has an effect in some of the designs, but not in others. See Brandts and Charness (2011).

2

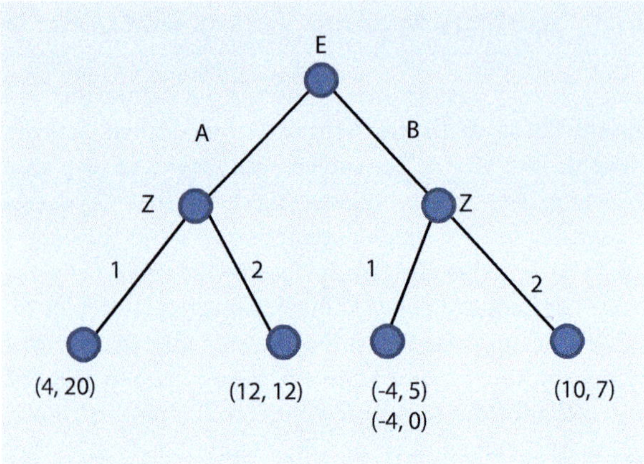

Fig. 2.2 Game tree in the experiment by Brosig et al. (2004, p. 579)

Brosig et al. (2004) use a simple sequential game with the following game tree in their experiment to investigate the strategy method (Fig. 2.2):

The game was played in two variants, which differed only in the payoff to the second mover at the end node which follows B and 1 – (−4, 5) as opposed to (−4, 0). The prediction of the subgame perfect equilibrium assuming pure selfishness is the same in both variants: the first mover E will choose B and the second move Z will select 2. The payoff difference should therefore be completely insignificant, in theory. However, this equilibrium is inefficient because both players can be better off if player E selects A and player Z chooses 2. However, if E plays A, the best answer from Z is to exploit this move and play 1. On the other hand, if E plays B, Z has the possibility to punish the first mover by choosing 1. The punishment costs of 2 or 7 incurred by player Z contrast with a loss of 14 for player E. Thus the two variants of the game differ in the effectiveness with which Z can punish the E.

Both experiments were carried out with direct elicitation ("hot") and with the strategy method ("cold"). It was found that for the most part the second movers acted as theoretically predicted, i.e. almost they always played their best response, with one exception. When the game was played hot and the possibilities to punish were good, over 40% of the E's B moves were punished by the Z players. In the cold version of the same game, only 4% of the B moves were punished. The interpretation of this result is clear: if you are immediately confronted with an unfriendly move, the willingness to punish at your own expense is much more pronounced than if you are confronted with the hypothetical possibility that the other player could be unfriendly.

The survey article of Brandts and Charness (2011) shows that the strategy method is not always neutral when it comes to punishment decisions. The pattern of behavior similar to that in Brosig et al. (2004) can also be found in Brandts and Charness (2003) and Oxoby and McLeish (2004). It is therefore reasonable to assume that in experiments in which subjects can decide whether they want to punish other subjects, the strategy method has a kind of "cooling effect" that results in punishments being carried out less frequently. Grimm and Mengel (2011) observed a very similar effect without using the strategy method. In their ultimatum game experiment, the responders had to complete a questionnaire after making their decision. They were then asked whether they wanted to stick to their original decision or change it. Some responders did in fact change their decision and it became apparent that in the treatment with the "second chance" significantly fewer proposers' offers were rejected than in a treatment in which the responders did not have the opportunity to revise their decision. Apparently, the decision to punish the proposer is easier to make if immediately faced with the offer than if the opportunity to consider it for a while exists.

Brandts and Charness (2011) identify other factors that bring out a difference between the strategy method and the direct response. It appears that simple decisions where subjects have to choose from a small number of alternatives are more susceptible than complex decisions. Thinking about a situation where decisions are more difficult to take seems to have a cooling effect similar to that observed by Grimm and Mengel (2011). If the decision is made repeatedly in the experiment, the differences between hot and cold treatments are also reduced.

However, this probably does not cover all the effects that the use of the strategy method can entail. This is reflected, for example, in Brosig et al. (2004) by the fact that not only the second movers respond to the use of the strategy method, but that the first movers also do so. As a reminder, the first movers had the choice between A and B. The latter was the equilibrium move, the former a move that puts the second mover in the role of a dictator who can divide 24 euros either fairly (12, 12) or in his own favor (4, 20). It was found that the proportion of A-moves in the two treatments involving the strategy method was significantly higher than in the corresponding treatments with direct response. At first sight, it comes as no surprise that the first movers see a difference between the hot and the cold variants, seeing that they could anticipate that a B-move in the hot variant is punished more often than in the cold variant. They should then, however, play A more frequently in the hot treatments and not in the cold treatments. Since the A-move was almost always exploited in all the treatments, the decision of the first mover in the experiments with the strategy method must have been based on a false belief as to the behavior of the second mover. Brosig et al. (2004) assume, among other things, that in the strategy method, the second mover must think through the entire game, i.e. he must look at the payoffs at all the end nodes and make a decision for every game outcome. This could – according to the possible expectation of the first mover – make the nice gesture associated with an A move of E more obvious and thus increase the probability of choosing 1 (reward).

Question

Can you think of real situations that you can design either "hot" or "cold"? What lessons would you draw for such situations from the experimental findings reported here?

In summary, using the strategy method has considerable advantages. It often allows the scarce resources available for experiments to be used in such a way that maximum information can be obtained. In many cases, it is very likely that the strategy method will not yield results different from eliciting direct responses. However, it may also be the case that the way the decision is elicited triggers differences in behavior. In particular, when there is punishment involved, caution is advisable. If there are doubts as to whether or not the strategy method is neutral in a specific case, a control experiment in which the methods are compared is helpful.

2.7.2 Experiments with Real Effort

Economic experiments almost always involve decisions in which costs play a role, whether it is a case of the subjects being faced with an allocation task in which every amount they give is at the expense of their payoff, purchasing goods or making a contribution to the production of goods. Occasionally, the work efforts that are exerted to fulfill a task are also represented by appropriately designed cost functions (for example, in the minimum effort coordination game). A two-stage procedure is usually used to implement costs in the laboratory. The first stage consists of giving the subjects an income in the form of an initial endowment (house money). This income can then be used to cover the costs incurred. In the second stage, the costs are specified in the form of a mathematical function, with there being considerable room for creativity. For example, the cost function can be convex to represent that it becomes increasingly difficult to exert the effort.

Inducing costs in this way has considerable advantages, especially in view of the fact that the experimenter retains complete control. Since the costs are part of the payoff function, it is indisputable to what extent they are actually incurred. However, this high degree of control comes at a price. Both stages of the procedure are certainly linked to problems. In ► Sect. 2.2.3, we dealt with the house money effect. People may treat the money they are given differently from the money they have earned from work. It is therefore not entirely unproblematic to first give subjects money that they can then use to cover costs. It must be assumed that the use of house money can adversely affect the external validity of experiments. The same applies to the specification of a cost function. We do not generally encounter costs – of any kind – in the real world in the form of a mathematical function. This is especially true when it comes to the physical or mental efforts we expend, for example, when doing work.

Now there is no avoiding both of these issues: it is necessary to provide the subjects with money, otherwise they would have to play with their own money and they would then probably not participate in the experiment. And, after all, it is necessary to be able to implement costs in the experiment. An alternative to issuing house money is to have the subjects work for the money they receive by introducing real effort. This increases external validity and avoids the house money effect, but has the disadvantage that the control over costs is lost. If subjects are allowed to "work" in order to impose costs on them, the actual level of costs that the subjects incur depends on the burden of the work they have to bear – and that cannot be observed!

The literature distinguishes between "real effort" designs and "chosen effort" designs (Gill and Prowse 2012). For the former, real efforts have to be made; for the latter, a level of effort is chosen that involves costs specified by a corresponding monetary cost function. The question is, under which conditions a real effort design is appropriate and, above all, how it can be designed in concrete terms. In order to answer the latter question, it must be borne in mind that the use of real effort in the experiment leads to a loss of control in two ways. First, the skills and preferences of the subjects are not observable with regard to the task assigned to them, so that a priori the actual work burden is not observable. Second, learning effects can scarcely be avoided if a task is repeated, i.e. the actual efforts and the associated actual costs can also change during the course of the experiment.

Which prerequisites should the work to be performed in an experiment reasonably fulfill? An important requirement is that it be structured in such a way that it can be assumed that at least at the beginning of the experiment all the subjects are equally good at achieving this performance. Therefore, no prior knowledge that may exist to varying degrees should be required and personal aptitude should not play an important role. It is also clear that the task should be easy to explain so that the subjects understand what is involved. Furthermore, the work outcome should be easily and reliably measurable and allow a comparison between the subjects. Finally, the task should be designed in such a way that possible learning effects are minimized and quantifiable, so that these effects can be corrected if necessary.

A central question, of course, is whether the use of real effort leads to results different to those obtained with a chosen level of effort that incurs monetary costs. To our knowledge, there are only a few studies in which a direct comparison is made. One is presented in more detail here because it provides a good demonstration of some important methodological aspects of experiments with real effort. Brüggen and Strobel (2007) carry out the comparison to answer a very specific research question. It concerns a central finding from what are known as gift-exchange experiments. In the gift-exchange game, companies offer workers wages. Once the workers have accepted a wage offer, they decide how hard they want to work. In subgame perfect equilibrium, the companies anticipate that the workers' best response is always to choose the least possible effort and the companies therefore offer the lowest possible wage. However, the experiment shows that companies offer higher wages than in equilibrium and that workers respond with greater effort. This behavior in the gift-exchange game is considered a prime example of reciprocity and has gained great importance, not least in the labor market literature (Fehr et al. 1993, 1998).

The question Brüggen and Strobel seek to answer is whether reciprocity can still be observed in the gift-exchange experiment when the workers have to do real work and cannot simply choose their level of effort. The work to be done in their experiment consisted of solving as many tasks involving the multiplication of two-digit numbers within 5 minutes. On the face of it, this is not a very suitable task because, first, the subjects can do mental arithmetic well to a greater or lesser extent and, second, learning effects are likely to occur if the task has to be performed several times (which was the case in the experiment). Brüggen and Strobel find a clever way to control for both effects, however.

The subjects were invited 1 week before the experiment and were required to solve the multiplication tasks. They were paid a fixed amount for each correct task, making it worthwhile to concentrate for 5 minutes. In this way, the experimenters learned about how good the individual subjects were in this discipline. The performance in the actual experiment was then evaluated as a percentage of the performance in the first part of the experiment. So if someone solved 20 tasks in the first part of the experiment and only 10 in the gift-exchange experiment, then that person had delivered an effort level of 50%. In order to detect possible learning effects, a group of subjects was subjected to the first part of the experiment twice, with a break of 1 week between. It was found that the second time round the performance was about 20% higher. This factor was used to correct for learning effects in the gift-exchange experiment.

Finally, at the end of the experiment, the subjects were required to complete a questionnaire to determine whether they enjoyed their work. This is a very important question, bearing in mind that the subjects who had the role of workers were given 5 minutes to calculate after accepting the contract. The alternative to calculating was simply to spend the 5 minutes waiting, which is comparatively boring. It actually turned out that the subjects enjoyed their work to a certain extent, meaning that the calculation tasks were not necessarily seen as a burden.

Brüggen and Strobel (2007) observed that both treatments, i.e. real efforts and chosen efforts, revealed the same response to the wage offers, with the workers increasing their work effort when their wages were increased. Thus reciprocity was shown to be present in both cases. There were also remarkable differences, however. For example, the variance in effort for real work performance was significantly higher than for simply choosing the level of effort. It was also astonishing that the average effort level of the multiplication tasks was four times higher than the average chosen-effort level.[40]

The experiment of Brüggen and Strobel (2007) shows that extensive corrections are necessary to take talent and learning effects into account when choosing the appropriate real effort task. In the literature a large number of real effort tasks can be found, the majority of which rely less on talent than the mental arithmetic of Brüggen and Strobel. We provide a small selection in the following. Gneezy et al. (2003) had the subjects find ways out of mazes, at Hoffman et al. (1994) they had to answer general quiz questions, Sutter and Weck-Hannemann (2003) presented the subjects with mathematical problems and Fahr and Irlenbusch (2000) had their subjects crack walnuts. Folding and putting letters into envelopes is also popular, as was practiced by Blaufuß et al. (2013) and Fochmann and Weimann (2013), for example.[41]

Two methods for generating real efforts are presented here since they have the advantage of meeting most of the above requirements relatively well. Gill and Prowse (2012) propose what they call a slider task, using sliders that can be moved with the computer mouse to any integer location in an interval from 0 to 100. Forty-eight of them are on one side and all are initially set to 0, with the task being to set as many sliders as possible to the exact value of 50 within 2 minutes. The task is easy to learn and use because the sliders were created with z-tree, a programming language widely used in experimental laboratories. The authors offer the program free of charge. Unfortunately, not even the slider task can eliminate the possibility of learning effects if it is used repeatedly. This is where the second method comes in. Benndorf et al. (2018) follow a method introduced by Erkal et al. (2011). The subjects are provided with a table in which letters are assigned a three-digit code. They are then presented with simple words and they must assign the corresponding code numbers to the letters that make up the word. The ingenious idea proposed by Benndorf et al. is to change the table in two ways after each pass – both the order of the letters and the numerical codes are randomly varied. The effect is that the codes cannot be learned, and this means that the overall learning effects are very small. In the Benndorf et al. experiment, when the task was repeated ten times, the real effort increased by only 8%. For other tasks it often increases by 20% or more. Both methods cannot exclude that subjects just enjoy solving the specific tasks.

40 The intrinsic value of the work has, of course, contributed to this, but the effort involved in the "chosen efforts" is surprisingly low at 23%, which results in costs of 1.3 monetary units (0.13 euros). Measured in the payoff space, this is close to the equilibrium.

41 To provide an idea of the subjects' work performance, in the second experiment (Fochmann and Weimann 2013) 43,300 letters were folded and placed in envelopes. On average, the subjects spent 72 minutes in the laboratory, with the duration of their stay being of their choice.

Finally, it should be noted that it is an unresolved question as to whether external validity can really be improved with real effort tasks as described here. The argument against this is that the tasks the subjects face are very far removed from what is required in real life. Behind this is a difficult trade-off: the closer the real effort task is to reality, the less control there will be over the actual costs borne by the subjects. If this loss of control is accepted, a large number of subjects could be employed in the individual treatments. This would then help to keep the probability that differences between the treatments are due to the spread of talents and preferences concerning the real effort task commensurately small.

2.7.3 Within- Versus Between-Subject Design

At the core of experimental research stands the comparison of different experimental treatments under controlled conditions. An experiment that consists of only one treatment makes relatively little sense. It is almost always a case of subjects making decisions under different conditions, with the treatments that are being compared as far as possible differing in only one parameter, thus enabling conclusions in relation to causality to be made. A fundamental issue of experimental design in this regard is whether each individual subject participates in a number of different treatments or whether every treatment involves different subjects, with each subject participating in only one treatment. The first case is described as a "within-subject design", since the comparison takes place *within* one and the same subject, while the latter case is called "between-subject design" due to the comparison *between* the subjects.

Both designs are employed in experimental research and each has its specific advantages and disadvantages, which we will discuss below.

Advantages of the Within-Subject Design

A very obvious advantage is that the number of observations per subject is greater when each subject participates in several treatments than when new subjects are invited to each treatment. Charness et al. (2012) use a very simple example to illustrate this. Suppose an experiment is being conducted with the intention of comparing the willingness to pay for a sandwich from the bakery around the corner with that for a sandwich from a food stand at the departure hall of a major airport. In the within-subject design, each subject would indicate her willingness to pay at the first location and then at the next location, resulting in two statements per person and thus twice as many statements as in the between-subjects design. However, it is also obvious that this advantage goes hand in hand with the fact that the two observations may not be independent of each other.

As compared to the between-subject design (henceforth, between design), the within-subject design (henceforth, within design) has the advantage that the internal validity of the experiment does not require successful randomization to have been carried out (Charness et al. 2012). In the former, if different people are asked at the airport and in the bakery about their willingness to pay, it must be ensured that the assignment to the two groups was purely random and that there was no selection effect. This is unnecessary with a within design.

Charness et al. (2012) see another advantage of the within design in its closer proximity to theory. For example, if the demand function of a household is formulated, then the idea behind it is that the household is faced with different prices and assigns different demand decisions to them – that is, there is a "within" connection between price and demand quantity. In addition, in the real world it is also important to describe the reaction of actors to parameter changes. Therefore, within design can be credited not only with higher internal validity but also higher external validity.

A within-subject design is ideal if one wants to test the behavioral effects arising from the introduction of a certain scheme (e.g. a new remuneration system) in the laboratory. Here the aim is to determine how subjects familiar with the old scheme react to the new scheme – and this can be observed very successfully in a within-subject design.

Disadvantages of the Within-Subject Design

As already mentioned, in the within design there is no avoiding that dependencies arise between the individual observations in the different treatments. This may well be desired, as described at the end of the last section. Let us go back to our sandwich example. If one and the same person provides the statements on the willingness to pay for the sandwiches from both locations, the first statement may set an anchor for the second. In order to be able to interpret the results, it is necessary to correct for this anchoring effect. One possibility, of course, is to send half of the subjects to the airport first and the other half to the bakery first. The variation of the order gives information about whether there is an anchoring effect, which would manifest itself in the form of an order effect. If this is detected, we are faced with a dilemma. It is clearly possible to consider only the first of the two statements in each respective case, since the anchoring effect does not exist in these; but then we would effectively be looking at the between design. The other possibility is to average the observations. However, it is then no longer entirely clear how the result should be interpreted. The average difference between the statements of willingness to pay will only correspond to the true difference if the distortions caused by the sequence are symmetrical. Whether this really is the case, however, is by no means certain.

The fact that the observations may be dependent also has direct implications for the analysis of the data. Since one and the same person is providing different information, a panel structure of the data is generated. This requires the use of econometric and statistical methods that take into account the fact that the observations made by one person are subject to effects stemming from the characteristics of that person.

The degree of dependence between the individual observations of a within design depends on the particular experiment. It is particularly noticeable if a practice effect occurs within the individual treatments. This happens, for example, when a parameter value is varied between the treatments, but the basic setup of the experiment is identical and has been designed to make the subjects better at making a decision over time. This point can be generalized: whenever individual treatments continue to have an effect on the subjects in any way, within-subject designs are particularly susceptible to bias.

What is probably the biggest issue that can cause trouble with a within design is that presenting the subjects with different treatments can lead to an experimenter demand effect. If the treatments differ only in that a single parameter of the experimental design has been changed, then this change very clearly indicates to the subject what is at issue in the experiment. In extreme cases, the experimenter's experimental setup fully reveals the research question. This may, but does not have to, lead to subjects seeking to behave in a manner they believe fulfills the experimenter's expectations. For example, if a price were being varied, the expectation would be that a price increase would probably lead to a decline in demand. Everything we said about the experimenter demand effect in ▶ Sect. 2.5.1 must be given particular consideration in within designs.

Advantages of the Between-Subject Design

A clear advantage of between designs is that they are very easy to handle. All the considerations that have to be made in connection with transfer effects between treatments in within designs can be ignored here. The only condition to be met is that the subjects are randomly assigned to the different treatments. This means not only letting chance decide who comes into which group, but also checking whether randomization "accidentally" led to a particular selection. For example, it is important to check whether the gender distribution in the groups is reasonably even. A dangerous selection could also emerge if the distribution of the students' subjects of study were to be very uneven. For example, if treatment A were mainly played by economics students and treatment B were played predominantly by humanities students, this random selection could lead to bias. If, however, randomization is successful, then it is absolutely true that "*random assignment is a powerful tool*" (Charness et al. 2012, p .3). It is true because successful randomization makes it possible to prove causal effects very reliably. If a treatment effect can be proven to be highly significant in a between design, then this is strong evidence that the parameter that was changed is responsible for the differences in behavior.

Another advantage is that the statistical analysis of between data is easier than within data, as it is not necessary to correct for dependencies between data elements. Charness et al. (2012) assume that between designs tend to lead to conservative results as compared to within designs. This also suggests that if there is significant evidence of an effect in between-subject design, there is a relatively high certainty that this finding is revealing a causality.

Disadvantages of the Between-Subject Design

The disadvantages are in a sense a mirror image of the advantages of the within design. For example, in some circumstances considerably more resources (time, money, subjects) may be required to obtain statistically meaningful data than with the within design approach. In other words, with the same use of resources, less statistical "power" is likely to be achieved with a between design than with a within design. Moreover, the external validity is not as direct as with a within design. However, it can be argued that this disadvantage is more than compensated for by the high degree of certainty that is gained in detecting causal effects.

As we can see, both methods have their advantages and disadvantages. The problems that arise in within designs due to transfer effects between the treatments can be mitigated somewhat by interspersing the treatments that matter in the experiment between those that have nothing to do with the actual research question and which are only carried out in order to distract the subjects and also to confuse them a little, so that they cannot so easily deduce the research question from the treatments. However, such an experiment requires more time and resources. In addition, even here one cannot exclude learning effects. It can also make sense to combine both designs. Let us assume that two treatments A and B are played in an experiment. Half the subjects undergo A, the other half undergo B. The result is a classic between-subject design, but there is no reason not to subsequently let those who played A play B and present A to those who played B. This does not change the quality of the data of the first run-through at all, but the result is the added advantage of gaining a within design with control over any order effects. In this way, the benefits of both these approaches can be used in one experiment.

Important

When it comes to the question of which issues the subjects should decide on, experimenters have to make three fundamental decisions. First, they must decide whether the subjects should respond to a specific decision of another player or whether they are to specify a complete strategy. The strategy method involves providing a response for all the possible moves of the other players. We have seen that the strategy method has substantial advantages because it makes it possible to generate many decisions with relatively little effort. However, it is important to be aware that it can also lead to a different response from players than a direct response to an opponent's decision.

Second, the experimenters have to decide whether they will give money to the subjects without the subjects having to perform some task for it or whether they will require the subjects do some prior task in order to receive the money. If they decide to have the subjects "work" for their money in order to avoid a house money effect, a decision has to be reached as to what kind of work is required. The main problem here is any learning effects that lead to the real effort being non-linear.

The third basic decision is to choose between a within-subject design and a between-subject design. Both have advantages and disadvantages that need careful consideration. A rule of thumb is that a between design is easier to handle statistically, but in many cases a within design may offer some extra benefit in terms of external validity.

2.8 The Repetition of Games

There are only very few decisions of economic interest that we take once only and are never faced with again. Normally we have to make decisions again and again. In fact, we may even make some of them very frequently. In a certain sense, this is a good thing because it gives us the opportunity to learn and adapt our behavior to experience. This is also the reason why many games are played repeatedly, meaning that the subjects make the same decision several times within one experiment. The methodological implications that this has depend on how the repetitions are designed. For example, for experiments in which strategic interactions occur, it makes a significant difference whether this interaction takes the form of a repeated "one-shot" game – i.e. with a new partner in each round – or whether it is a repeated interaction with one and the same partner. It is possible – although rarely done in practice – to repeat experimental sessions with the same subjects. Here, too, a few things have to be taken into account so that the data obtained can still be meaningfully interpreted.

2.8.1 Repetition Within a Session

The majority of games tested in economic experiments are played over several rounds, i.e. the respective game is played repeatedly with the same subjects within one session. The main reason for these repetitions is to give the players the opportunity to gain experience and learn the game. To a lesser extent, it is also to investigate whether and how behavior changes when one and the same task has to be solved repeatedly. For example, public good experiments are usually played over 10 rounds. A stylized fact that can be deduced from hundreds of such experiments is that contributions to the establishment of the public good decline when the game is repeated. Fischbacher and Gächter (2010) show that there is a high probability of social learning behind this phenomenon.

Learning involves subjects receiving information or gaining experience from which they can draw conclusions. Which information the subjects receive, however, is again a question of experimental design. This means that this design also determines the opportunities for learning in an experiment.

Before dealing with learning in experiments, we first revisit a point from the previous section. It is of great relevance for the strategic situation in which the subjects find themselves whether they are always put together with new partners during the game repetitions or whether they interact with one and the same person several times. It is not crucial that they know for sure that they are always playing with the same partner. It suffices that there is a sufficiently high probability of meeting the same person again. This point is so important because reputation effects, which are not possible in a one-time interaction, can play a role in a repeated interaction. In many experiments found in literature, this distinction is not discernible in the clarity and accuracy that would actually be desirable.

For example, a distinction is often made between partner and stranger design. The first means that a repeated experiment is performed in fixed groups. The subjects therefore know that they are playing with the same person(s) in each round and they receive information about the actions of the other person(s) in the individual rounds. With a partner design, it is clear that we are dealing with a repeated interaction and that reputation and learning effects that are not possible with a one-time interaction may occur. For example, the subjects learn something about how the other person(s) have behaved and they know that the other person(s) can observe what they themselves have done. It should be clear that both could have an impact on future rounds.

However, a completely different strategic situation exists if the subjects know that they will only interact with every other subject once.[42] Although it is possible to learn something about the behavior of the subjects in general, this does not reveal precisely how the game partners in the next round will behave, since their behavior in the preliminary rounds cannot be observed and no direct "messages" can be sent to them through one's own behavior. In order to clearly delineate direct and indirect reputation and learning effects, stranger designs should therefore be structured in such a way that

42 Cooper et al. (1996) experimentally show that the difference between repeated one-shot games and a repeated game can be substantial.

Fig. 2.3 Round Robin Design

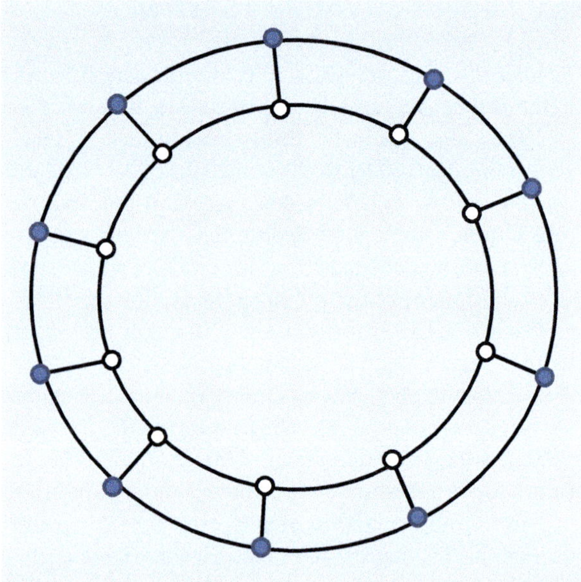

the subjects know that they will definitely not meet subjects of the current round again in future rounds. Indirect effects are only ruled out if it is also certain that the subjects with whom the player will interact in round t also had no indirect contact with subjects with whom the player interacted in $t - i$. To ensure this, what is known as the round-robin design has proved its worth. Suppose a two-person game is to be repeated ten times one-shot. Then 20 subjects are needed to set up a round-robin design. These are divided into two groups of 10 each. To illustrate the method, imagine these 10 subjects in two circles arranged concentrically, resulting in 10 pairs (Fig. 2.3):

After one round, one of the two circles is turned one position further, resulting in 10 new pairs. This rotation is repeated in the same direction until 10 rounds have been completed. This method of creating repeated one-shot games was proposed by Cooper et al. (1996). Kamecke (1997) has shown that it is the only way to avoid direct or indirect reputational effects.

Simply forming new pairs randomly in each game round without using the round-robin design can lead to a problem. Suppose there are two roles in a two-person experiment. Five subjects are assigned role A and five subjects are assigned role B. Then the game is played ten times and five new pairs are formed in each round. Such designs are also sometimes referred to as stranger designs. But strictly speaking, it is a repeated game because the probability of interacting multiple times with at least one of the subjects in the other group is 1, which means that reputation effects can play a role.

What role does learning play in repeated experiments? First of all, it is important to realize that there are different things that can be learned in an experiment. It is useful to differentiate between things that can only be learned if the experimenter provides appropriate information (feedback) and things that can be learned without any feedback. The former includes learning about other subjects' behavior. Different conclusions can be drawn from this behavior. In strategic situations, the information helps to form better expectations about future moves of other subjects. But the information can also be used to find out which "social norm" is currently at work, for example in experiments in which subjects can act more or less socially.

In more complex decision situations, the repetition of the game undoubtedly also leads to the subjects learning to understand the game better. This does not necessarily mean that they will converge toward the theoretically predicted behavior, but they could. For this, it is essential that subjects receive feedback after each round, even if sometimes only in the form of their own payoff.

It is also possible, however, for the subjects to learn without receiving feedback. Weber (2003) demonstrated that subjects were even able to learn about other people's behavior without being told how they behaved. The experiment in which he observed such "learning without feedback" used a so-called guessing game. The object of this game is that all the members of a group specify a number between 0 and 100. The winner is the group member whose number is closest to 2/3 of the average of all the numbers mentioned. Research has shown that the behavior observed in such experiments can be readily explained by what is known as k-level reasoning. A level-0 player randomly selects a number. A level-1 player assumes that the other players are level-0 players and therefore chooses 2/3 of 50 as number (33). A level-2 player assumes that the other players are level-1 players and selects 2/3 of 33 (22), and so on. In the Nash equilibrium of the game, all the players choose zero. Weber was able to observe the subjects converging toward the equilibrium when repeating the game, although they received no feedback. The explanation for this is that a subject who behaved like a level-1 player in round 1 could then assume that in round 2 the others will behave like level-1 players and he will therefore become a level-2 player. All he has to keep in mind is that the others may have thought the same way he did.

Yet there is still more for a person to learn without feedback, especially about himself. He learns how he feels when he plays a certain strategy. This is especially important when the experiment brings subjects into conflict, for instance, between selfish and social behavior. The subject who acted selfishly may have learned that this behavior was not sanctioned – neither by the experimenter nor by his own conscience. The player who behaved socially might learn that she did enough for others in the last round and can now also think of herself. In addition, subjects generally learn what it is like to go through the experiment. They get used to the environment, to the processes and to the presence of an experimenter. The situation in which they find themselves loses its unfamiliarity. All of this can lead to changes in behavior over time. If such changes can be observed, it may be worth investigating their cause. Sometimes this is possible by varying learning opportunities and monitoring whether this changes behavior. As a matter of principle, it is vital to control the learning opportunities as effectively as possible in a repeated experiment.

A particular problem in relation to the repetition of experiments is the endeavor to perform games that are repeated "infinitely often" in the laboratory. The reason for this is that such games have an important role to play from a game-theoretical point of view. The best example of this is the fact that cooperative behavior – provided that the players' self-interest and rationality are common knowledge – can only be modeled in prisoner's dilemma situations in game theory if the game is repeated infinitely often. Now it is clear that an experiment cannot really be repeated indefinitely. The most common method of implementing something similar to "infinity" in the laboratory is to repeat experiments indefinitely and often. This is achieved by performing a next round only with a predefined probability after a certain number of definite repetitions, so that the experiment is terminated in finite time, but the subjects do not know when this will happen. This method, admittedly, quickly comes up against natural limits. Bruttel and Kamecke (2012) point out that laboratory experiments can only take a certain number of hours and that this means that the probability of termination usually either has to be very high or cannot be stationary, because it must approach a value of one at some point. As a result, what is termed the final-round effect, which normally occurs in experiments with a fixed number of rounds, can even be found in experiments with random termination. In cooperation experiments, for example, the willingness to cooperate diminishes in the last few rounds.

Bruttel and Kamecke (2012) propose a procedure that avoids this effect, but has the disadvantage of being comparatively complicated. They have the subjects make a series of consecutive decisions. This happens until (in a two-person game with two pure strategies each) one of the four possible strategy combinations is played twice in a row. If this happens, this combination is counted as what is played to infinity, and the computer calculates the resulting payoffs using a discount factor of 0.8. Alternatively, after the introductory rounds, the subjects can also specify their own future strategy by indicating how they will play in the next round if one of the four possible strategy combinations occurs in the current round. Bruttel & Kamecke's experiment showed that the first method in particular had a stabilizing effect and was able to mitigate the final-round effect.

Question

Can you think of any other reasons (apart from trying to create "infinity") why you should play experiments that do not have a definite end-point?

The subjects' making the same decision several times in a row in an experiment has an impact on the statistical analysis of the data, since these then have the character of panel data, i.e. there are several temporally different observations for one and the same person. This necessitates taking into account the conditions that are specific to the individual subject (for example, their preferences, their personality) and that do not change during the various runs of the experiment. The procedures available for this will be dealt with in a separate section in ▶ Chap. 4, which deals with statistical analysis.

Important

Repeating games within a session can create different learning effects. The experimenter determines which are possible, at least in part, through the design of the experiment. For example, if the same partners always play the game together, the learning opportunities that arise are different from those that occur when playing with new subjects in each round. If the experiment is to be designed in such a way that one-shot interactions are to be repeated, then a round-robin design should be chosen because this is the only way to avoid direct and indirect reputation effects.

Which learning effects occur also depends on the complexity and type of the game. In more complex games, repetition can lead to a better understanding of the game.

In repeated games, it is important whether the test subjects know how often they play or whether they are unaware of it. Games played with random termination can be used as a (more or less close) substitute for unending games.

2.8.2 The Repetition of Sessions

It is very common for games to be repeated within a session. This is probably the case in the majority of experiments. It is very rare, however, that entire sessions are repeated identically with the same subjects (exceptions are Volk et al. 2012; Carlsson et al. 2014; Sass and Weimann 2015; and Brosig-Koch et al. 2017; and Sass et al. 2018). Yet such repetitions make perfect sense. As we have already said, the vast majority of decisions that people make are not made just once. They are repeated. The same applies to many interactions we have with our fellow human beings. Whether to be cooperative or not, whether to voluntarily sacrifice income to help others, whether to tell the truth when a lie would benefit us more – all these are issues that every human being faces very often and in a very similar way. The question is whether we can learn how people behave in such recurring situations from experiments that only take place once in their own very unique way. This question seems all the more justified because the experimental situation is a special one – one that inevitably differs from real-world situations. Would it not make sense to repeat sessions to make them more similar to real-world decision-making situations?

There is a second point to add. Laboratory experiments present the experimental subjects with a very special and necessarily artificial situation. We have pointed out in many places in this book that this particular laboratory situation can have an impact on the behavior of the subjects. These effects are likely to be all the stronger, the more unfamiliar the laboratory environment is to the subjects. By repeating sessions with a certain amount of time between them, it can be possible to accustom the subjects to the experimental situation and thus eliminate the unfamiliarity. It is reasonable to assume, for example, that experimenter demand effects can be mitigated. The subjects are under observation and this can influence their behavior. It is possible to become accustomed to being observed, however, and when this happens, repeated sessions are more likely to reveal "true" behavior than those that are performed only once.

Yet why shouldn't it be possible to achieve these effects by repeating the game within one session? There are at least two arguments against it. First, it could be that subjects view a repeated one-shot game within a session as *one* game. An example may help to clarify this point. The phenomenon of moral self-licensing has been described in the literature.[43] This means that after doing something "good" or "social", people grant themselves the right to think more about themselves and act more selfishly at the next opportunity. This phenomenon is more likely to be observed in repeated sessions than in repeated games in a session because in the latter case the decisions could be viewed as a whole and therefore there is no real repetition of the situation.

The second reason is that when deciding how to behave in the experiment, the subjects are also likely to be guided by the opportunity costs of participating in the experiment. If an experiment is played over several rounds and these rounds are paid off, the average opportunity costs per round in the course of the experiment will decline.

43 Merritt et al. (2010) provide a survey of the literature on "moral licensing". More recent experimental studies relating to this effect have been conducted by Brañas-Garza et al. (2013), Ploner and Regner (2013), Clot et al. (2014), Cojoc and Stoian (2014) and Brosig-Koch et al. (2017), for example.

This means the individual rounds are no longer identical. Subjects who go into the experiment with the goal that they definitely want to recoup their opportunity costs, for example, may change their behavior once this goal has been achieved.

Hence, there are good reasons for repeating sessions identically. Seeing as this is the case, why are experiments that do just that such a big exception? Any answer to this question is, of course, speculative. One theory is that experimental economists refrain from such experiments because they are accompanied by an unavoidable loss of control. It is not possible to control everything that the subjects do between sessions, and it cannot be ruled out that things may happen that might influence their behavior in the experiment. This places a constraint on one advantage of laboratory experiments – their high internal validity.

An obvious complication of such experiments would be that the subjects talk to each other between the experiments, with all things being possible. For instance, they might discuss what behavior is appropriate in the experiment, or coordinate their actions and so on. In such a case, the experimenter would have no control whatsoever over the communication content and thus no control over the experiment. Uncontrolled conversations between the subjects are in a sense the worst-case scenario in experiments with repeated sessions. However, they are relatively easy to prevent. Strict adherence to the anonymity of the subjects is all that is necessary, beginning with subject selection. If they are students (which is generally the case), the subjects should not all be from the same faculty and, if possible, not from the same academic year. This already reduces the probability of meeting each other by chance after the experiment. It is, of course, imperative that the subjects have no contact with each other before, during or after the experiment. They should therefore be individually invited to different meeting points and picked up from there. During the experiment they should sit in soundproof booths that prevent them from seeing other subjects, thus making it impossible to make contact with them. After the experiment, they are to be led out of the booths individually, so that no contact can be made at this point either. If all these rules are observed, at least the probability of direct communication between the subjects of the experiment is very small.

Of course, the subjects cannot be prevented from talking to other people about the experiment and gaining experience over which the experimenter has no control. Even if the subjects cannot be relied upon to comply, the instructions should in any case include the statement that such conversations are undesirable. It also cannot be avoided that subjects inform themselves about the experiment, for example, by doing research on the Internet. Although such behavior may be the exception, it cannot be ruled out. All these factors lead to a loss of control on the part of the experimenter. However, this does not necessarily mean that the results of repeated sessions cannot be interpreted. First of all, it must be conceded that even between two more or less identical decision-making situations in real life, people gather new information and gain new insights. This means that the restriction of internal validity is accompanied by an increase in external validity. In reality, the gain in information and knowledge will not be very systematic, i.e. the experiences gathered are likely to be randomly distributed. One might argue that the same is likely to happen between experimental sessions. In other words, the influences to which the subjects are exposed between two sessions are likely to go in very different directions. If the experiment shows that there is a uniform change in behavior throughout the sessions, it is therefore unlikely that this will have been generated by experiences made between the sessions.

Another problem with repeated sessions is unreliable subjects who do not attend all the sessions. Subjects failing to show up is not only annoying, but also reduces the interpretability of the experimental results because it cannot be ruled out that a selection process is associated with their absence. For example, it could be precisely the subjects who had certain kinds of experience in previous sessions who are missing in later sessions. A very effective means against such absences is to postpone payments to the end of the series. The threat of going away empty-handed if one does not show up for all the sessions is quite credible and should not fail to have the desired effect.

Furthermore, careful consideration should be given to the payoff design in repeated sessions. For example, if a session is played four times, the payoffs in the first sessions may result in income effects that could influence behavior in subsequent sessions. In addition, subjects may carry out a kind of portfolio planning to manage their payoffs in one way or another over the four sessions. Both effects can be eliminated by randomly selecting one of the four sessions at the end of the last session to be the payoff-relevant session. If it is desirable to prevent the incentives from becoming too weak, the proceeds from this session can be paid out four times. This payoff method has another advantage that might play a role in certain experiments. It can prevent information about other subjects' behavior that the payoffs might indirectly convey from reaching the subjects. This could be important, for example, if the experiment aims to investigate the pure repetition effect and it is desirable to avoid learning and reputation effects as far as possible.

Although the repetition of sessions raises a number of additional methodological problems, it is able to solve some problems that are frequently encountered in connection with conventional experiments[44] and that mainly concern the external validity of experiments. For example, it is criticized that subjects have too little opportunity to become accustomed to the artificial experimental situation. This is combined with the criticism that experiments do not give the subjects enough scope for learning processes and gaining experience. Having the subjects repeat sessions obviously resolves these issues. There is no denying the loss of control that goes hand-in-hand with this, but it is also quite certain that repeated sessions lead to "more mature" behavior being observed in the experiment, i.e. behavior that might have been reflected on more and that therefore might approach behavior in real contexts more closely than does behavior in one-off sessions. It is important to weigh this advantage against the disadvantage associated with loss of control.

Important

The repetition of sessions is comparatively rare. One reason for this may well be the fact that it is not possible to control what the subjects do between sessions, thus implying a loss of control. On the other hand, repetition may lead to an increase in external validity and allow mature behavior to be observed in the laboratory. The loss of control can, in particular, be minimized by ensuring that the probability of contact between the subjects is as low as possible. This is achieved first and foremost by maintaining strict anonymity during the experiment.

44 See, for example, the discussion initiated mainly by Levitt and List (2007).

2.9 The Reproducibility of Experiments

Scientific research aims to make general statements about causal relationships whose validity can be verified intersubjectively. An experiment conducted in a particular laboratory at a particular time with particular subjects cannot be the basis for such a statement. Its results are in the first instance no more than a single observation. If a causal relationship is established, it applies specifically to this experiment, and it cannot easily be generalized. Experimental findings become usable – to put it more precisely – actually only when they have been proven several times and it has been shown that they apply irrespective of time and place. Of course, experimental methods cannot prove that a certain causality actually exists. This applies to all experimental and empirical disciplines. To describe it using a famous example, even if one has observed so many white swans, one cannot logically derive the statement "All swans are white" from this. That would be invalid inductive reasoning.

In the final analysis, however, it is not the aim of experimental research to produce "true" statements in the above sense. Rather, it is about creating empirical, experimental evidence. This means gaining insight into which causal relationships are likely to be encountered under which circumstances. This is not possible with a single experiment, but requires the experience that observations can be reproduced relatively reliably. This has some methodological implications, also for the design of experiments. The most important implication is that experiments must always be designed in such a way that they are reproducible.

The requirement of reproducibility is essential for scientific work. If it is not fulfilled, a basic prerequisite for the scientificality of results cannot be fulfilled: the intersubjective verifiability of scientific statements. In principle, experiments lend themselves very well to being reproduced. The prerequisite for this is that the design of the experiment does not contain any elements that are bound to a specific location or to a specific person. An experimental result that can only be observed if the experiment is carried out by assistant X but not if it is supervised by someone else is worthless. The same applies if the result can only be observed in a particular laboratory. Experiments must therefore be designed in such a way that they can be reproduced in any laboratory by any experimenter.

However, it is not enough to simply create a design that allows an experiment to be reproduced. So that a replication can actually be conducted, this design and all its elements must also be well documented. Comprehensive documentation must guarantee that the experiment can be conducted identically by other people at a different location. This means that details of all the auxiliary materials used must be supplied. Not only the instructions that the subjects received, for instance, but also the software used in the experiment should be available to those who want to reproduce it. In addition, the procedure of the experiment must be documented very precisely. This includes, for example, the way in which the subjects were invited and received in the laboratory, whether and in what form they had contact with each other, how the instructions were distributed, whether they were read aloud and how questions of understanding were dealt with. Every single detail could be important. Of course, the raw data collected in an experiment are also among the things that have to be documented. This is not absolutely necessary for reproducing the experiment, but reproducibility refers not only to the experiment, but also to the subsequent statistical analysis. There are several reasons for this. On the one hand, it is then possible to check that the analysis is free of errors. On the other hand, the number of possible statistical procedures that can be applied is very large. Sometimes the question arises whether the results will change if other statistical methods are used. This is also a kind of robustness check which can be carried out and which says something about the stability of the results.

Documenting an experiment properly and comprehensively is a necessary condition for it to be replicable, but it is not sufficient. For the experiment to be replicated identically, all these documents must also be accessible. This is a problem that should not be underestimated. At present, the situation in experimental economics means that it is more or less a matter of chance whether the information needed to replicate a particular experiment can be obtained. In practice, this is done by writing an email to the relevant colleague and asking him or her to provide the documentation. Success in this depends on how well organized the documentation department in the laboratory concerned is. It does sometimes happen that things go missing. This risk is especially high when many authors have worked on the experiment, all of whom have relied on each other to document it. Recently there have been various initiatives aimed at creating central repositories where experimental data is collected, processed and made available in such a way that the replication of experiments is assured. At the time of writing, for example, the *Deutsche Forschungsgemeinschaft* (German Research Foundation) is funding a project that aims to create a database that will collect experimental data from all the social sciences and provide access to it in a repository. Ideally, such initiatives will result in the establishment of a central collection point that will provide clear guidelines for documenting experiments and provide access to all the experiments that are appropriately prepared and stored to all researchers worldwide. In this case, it is easy to imagine that a routine of sorts will arise, with every experiment carried out anywhere in the world being documented accordingly. Routines are very helpful here, because if they exist, an experiment can from the outset be designed in such a way that it can easily be documented according to the appropriate standard.

In order for an experiment to be replicated identically, it is necessary to provide the experimental design in detail. This does not mean, however, that variations in design should not be allowed when reproducing experiments. As long as the experiment is reproduced in all its essential aspects, it can even be very helpful if minor deviations from the original occur. It is ultimately a matter of obtaining *robust* results, and an experimental observation that disappears when the experiment is performed a little differently is not robust.

The reproducibility of experiments is important not only because it is the prerequisite for generalizable, robust causal statements. It has a second and equally important function: it is an indispensable tool in ensuring good scientific practice. Behind this lies the fact that in the very competitive scientific world there are substantial incentives to deviate from this practice. The more unexpected and surprising an experimental result is, the greater the chance of publishing it in a good journal. Such results are easy to achieve – by manipulating the data. The calculation is very simple. The easier it is to replicate an experiment, the less it pays to manipulate the data. If someone invents sensational results, he can be pretty sure that there will be people who will make the effort to reconstruct the experiment. Spectacular examples from the natural sciences show how great the danger of data being manipulated is. Experiments there are sometimes much more difficult to reproduce because they may involve substantial technical and human resources.

Experimental economists have a huge interest in everything being done legitimately in their scientific community. If manipulations were to occur, the credibility of the entire profession would be massively shaken. By ensuring that experiments can be easily replicated, the scientific community is very effectively protected against this danger. The biggest problem with reproducing experiments is often not the experiment itself, but the fact that there are far too few incentives in the scientific world to carry it out. Imitating the experiment of one or the other scientist – possibly with the same results – is not very creative and satisfies scientific curiosity only to a very limited extent. In addition, the chances of publication of a straightforward replication are extremely meager. The information that a particular result could be reproduced can in principle be summarized in one sentence: "The results of the experiment that Mr. X or Ms. Y published in Z could be reproduced." This is not something that will boost the impact of a journal. But if nobody wants to publish simple replications, why even do them?

Replications are of course interesting if they show that the result of an experiment cannot be confirmed. Whether this is the case, however, is only known after repeating the experiment. For this reason, this incentive does not have a special appeal either since there is a very high risk of obtaining the same results as in the original experiment. The fact that many experiments have been replicated despite these problems is due to the fact that many experiments have not been completely redesigned, but are rather variations of already known designs. As a result, the baseline treatment used in the respective type of experiment is also played in the variations. For example, the ultimatum game experiment has been carried out in many variations. In most of these variants, the original ultimatum game is needed as a baseline against which the effects of a new variation of the experiment are measured. In this way, replications of the baseline treatment are created as a by-product, as it were. The same applies to all the major experimental designs, with all of them having been investigated in ever new variations.

Nevertheless, it remains a fundamental problem of experimental economic research that there is such a lack of incentive to carry out the important task of reproducing experiments.

Important

Experiments must be reproducible, for only then can stylized facts, i.e. observations that can be reproduced again and again, be derived from individual observations that are tied to a time and a place. Moreover, reproducibility ensures the reliability of experimental results and protects the scientific community from manipulation. Although reproducing experiments is vital, there is little incentive to do so.

References

Abbink, K., & Hennig-Schmidt, H. (2006). Neutral versus loaded instructions in a bribery experiment. *Experimental Economics, 9*(2), 103–121.

Abeler, J., & Nosenzo, D. (2015). Self-selection into laboratory experiments: Pro-social motives versus monetary incentives. *Experimental Economics, 18*(2), 195–214.

Alberti, F., & Güth, W. (2013). Studying deception without deceiving participants: An experiment of deception experiments. *Journal of Economic Behavior & Organization, 93*, 196–204.

Anderson, J., Burks, S., Carpenter, J., Gotte, L., Maurer, K., Nosenzo, D., Potter, R., Rocha, K., & Rustichini, A. (2013). Self-selection and variations in the laboratory measurement of other-regarding preferences across subject pools: Evidence from one college student and two adult samples. *Experimental Economics, 16*(2), 170–189.

Anderson, L. R., Mellor, J. M., & Milyo, J. (2008). Inequality and public good provision: An experimental analysis. *Journal of Socio-Economics, 37*, 1010–1028.

Andreoni, J. (1988). Why free ride?: Strategies and learning in public goods experiments. *Journal of Public Economics, 37*, 291–304.

Bardsley, N. (2008). Dictator game giving: Altruism or artefact? *Experimental Economics, 11*(2), 122–133.

Barrera, D., & Simpson, B. (2012). Much ado about deception: Consequences of deceiving research participants in the social sciences. *Sociological Methods Research, 41*(3), 383–413.

Barmettler, F., Fehr, E., & Zehnder, C. (2012). Big experimenter is watching you! An-onymity and prosocial behavior in the laboratory. *Games and Economic Behavior, 75*(1), 17–34.

Becker, G. M., DeGroot, M. H., & Marschak, J. (1964). Measuring utility by a single-response sequential method. *Systems Research and Behavioral Science, 9*(3), 226–232.

Belot, M., Duch, R., & Miller, L. (2015). A comprehensive comparison of students and non-students in classic experimental games. *Journal of Economic Behavior & Organization, 113*, 26–33.

Benndorf, V., Rau, H., Sölch, C. (2018). Minimizing learning behavior in experiments with repeated real-effort tasks. SSRN: https://ssrn.com/abstract=2503029 or https://doi.org/10.2139/ssrn.2503029

Ben-Ner, A., Putterman, L., & Ren, T. (2011). Lavish returns on cheap talk: Two-way communication in trust games. *Journal of Socio-Economics, 40*, 1–13.

Binmore, K., Shaked, A., & Sutton, J. (1985). Testing noncooperative bargaining theory: A preliminary study. *American Economic Review, 75*(5), 1178–1180.

Binswanger, H. P. (1980). Attitudes toward risk: Experimental measurement in rural India. *American Journal of Agricultural Economics, 62*(3), 395–407.

Binswanger, H. P. (1981). Attitudes toward risk: Theoretical implications of an experiment in rural India. *The Economic Journal, 91*(364), 867–890.

Blanco, M., Engelmann, D., Koch, A. K., & Normann, H. T. (2010). Belief elicitation in experiments: Is there a hedging problem? *Experimental Economics, 13*(4), 412–438.

Blaufuß, K., Fochmann, M., Hundsdoerfer, J., Kiesewetter, D., & Weimann, J. (2013). Net wage illusion in a real-effort experiment. *The Scandinavian Journal of Economics, 115*(2), 476–484.

Blume, A., & Ortmann, A. (2000). The effect of costless pre-play communication: Experimental evidence from a game with Pareto-ranked equilibria. *Journal of Economic Theory, 132*, 274–290.

Bortolotti, S., Casari, M., & Pancotto, F. (2015). Norms of punishment: Experiments with students and the general population. *Economic Enquiry, 53*, 1207–1223.

Bohnet, I., & Frey, B. S. (1999). The sound of silence in prisoner's dilemma and dictator games. *Journal of Economic Behavior & Organization, 38*(1), 43–57.

Bonetti, S. (1998). Experimental economics and deception. *Journal of Economic Psychology, 19*, 377–395.

Brañas-Garza, P. (2007). Promoting helping behavior with framing in dictator games. *Journal of Economic Psychology, 28*, 477–486.

Brañas-Garza, P., Bucheli, M., Espinosa, M. P., & García-Muñoz, T. (2013). Moral cleansing and moral licenses: Experimental evidence. *Economics & Philosophy, 29*(2), 199–212.

Brandts, J., & Charness, G. (2003). Truth or consequences: An experiment. *Management Science, 49*(1), 116–130.

Brandts, J., & Charness, G. (2011). The strategy versus the direct-response method: A first survey of experimental comparisons. *Experimental Economics, 14*, 375–398.

Brosig, J., Weimann, J., & Ockenfels, A. (2003). The effect of communication media on cooperation. *German Economic Review, 4*(2), 217–241.

Brosig, J., Weimann, J., & Yang, C. L. (2004). The hot versus cold effect in a simple bargaining experiment. *Experimental Economics, 6*(1), 75–90.

Brosig, J., Heinrich, T., Riechmann, T., Schöb, R., & Weimann, J. (2010). Laying off or not? The influence of framing and economics education. *International Review of Economics Education, 9*, 44–55.

2

Brosig-Koch, J., & Heinrich, T. (2018). The role of communication content and reputation in the choice of transaction partners: A study based on field and laboratory data. *Games and Economic Behavior, 112*, 49–66.

Brosig-Koch, J., Helbach, C., Ockenfels, A., & Weimann, J. (2011). Still different after all these years: Solidarity behavior in East and West Germany. *Journal of Public Economics, 95*, 1373–1376.

Brosig-Koch, J., Riechmann, T., & Weimann, J. (2017). The dynamics of behavior in modified dictator games. *PLoS One, 12*(4), e0176199.

Brüggen, A., & Strobel, M. (2007). Real effort versus chosen effort in experiment. *Economics Letters, 96*, 232–236.

Bruttel, L., & Kamecke, U. (2012). Infinity in the lab. How do people play repeates games. *Theory and Decision, 72*, 205–219.

Buchan, N. R., Johnson, E. J., & Croson, R. T. (2006). Let's get personal: An international examination of the influence of communication, culture and social distance on other regarding preferences. *Journal of Economic Behavior & Organization, 60*(3), 373–398.

Burnham, T., McCabe, K., & Smith, V. L. (2000). Friend-or-foe intentionality priming in an extensive form trust game. *Journal of Economic Behavior & Organization, 43*(1), 57–73.

Cachon, G., & Camerer, C. (1996). Loss avoidance and forward induction in experimental coordination games. *Quarterly Journal of Economics, 111*(1), 166–194.

Cappelen, A. W., Nygaard, K., Sørensen, E., & Tungodden, B. (2015). Social preferences in the lab: A comparison of students and a representative population. *Scandinavian Journal of Economics, 117*(4), 1306–1326.

Camerer, C. F., & Hogarth, R. M. (1999). The effects of financial incentives in experiments: A review and capital-labor-production framework. *Journal of Risk and Uncertainty, 19*, 7–42.

Carlsson, F., He, H., & Martinsson, P. (2013). Easy come, easy go. *Experimental Economics, 16*(2), 190–207.

Carlsson, F., Johansson-Stenman, O., & Nam, P. K. (2014). Social preferences are stable over long periods of time. *Journal of Public Economics, 117*, 104–114.

Charness, G., & Dufwenberg, M. (2006). Promises and partnership. *Econometrica, 74*, 1579–1601.

Charness, G., & Gneezy, U. (2008). What's in a name? Anonymity and social distance in dictator and ultimatum games. *Journal of Economic Behavior & Organization, 68*(1), 29–35.

Charness, G., Gneezy, U., & Kuhn, M. A. (2012). Experimental methods: Between-subject and within-subject design. *Journal of Economic Behavior & Organization, 81*(1), 1–8.

Charness, G., Haruvy, E., & Sonsino, D. (2007). Social distance and reciprocity: An internet experiment. *Journal of Economic Behavior & Organization, 63*(1), 88–103.

Cherry, T., Frykblom, P., & Shogren, J. (2002). Hardnose the dictator. *American Economic Review, 92*(4), 1218–1221.

Cherry, T. L., Kroll, S., & Shogren, J. F. (2005). The impact of endowment heterogeneity and origin on public good contributions: Evidence from the lab. *Journal of Economic Behavior & Organization, 57*(3), 357–365.

Cherry, T. L., & Shogren, J. F. (2008). Self-interest, sympathy and the origin of endowments. *Economics Letters, 101*(1), 69–72.

Clark, J. (2002). House money effects in public good experiments. *Experimental Economics, 5*(3), 223–231.

Cleave, B. L., Nikiforakis, M., & Slonim, R. (2013). Is there selection bias in laboratory experiments? The case of social and risk preferences. *Experimental Economics, 16*(3), 372–382.

Clot, S., Grolleau, G., & Ibanez, L. (2014). Smug alert! Exploring self-licensing behavior in a cheating game. *Economics Letters, 123*(2), 191–194.

Cojoc, D., & Stoian, A. (2014). Dishonesty and charitable behavior. *Experimental Economics, 17*(4), 717–732.

Cooper, D. (2014). A note on deception in economic experiments. *Journal of Wine Economics, 9*(2), 111–114.

Cooper, R., DeJong, D. V., Forsythe, R., & Ross, T. W. (1996). Cooperation without reputation: Experimental evidence from prisoner's dilemma games. *Games and Economic Behavior, 12*, 187–218.

Costa-Gomes, M. A., & Weizsäcker, G. (2008). Stated beliefs and play in normal-form games. *Review of Economic Studies, 75*(3), 729–762.

Cox, J. C., & Deck, A. A. (2005). On the nature of reciprocal motives. *Economic Inquiry, 43*(3), 623–635.

Cox, J. C., Robertson, B., & Smith, V. L. (1982). Theory and behavior of single object auctions. In V. L. Smith (Ed.), *Research in experimental economics* (Vol. 2, pp. 1–43). Greenwich: JAI Press.

Cox, J. C., Sadiraj, V., & Schmidt, U. (2015). Paradoxes and mechanisms for choice under risk. *Experimental Economics, 18*(2), 215–250.

Cox, J. C., Smith, V. L., & Walker, J. M. (1992). Theory and misbehavior of first-price auctions: Comment. *American Economic Review, 82*(5), 1392–1412.

Crawford, V. (1998). A survey of experiments on communication via cheap talk. *Journal of Economic Theory, 78*(2), 286–298.

Csukás, C., Fracalanza, P., Kovács, T., & Willinger, M. (2008). The determinants of trusting and reciprocal behaviour: Evidence from an intercultural experiment. *Journal of Economic Development, 33*(1), 71–95.

Dawes, R. M., McTavish, J., & Shaklee, H. (1977). Behavior, communication, and assumptions about other people's behavior in a commons dilemma situation. *Journal of Personality and Social Psychology, 35*(1), 1–11.

Dreber, A., Ellingsen, T., Johannesson, M., & Rand, D. G. (2013). Do people care about social context? Framing effects in dictator games. *Experimental Economics, 16*(3), 349–371.

Dufwenberg, M., Gächter, S., & Hennig-Schmidt, H. (2011). The framing of games and the psychology of play. *Games and Economic Behavior, 73*, 459–478.

Eckel, C. C., & Grossman, P. J. (1996). Altruism in anonymous dictator games. *Games and Economic Behavior, 16*(2), 181–191.

Ellingsen, T., & Johannesson, M. (2004). Promises, threats and fairness. *The Economic Journal, 114*(495), 397–420.

Erkal, N., Gangadharan, L., & Nikiforakis, N. (2011). Relative earnings and giving in a real-effort experiment. *American Economic Review, 101*(7), 3330–3348.

Erhard, K.-A., & Keser, C. (1999). *Mobility and cooperation: On the run* (Scientific series). Montreal: CIRANO, 99s-24.

Exadaktylos, F., Espín, A. M., & Branas-Garza, P. (2013). Experimental subjects are not different. *Scientific Reports, 3*, 1213.

Fahr, R., & Irlenbusch, B. (2000). Fairness as a constraint on trust in reciprocity: Earned property rights in a reciprocal exchange experiment. *Economics Letters, 66*, 275–282.

Falk, A., Meier, S., & Zehnder, C. (2013). Do lab experiments misrepresent social preferences? The case of self selected student samples. *Journal of the European Economic Association, 11*(4), 839–852.

Farrell, J., & Rabin, M. (1996). Cheap talk. *Journal of Economic Perspectives, 10*(3), 103–118.

Fehr, E., Kirchler, E., Weichbold, A., & Gächter, S. (1998). When social norms over-power competition: Gift exchange in experimental labor markets. *Journal of Labor Economics, 16*(2), 324–351.

Fehr, E., Kirchsteiger, G., & Riedl, A. (1993). Does fairness prevent market clearing? An experimental investigation. *Quarterly Journal of Economics, 108*, 437–460.

Fehr, E., & Schmidt, K. M. (2006). The economics of fairness, reciprocity and altruism–experimental evidence and new theories. In S.-C. Kolm & J. M. Ythier (Eds.), *Handbook of the economics of giving, altruism and reciprocity* (Vol. 1, pp. 615–691). Amsterdam: North-Holland.

Feltovich, N. (2011). What's to know about laboratory experimentation in economics. *Journal of Economic Surveys, 25*, 371–379.

Fischbacher, U., & Föllmi-Heusi, F. (2013). Lies in disguise—An experimental study on cheating. *Journal of the European Economic Association, 11*(3), 525–547.

Fischbacher, U., & Gächter, S. (2010). Social preferences, beliefs, and the dynamics of free riding in public goods experiments. *American Economic Review, 100*(1), 541–556.

Fischbacher, U., Gächter, S., & Fehr, E. (2001). Are people conditionally cooperative? Evidence from a public goods experiment. *Economics Letters, 71*(3), 397–404.

Fochmann, M., & Weimann, J. (2013). The effects of tax salience and tax experience on individual work efforts in a framed field experiment. *Public Finance Analysis, 69*, 1–32.

Fonseca, M. A., & Normann, H. T. (2008). Mergers, asymmetries and collusion: Experimental evidence. *The Economic Journal, 118*(527), 387–400.

Fonseca, M. A., & Normann, H. T. (2012). Explicit vs. Tacit collusion – The impact of communication in oligogoply experiments. *European Economic Review, 56*, 1759–1772.

Forsythe, R., Horowitz, J. L., Savin, N. E., & Sefton, M. (1994). Fairness in simple bar-gaining experiments. *Games and Economic Behavior, 6*(3), 347–369.

Frank, R. H., Gilovich, T., & Regan, D. T. (1993). Does studying economics inhibit co-operation? *Journal of Economic Perspectives, 7*(2), 159–171.

Fréchette, G. R., & Schotter, A. (2015). *Handbook of experimental economic methodology*. Oxford: Oxford University Press.

Frey, B. S., & Oberholzer-Gee, F. (1997). The cost of price incentives: An empirical analysis of motivation crowding-out. *American Economic Review, 87*(4), 746–755.

Friedman, D. (1992). Theory and misbehavior of first-price auctions: Comment. *American Economic Review, 82*(5), 1374–1378.

Fryer, R. G. (2013). Teacher incentives and student achievement: Evidence from New York City public schools. *Journal of Labor Economics, 31*(2), 373–407.

Gächter, S., & Fehr, E. (2000). Cooperation and punishment in public goods experiments. *American Economic Review, 90*(4), 980–994.

Geanakoplos, J. D., Pearce, J. D., & Stacchetti, E. (1989). Psychological games and sequential rationality. *Games and Economic Behavior, 1*(1), 60–79.

Gill, D., & Prowse, V. (2012). A structural analysis of disappointment aversion in a real effort competition. *American Economic Review, 102*(1), 469–503.

Gneezy, U., Niederle, M., & Rustichini, A. (2003). Performance in competitive environments: Gender differences. *Quarterly Journal of Economics, 118*, 1049–1074.

Gneezy, U., & Rustichini, A. (2000). Pay enough or don't pay at all. *Quarterly Journal of Economics, 115*(3), 791–810.

Goerg, S. J., Hennig-Schmidt, H., Walkowitz, G., & Winter, E. (2016). In wrong anticipation-miscalibrated beliefs between Germans, Israelis, and Palestinians. *PLoS One, 11*(6), e0156998.

Greiner, B., Güth, W., & Zultan, R. (2012). Social communication and discrimination: A video experiment. *Experimental Economics, 15*, 398–417.

Grimm, V., & Mengel, F. (2011). Let me sleep on it: Delay reduces rejection rates in ultimatum games. *Economics Letters, 111*(2), 113–115.

Güth, W., & Kocher, M. G. (2014). More than thirty years of ultimatum bargaining experiments: Motives, variations, and a survey of the recent literature. *Journal of Economic Behavior & Organization, 108*, 396–409.

Güth, W., Schmittberger, R., & Schwarze, B. (1982). An experimental analysis of ultimatum bargaining. *Journal of Economic Behavior & Organization, 3*, 367–388.

Harbring, C. (2006). The effect of communication in incentive systems – An experimental study. *Managerial Decision Economics, 27*, 333–353.

Harrison, G. W. (1989). Theory and misbehavior of first-price auctions. *American Economic Review, 79*(4), 749–762.

Harrison, G. W. (1992). Theory and misbehavior of first-price auctions: Reply. *American Economic Review, 82*(5), 1426–1443.

Harrison, G. W. (2007). House money effects in public good experiments: Comment. *Experimental Economics, 10*(4), 429–437.

Harrison, G. W., Lau, M. I., & Rutström, E. E. (2009). Risk attitudes, randomization to treatment, and self-selection into experiments. *Journal of Economic Behavior & Organization, 70*(3), 498–507.

Harrison, G. W., & List, J. A. (2004). Field experiments. *Journal of Economic Literature, 42*(4), 1009–1055.

Harrison, G. W., Martínez-Correa, J., & Swarthout, J. T. (2013). Inducing risk neutral preferences with binary lotteries: A reconsideration. *Journal of Economic Behavior & Organization, 94*, 145–159.

Harrison, G. W., Martínez-Correa, J., & Swarthout, J. T. (2015). Reduction of compound lotteries with objective probabilities: Theory and evidence. *Journal of Economic Behavior & Organization, 119*, 32–55.

Harrison, G. W., & Rutström, E. E. (2008). Risk aversion in the laboratory. In J. C. Cox & G. W. Harrison (Eds.), *Risk aversion in experiments* (Research in Experimental Economics) (Vol. 12, pp. 41–196). Bradford: Emerald Group Publishing Limited.

He, S., Offerman, T., & van de Ven, J. (2016). The sources of the communication gap. *Management Science, 63*, 2832–2846.

Hey, J. D. (1998). Experimental economics and deception: A comment. *Journal of Economic Psychology, 19*, 397–401.

Hey, J. D., & Orme, C. (1994). Investigating generalizations of expected utility theory using experimental data. *Econometrica, 62*, 1291–1326.
Hoffman, E., McCabe, K., Shachat, K., & Smith, V. L. (1994). Preferences, property rights, and anonymity in bargaining games. *Games and Economic Behavior, 7*(3), 346–380.
Hoffman, E., McCabe, K., & Smith, V. L. (1996). On expectations and the monetary stakes in ultimatum games. *International Journal of Game Theory, 25*, 289–300.
Holt, C. A., & Laury, S. K. (2002). Risk aversion and incentive effects. *American Economic Review, 92*(5), 1644–1655.
Holt, C. A., & Smith, A. M. (2009). An update on Bayesian updating. *Journal of Economic Behavior & Organization, 69*(2), 125–134.
Irlenbusch, B., & Sutter, M. (2006). An experimental analysis of voting in the Sta-bility and Growth Pact in the European Monetary Union. *Public Choice, 129*(3–4), 417–434.
Isaac, R. M., Walker, J. M., & Thomas, S. H. (1984). Divergent evidence on free riding: An experimental examination of possible explanations. *Public Choice, 43*(2), 113–149.
Isaac, R. M., & Walker, J. M. (1988). Group size effects in public goods provision: The voluntary contributions mechanism. *Quarterly Journal of Economics, 103*(1), 179–199.
Ivanov, A. (2011). Attitudes to ambiguity in one-shot normal-form games: An experimental study. *Games and Economic Behavior, 71*(2), 366–394.
Jamison, J., Karlan, D., & Schechter, L. (2008). To deceive or not to deceive: The effect of deception on behavior in future laboratory experiments. *Journal of Economic Behavior & Organization, 68*, 477–488.
Kagel, J. H., & Roth, A. E. (1992). Theory and misbehavior in first-price auctions: Comment. *American Economic Review, 82*(5), 1379–1391.
Kamecke, U. (1997). Rotations: Matching schemes that efficiently preserve the best reply structure of a one shot game. *International Journal of Game Theory, 26*(3), 409–417.
Karni, E. (2009). A mechanism for eliciting probabilities. *Econometrica, 77*(2), 603–606.
Kartik, N. (2009). Strategic communication with lying costs. *The Review of Eonomic Studies, 76*, 1359–1395.
Kimball, M. S. (1990). Precautionary saving in the small and in the large. *Econometrica, 58*(1), 53–73.
Kimball, M. S. (1992). Precautionary motives for holding assets. In P. Newman, M. Milgate, & J. Falwell (Eds.), *The new Palgrave dictionary of money and finance* (Vol. 3, pp. 158–161). London: MacMillan.
Kroll, E., Morgenstern, R., Neumann, T., Schosser, S., & Vogt, B. (2014). Bargaining power does not matter when sharing losses – Experimental evidence of equal split in the Nash bargaining game. *Journal of Economic Behavior & Organization, 108*, 261–272.
Kroll, S., Cherry, T. L., & Shogren, J. F. (2007). The impact of endowment heterogeneity and origin on contributions in best-shot public good games. *Experimental Economics, 10*(4), 411–428.
Krupka, E. L., & Weber, R. A. (2013). Identifying social norms using coordination games: Why does dictator game sharing vary? *Journal of the European Economic Association, 11*(3), 495–524.
Laury, S. K., Walker, J. M., & Williams, A. W. (1995). Anonymity and the voluntary provision of public goods. *Journal of Economic Behavior and Organization, 27*, 365–380.
Levitt, S. D., & List, J. A. (2007). What do laboratory experiments measuring social preferences reveal about the real world. *Journal of Economic Perspectives, 21*, 153–174.
Liberman, V., Samuels, S. M., & Ross, L. (2004). The name of the game: Predictive power of reputations versus situational labels in determining prisoner's dilemma game moves. *Personality and Social Psychology Bulletin, 30*(9), 1175–1185.
Lichtenstein, S., & Slovic, P. (1971). Reversals of preference between bids and choices in gambling decisions. *Journal of Experimental Psychology, 89*(1), 46–55.
List, J. A. (2007). On the interpretation of giving in dictator games. *Journal of Political Economy, 115*(3), 482–493.
Loewenstein, G., & Issacharoff, S. (1994). Source dependence in the valuation of objects. *Journal of Behavioral Decision Making, 7*(3), 157–168.
Loomes, G. (1999). Some lessons from past experiments and some challenges for the future. *The Economic Journal, 109*(453), 35–45.
Manski, C. F. (2002). Identification of decision rules in experiments on simple games of proposal and response. *European Economic Review, 46*(4), 880–891.

2

Marwell, G., & Ames, R. E. (1981). Economists free ride, does anyone else? Experiments on the provision of public goods. *Journal of Public Economics, 15*(3), 295–310.

McDaniel, T., & Starmer, C. (1998). Experimental economics and deception: A comment. *Journal of Economic Psychology, 19*, 403–440.

Mellström, C., & Johannesson, M. (2008). Crowding out in blood donation: Was Titmuss right? *Journal of the European Economic Association, 6*(4), 845–863.

Merlo, A., & Schotter, A. (1992). Theory and misbehavior of first-price auctions: Comment. *American Economic Review, 82*(5), 1413–1425.

Merritt, A. C., Effron, D. A., & Monin, B. (2010). Moral self-licensing: When being good frees us to be bad. *Social and Personality Psychology Compass, 4*(5), 344–357.

Muehlbacher, S., & Kirchler, E. (2009). Origin of endowments in public good games: The impact of effort on contributions. *Journal of Neuroscience, Psychology, and Economics, 2*(1), 59–67.

Niederle, M. (2015). Intelligent design: The relationship of economic theory to experiments: Treatment driven experiments. In G. R. Fréchette & A. Schotter (Eds.), *Handbook of experimental economic methodology* (pp. 104–131). Oxford: Oxford University Press.

Nyarko, Y., & Schotter, A. (2002). An experimental study of belief learning using elicited beliefs. *Econometrica, 70*(3), 971–1005.

Ockenfels, A., & Weimann, J. (1999). Types and patterns: An experimental east-west comparison of cooperation and solidarity. *Journal of Public Economics, 71*, 275–287.

Offerman, T. J. S., & Sonnemans, J. H. (2001). Is the quadratic scoring rule behaviorally incentive compatible?. CREED Working Paper.

Offerman, T., Sonnemans, J., Van de Kuilen, G., & Wakker, P. P. (2009). A truth serum for non-bayesians: Correcting proper Scoring Rules for risk attitudes. *Review of Economic Studies, 76*(4), 1461–1489.

Ortmann, A., & Hertwig, R. (2002). The costs of deception: Evidence from psychology. *Experimental Economics, 5*, 111–131.

Oxoby, R. J., & McLeish, K. N. (2004). Sequential decision and strategy vector methods in ultimatum bargaining: Evidence on the strength of other-regarding behavior. *Economics Letters, 84*(3), 399–405.

Oxoby, R. J., & Spraggon, J. (2008). Mine and yours: Property rights in dictator games. *Journal of Economic Behavior & Organization, 65*, 703–713.

Ploner, M., & Regner, T. (2013). Self-image and moral balancing: An experimental analysis. *Journal of Economic Behavior & Organization, 93*, 374–383.

Pruitt, D. G. (1967). Reward structure and cooperation: The decomposed Prisoner's dilemma game. *Journal of Personality and Social Psychology, 7*, 21–25.

Rankin, F. W. (2003). Communication in ultimatum games. *Economics Letters, 81*(2), 267–271.

Riechmann, T., & Weimann, J. (2008). Competition as a coordination device: Experimental evidence from a minimum effort coordination game. *European Journal of Political Economy, 24*(2), 437–454.

Roth, A. E. (1995). Bargaining experiments. In J. H. Kagel & A. E. Roth (Eds.), *Handbook of experimental economics* (pp. 253–348). Princeton: Princeton University Press.

Roth, A. E., & Malouf, M. W. (1979). Information in bargaining. *Psychological Review, 86*(6), 574–594.

Roux, C., & Thöni, C. (2015). Do control questions influence behavior in experiments. *Experimental Economics, 18*(2), 185–194.

Rubinstein, A. (2006). A sceptic's comment on the study of economics. *The Economic Journal, 116*, 1–9.

Rydval, O., & Ortmann, A. (2005). Loss avoidance as selection principle: Evidence from simple stag-hunt games. *Economics Letters, 88*(1), 101–107.

Sass, M., Timme, F., & Weimann, J. (2018). The Cooperation of Pairs. *Games, 9*, 68. https://doi.org/10.3390/g9030068.

Sass, M., & Weimann, J. (2015). Moral self-licensing and the direct touch effect, Cesifo Working Paper 5174.

Schotter, A., & Trevino, I. (2014). Belief elicitation in the laboratory. *Annual Review of Economics, 6*(1), 103–128.

Selten, R. (1967). Die Strategiemethode zur Erforschung des eingeschränkt rationalen Verhaltens im Rahmen eines Oligopolexperiments. In H. Sauermann (Hrsg.): Beiträge zur Experimentellen Wirtschaftsforschung (pp. 136–168). Tübingen: JCB Mohr (Paul Siebeck).

Selten, R., Sadrieh, A., & Abbink, K. (1999). Money does not induce risk neutral behavior, but binary lotteries do even worse. *Theory and Decision, 46*, 211–249.

Smith, V. L. (1976). Experimental economics: Induced value theory. *American Economic Review, 66*(2), 274–279.

Sturm, B., & Weimann, J. (2007). Unilateral emissions abatement: An experiment. In T. L. Cherry, J.-s. F. Shogren, & S. Kroll (Eds.), *Experimental methods, environmental economics* (pp. 157–183). London: Routledge.

Sutter, M., & Weck-Hannemann, H. (2003). Taxation and the veil of ignorance - a real effort experiment on the Laffer curve. *Public Choice, 115*, 217–240.

Sutter, M., & Strassmair, C. (2009). Communication, cooperation and collusion in team tournaments – An experimental study. *Games and Economic Behavior, 66*(1), 506–525.

Trautmann, S. T., & van de Kuilen, G. (2015). Belief elicitation: A horse race among Thruth serums. *The Economic Journal, 125*, 2116–2135.

Vanberg, C. (2008). Why do people keep their promises? An experimental test of two explanations. *Econometrica, 76*(6), 1467–1480.

Van Huyck, J. B., Battalio, R. C., & Beil, R. O. (1990). Tacit coordination games, strategic uncertainty, and coordination failure. *American Economic Review, 80*(1), 234–248.

Vieider, F. M. (2011). Separating real incentives and accountability. *Experimental Economics, 14*(4), 507–518.

Volk, S., Thöni, C., & Ruigrok, W. (2012). Temporal stability and psychological foundations of cooperation preferences. *Journal of Economic Behavior & Organization, 81*(2), 664–676.

Wakker, P., & Deneffe, D. (1996). Eliciting von Neumann-Morgenstern utilities when probabilities are distorted or unknown. *Management Science, 42*(8), 1131–1150.

Weber, R. A. (2003). Learning'with no feedback in a competitive guessing game. *Games and Economic Behavior, 44*(1), 134–144.

Zizzo, D. J. (2010). Experimenter demand effects in economic experiments. *Experimental Economics, 13*(1), 75–98.

Experimental Practice

J. Weimann, J. Brosig-Koch, *Methods in Experimental Economics*, Springer Texts in Business and Economics, https://doi.org/10.1007/978-3-319-93363-4_3

3

Overview
In this chapter of the book we will examine the practical aspects of experimental research. The main question is how to proceed in concrete terms when setting up a laboratory, preparing an experiment and then conducting it. Of course, the methodological principles that we have dealt with in the second chapter play a central role. It will therefore not be possible to proceed without repeating a number of previously described issues, but we will try to reduce any unnecessary repetition by making appropriate references to what is necessary.

3.1 Setting Up an Experimental Laboratory

The basic arrangement of a laboratory consists of a series of computer workstations for the subjects and a workstation for the experimenter who manages and conducts the experiment. When planning to set up such a laboratory, the first thing needed is a kind of floor plan that shows how the subjects' workstations will be arranged and how the experimenter's workstation will be positioned in relation to them. There are two aspects which must be taken into account here and which are to a certain extent contradictory. On the one hand, it is necessary for the experimenter to be able to monitor the subjects of the experiment, for example, to prevent unwanted communication. On the other hand, it is important to avoid as far as possible the subjects feeling that they are under observation.

A very good solution to this problem is to place the subjects in soundproof booths (▫ Figs. 3.1 and 3.2). This prevents unauthorized communication between the subjects without the experimenter practically having to stand behind the them. A booth solution also has other advantages. If strict anonymity is required and the subjects are therefore not allowed to meet, they can be led individually into the booths and individually released from them again after the experiment. This also ensures that no contact can occur before and after the experiment.

▫ **Fig. 3.1** Soundproof booths and open workstations in the Magdeburg Experimental Laboratory of Economic Research (MaXLab)

Fig. 3.2 Soundproof booths and open workstations at the Essen Laboratory for Experimental Economics (elfe)

Another advantage of soundproof booths is that the *desired* communication between the subjects can be permitted in a controlled way. Since the workstations have to be equipped with computers anyway, it is not a problem to extend the system to allow any form of communication with the aid of cameras, headphones and microphones. From chat messages to conference calls to video conferencing, various types of communication can be set up and controlled effortlessly. Equipping the booths with two monitors facilitates viewing a videoconference and the display for the experimental software at the same time.

In order to be able to control communication, the experimenter's workstation must be equipped with technology that makes it possible, for example, to follow a videoconference. If all the communication is handled digitally, it is not a technical problem to store and archive it either. In this way, communication content and its effects can be investigated very effectively.

A further advantage of booths is that they ensure the subjects are not exposed to any distractions. This makes it easier for them to concentrate fully on the experiment. However, this is only possible if the computers in the booths do not provide Internet access and if the use of smartphones is banned during the experiment. The simplest way to achieve the latter is by having the devices handed in at the beginning of the experiment. Another variant is that the video cameras installed in the booths for communication purposes are used to monitor whether the subjects are using their smartphones. However, this has the drawback that the subjects are given a very clear understanding that they are constantly being observed.

Thus, using booths has some advantages, but it also has its drawbacks. Soundproof booths take up considerably more space than simple workspaces for the subjects. If the space available for a laboratory is limited, it may be necessary to accept limits on the number of seats if a booth solution is desired. A second disadvantage is that booths are significantly more expensive than simple laboratory space. In order to be able to use them effectively, they absolutely must be soundproof. They should also not be too compact and should provide sufficient space, lest the subjects feel uncomfortable in them. For the same reason, good ventilation, air conditioning and lighting must be provided. In addition, fire alarms and the like must be installed in the soundproof booths. All these features drive up the cost of a booth (Fig. 3.3).

Fig. 3.3 Soundproof booth in the elfe

The benefit of booths is that the subjects can be separated from each other very well and all communication can be perfectly controlled during the experiment. There may be experimental designs, however, in which this is not desired at all. For example, it is conceivable that groups have to make decisions in an experiment. Of course, this can also be solved in booths by allowing the group to hold a videoconference. However, if as realistic face-to-face communication as possible is wanted, then booths are rather a hindrance. Ideally, the laboratory should therefore have both a sufficient number of booths and "open" workstations.

Workstations for subjects that are not located in booths should also meet certain requirements. They should be designed to allow the subjects to be undisturbed by the other subjects if necessary, but at the same time to enable them to make contact with the other subjects as needed. It is therefore a good solution if the individual workstations – which of course should have appropriate basic technical equipment – are separated from each other laterally by walls and – depending on the location of the open cubicles – have a curtain which can also be used to provide privacy from behind. The purpose of the curtain is to eliminate the feeling of being permanently observed. With the curtain open, direct face-to-face communication between the subjects is possible. This is made easier if the workstations are equipped with office chairs on rollers (Fig. 3.4).

Fig. 3.4 Open workstation in MaXLab

It is highly advisable to place the experimenter's workstation near the laboratory entrance – if only because this is the best way for the experimenter to monitor who enters and who leaves the laboratory. Technically, the experimenter's workstation should be suitably equipped to monitor and control the experiment from there. It has proven to be very effective and time-saving to set up the experimental software in such a way that all the workstations can be started centrally from the experimenter's workstation. In communication experiments it is important that the experimenter can monitor the communication from his or her seat, for example, if the instructions stipulate that the communication content is limited. For this restriction to be monitored, the experimenter must be able to listen in. Of course, this must be done with headphones to ensure that the communication content remains confidential.

In addition to the experimenter's workstation, there should also be a place for the payment of the subjects at the end of the experiment. It helps considerably if the subjects still in the laboratory cannot see this area. Ideally, it should be accommodated in a separate room. If this is not possible, a kind of mobile payment station can be set up a in the corridor in front of the laboratory, where the transaction can take place quietly and undisturbed.

A fundamental question when setting up a laboratory is how many workstations it should have. Of course, in the majority of cases it is better to have too many than too few, but two restrictions must be kept in mind. First, the available space is usually limited, which already means a limited number of workstations. Second, it should be remembered that the utilization of the laboratory requires considerable resources. Apart from the fact that a sufficient number of experimental projects and experimenters are needed, each experiment requires a sufficiently large number of subjects and adequate financial resources to create the monetary incentives. The larger a laboratory is, the more demanding it is to ensure a satisfactory utilization of capacity. In view of the scarcity of resources, a common problem for economic laboratory research, and the competition for these resources, in which experimenters all too often find themselves, it would be difficult to justify a laboratory that is permanently underutilized.

Apart from the number of workstations and the size of the room, it is also worth considering spreading the laboratory over more than one room. This provides flexibility in deciding on the size of the laboratory and the number of laboratory workstations. Not having all the subjects in one room might seem impractical at first glance, but on closer inspection there are some advantages to having two rooms. Ideally, they should be next to each other. This opens up some additional options. For example, if the laboratory is very busy, the two-room solution can be used to carry out different tests in parallel. Two rooms are particularly advantageous, furthermore, if the experiment involves two groups of subjects who assume different roles in the experiment and who as far as possible should not come into contact with each other. It is significantly easier to do this by having one group located in the first lab room and the other in the second lab room than in the case where only one room is available.

So far we have only talked about the basic equipment in a laboratory, i.e. the subjects' workstations and the experimenter's workstation. And it is the overwhelming majority of laboratories that consist solely of this basic equipment. Yet it is quite possible to upgrade laboratories with new technologies. Eye tracking is possible, for example, with the assistance of the appropriate devices. One possible application for this is the question, which options offered on a screen are viewed, for how long and in which order. Eye tracking is still relatively rarely used in economic experiments, but there may be new experimental designs in the future that can benefit from this technique.

Psychology relatively often uses equipment that makes it possible to measure physiological states that allow conclusions to be drawn about emotional states. One such example is the skin's resistance to stress situations. The same applies to the pulse. Such devices can also be used in economic experiments. In neuroeconomics, in addition to the very complex fMRI method,[1] EEG[2] is still used as an imaging method in which electrical activities of the brain are visualized by recording voltage differences on the scalp. All these methods have so far only been used sporadically in economic laboratory research. An important reason for this may be that using them properly requires competences that economists do not normally possess. Against this background, it is very welcome that in the meantime a few scattered laboratory communities are forming. These are laboratories with a comprehensive range of technical equipment that are used by economists, psychologists and neuroscientists.[3] Of course, such laboratories also offer excellent conditions for interdisciplinary research. In view of discipline-specific methodological requirements, it may be useful to define common methodological standards that then apply to all the users of the laboratory.

---

1 Functional magnetic resonance imaging.
2 Electroencephalography.
3 For example, such a laboratory has recently been inaugurated at the University of Karlsruhe.

Box 3.1 The Right Software

Every laboratory needs at least two types of software. One that can be used to program the experiments so that they can be run over a computer network, and one that can be used for the purpose of administration and recruitment of the subjects. Different software solutions are available for both tasks. We cannot present all programs here and therefore concentrate on those that have become the standard so far. Having such standards is of some advantage because it greatly enhances the reproducibility of experiments. Being a standard, however, does not mean offering the best solution. For this reason, it should be left to each laboratory to choose the most suitable programs from the relatively large number now available to meet their specific requirements. However, it is always advisable that those working in a laboratory agree to use only one program at a time. This greatly simplifies the life of the laboratory manager and safeguards against avoidable errors.

z-Tree has become the global standard for experiment programming. The tool was developed by Urs Fischbacher and has been updated and further developed for many years.[4] z-Tree offers the possibility to program almost any experiment in a relatively simple way. Since it is precisely adapted to the needs of experimental economic research, it mainly contains elements that are frequently used there. This has the advantage that z-Tree is relatively streamlined and therefore easy to learn. The program is available for free and it is well documented. Many laboratories offer regular z-Tree workshops in which, amongst others, students planning to conduct an experiment as part of their master thesis learn how to use the software. As already mentioned, it has proven to be a great advantage that z-Tree is a program that is used all over the world. Experimental research is very often teamwork and it is not uncommon for teams to be composed of people from all over the world. It is therefore of great help if everyone is familiar with the same software.

In addition to the programming of experiments, laboratories need professional recruitment and supervision of subjects if these laboratories are to conduct experiments on a regular basis. Programs are also available for this purpose. ORSEE, developed by Ben Greiner,[5] played a similar role to z-Tree for quite some time. In contrast to the programming of the experiments, when it comes to the recruitment software it is not so important that many laboratories choose the same program, since the recruitment always takes place locally. This makes it a little easier for newcomers and is probably the reason why HROOT has in the meantime become a strong competitor for ORSEE. Both solutions have a similar scope of services. With the aid of the recruitment software, it is possible to manage all the potential subjects online. People who would like to participate in experiments can register for the database online. The most important characteristics of the respective person are recorded in the database. In addition to demographic data, this includes above all information on the experiments in which a particular person has already participated. It is very important to know this because, as a rule, researchers are interested in people who have not yet had any experience with the planned experimental setup. Sometimes, however, it is desirable to invite precisely those people who have already participated in a similar experiment to the laboratory.

The recruitment process is the same for both programs. Once the criteria for the selection of the subjects have been defined, a sufficiently large number of suitable potential participants are randomly selected and invited to the experiment by email. Those who want to participate can register online. As soon as enough applications are received, the recruitment process ends.

---

4 See Fischbacher (2007).
5 See Greiner (2015).

General rules for the use of the laboratory are not only useful when the laboratory is used by different disciplines. More importantly, a general code of conduct should be agreed which laboratory users are obliged to observe. Everything that happens in the laboratory affects the reputation of all the scientists working there. Therefore, steps should be taken to avoid that the behavior of individuals could damage this reputation. For example, the laboratory rules should determine how to deal with a shortage of subjects or a computer crash. Furthermore, it is imperative that the rules stipulate that subjects must not be manipulated. It is also likely to be of great benefit if negative payoffs to subjects are precluded as this could be particularly damaging to the reputation of the researchers. In general, those who use the laboratory should agree on their understanding of the rules of good (laboratory) science and then put this in writing in the form of a laboratory code. Nevertheless, the best laboratory rules are of no use if compliance with them is not monitored. This leads us to the question of whether a laboratory needs a laboratory manager.

Box 3.2 Is a Laboratory Manager Needed?

The answer to this question naturally depends on how intensively the laboratory is used and how well equipped it is. If it is a computer lab that is only occasionally used for an experiment, no one is needed to monitor the laboratory. However, if the laboratory is set up specifically for research and equipped with the appropriate technology, it is highly recommended that a laboratory manager be employed.

The tasks of the laboratory manager are manifold. The primary task is to ensure that the technical conditions for experiments are always at an optimum level. This requires someone who is very well acquainted with both the hardware and the software used in experiments. Both of these components are constantly evolving and the laboratory should always be at the cutting edge of technology. That is at least the goal. How far it can be achieved naturally depends on the resources available. Another important task of the laboratory manager is to regularly recruit new subjects for registration in the online database. In addition, the laboratory manager should of course also ensure that laboratory capacities are allocated fairly and that laboratory rules are actually observed. In order to fulfill all these tasks, it is of great advantage for the laboratory manager herself to be an active researcher. This ensures that she can communicate on an equal footing with other laboratory users and that she is familiar with the state of the art in these matters. This makes it much easier to determine which technology the laboratory should have and which is superfluous.

3.2 Preparing an Experiment

In this section, we look at the practical side of what happens before the first subject enters the laboratory. Until then, a large number of questions must be resolved and the methodological foundations that we laid in ► Chap. 2 will be taken into account in answering these questions.

3.2.1 Choosing the Design and the Treatments

The choice of the experimental design depends on the specific research question to be answered by the experiment. Of course, this presupposes that this research question has been given careful consideration. Once this question has been formulated, the experiment must be designed in such a way that it produces data that make it possible to decide how the research question can be answered. Ideally, this is achieved by deriving hypotheses from the research question. The hypotheses can then be either confirmed or rejected on the basis of the experimental data. Formulating hypotheses thus serves, above all, the purpose of determining the experimental design. All those who are conducting an experiment should also ask themselves what information is being provided by the data generated in the experiment. On the basis of this information, are they really in a position to decide whether or not a hypothesis should be rejected and whether it can be clearly separated from other hypotheses? Only if these questions can be answered with "yes" is the design tailored to the research question.

The role hypotheses play in experimental studies is controversial. Some disciplines adopt a very puristic attitude to this, which is based on the notion that only those findings of an experiment that refer to the previously defined hypotheses can be used. The reason for this restrictive attitude is the fear that hypotheses are formulated ex post, so to speak, after the experimental results have been obtained. This contradicts the ideal image of the scientist who works empirically and devises an experiment in order to answer a clearly formulated research question. The first step must be to derive the hypotheses from one or more theories, which must ultimately be verified. If this path is not followed, the experiment is not theory-based and is thus scientifically inferior – at least this is the purists' point of view. We are in favor of a rather more pragmatic view here. Theory, or at least a clear hypothesis, should of course guide the experiment and determine the experimental design. But should we ignore findings that do not fit the hypotheses? Reinhard Selten, Nobel Prize winner and pioneer of experimental research, shared the following story, which was passed on to us by word of mouth. Some astronauts are sent to Mars and told: "Our hypothesis is that there exist red rocks on Mars. Bring back rocks that are consistent with this hypothesis." The astronauts land on Mars and find red rocks, but they discover even more. Under every stone they pick up, they find little Martian worms. What are they supposed to do? Only take the red rocks to Earth because there was no hypothesis as to whether and in what form there is life on Mars? From a puristic point of view, they would have to do this and start a new expedition to Mars, this time equipped with the hypothesis that there is life on Mars.

We propose not to ignore findings that arise in an experiment and that lie outside the hypotheses that were formulated *ex ante*. The history of science is full of examples of important discoveries made by chance and not the result of targeted research. This does not mean, of course, that we should refrain from formulating the most accurate hypotheses possible. Random findings are more the exception than the rule and working inaccurately when formulating the hypotheses may ultimately reveal that the data generated by the experiment is not at all suitable for answering the research question. This should be avoided at all costs.

When formulating the research question, it is not only the creativity of the researcher that is decisive, but also a thorough investigation of the scientific world. Is there already empirical or experimental evidence on this issue? Only an intensive search of the literature can shed light on this. It is a good idea to proceed carefully and invest the necessary time. Nothing is more annoying than having an interesting experimental result and then discovering that you are not the first to do so. It should be remembered that experimental works are not only published in the relevant economic journals, but also increasingly find their way into general scientific journals.[6] Of course, it is not only necessary to clarify whether the specific experiment has already been carried out elsewhere, but also to find the literature that is closely related to one's own research question. In doing so, it is advisable not to limit oneself to experimental work (and also not only to works from one's own discipline).

Experimental research is directly related to economic theory. It is therefore also important to ascertain whether there are models in the literature that are relevant to the research question under consideration. For economists, this question is of particular significance because there is a reference point of sorts for the interpretation of experimental results that can rarely be avoided: What is the prediction that can be derived from rational (and self-interested) behavior? In order to be able to answer this question, a suitable model is required. Either this already exists in the theoretical literature or it needs to be developed and solved.

The situation is somewhat different when we are dealing with an experiment in which it is already clear that rational and self-interested behavior does not provide any useful predictions. This applies to many of the standard experiments used in experimental economic research. There is no way to explain the behavior in the ultimatum game experiment, in public good experiments or in the trust game experiment if we assume that the subjects behave strictly rationally and try to maximize their own payoff. This leaves us with two possibilities. Either we have a model that deviates from these assumptions and tries to organize the experimental findings, or we limit ourselves to a purely exploratory study that tries to gain information about individual behavior that might help to find an explanation for what happens in the experiment. In studies that do not examine a specific model and for which it has already been shown that rational and self-interested behavior does not provide a good explanation, it is sometimes difficult to come up with reasonable hypotheses. Sometimes there is simply no theory that could provide a clear prediction of what should happen. Even if the hypotheses that we can then use have a certain degree of arbitrariness, they should nevertheless be formulated *ex ante*.

Once the hypotheses have been determined, the experimental variables relevant for hypothesis testing must be defined. This is a very important point, and there are certainly degrees of freedom. Let us take an example. Suppose an experiment is planned with a two-person game in which player **A** can punish player **B** if **B** does something that **A** does not like. The punishment is that **A** reduces the amount paid to **B**. However, this sanction incurs costs of a to **A** for every euro he reduces the amount paid to **B**. If it is punishment behavior that is being investigated, what is the relevant variable, the *costs* **A** will incur to punish **B**, or the *punishment* **B** will have to bear? In any case, the information given in Box 2.1 should be observed: the payoff function is crucial and should not be too flat.

When it comes to the specific design of the experiment, the key question is what needs to be controlled and how this should be achieved in each individual case. Basically there are four things that can (and should) be controlled.

1. Preferences, Motives, Attitudes

The question of how preferences can be controlled was discussed in detail in ▶ Sects. 2.2 and 2.4. The induced value method replaces the utility function with a payoff function, assuming that the subjects always act according to the maxim "a higher payoff is

---

6 Journals such as Nature, PLOS ONE, PNAS or Science are meant.

better than a lower one". An obvious hypothesis9* in any experiment using monetary incentives is therefore that the subjects behave in such a way that their payoff is maximized. However, as already indicated, other motives can also play a role. People can behave unselfishly because they have altruistic motives, or generally include the well-being of other subjects in their considerations. On the other hand, they may also be willing to harm their fellow players, for example to punish them for certain types of behavior or because they simply want to do "better" than the other subjects in the experiment.

When analyzing behavioral motives that deviate from pure payoff maximization, it is important to realize that such motives cannot be observed directly. This means that their existence can only be concluded if they lead to deviations from payoff-maximizing behavior. This is an important point to consider when designing the experiment. If it is to be possible to deduce certain behavioral motives from the subjects' behavior, then the monetary incentives must be set in such a way that a specific motive can be deduced as clearly as possible from the deviations from purely selfish behavior. This is not always easy. Let us look at an example. In an ultimatum game experiment, if we observe that the proposers offer to split the amount of money to be allocated 50:50, then we cannot directly conclude the motive for this. It may be due to a strong sense of fairness, but it may also be that this behavior is driven solely by the fear that the responder will reject an offer that is less favorable for him than 50:50.

Question

When discussing payoff functions in ▶ Chap. 2, we pointed out that payoffs should be *salient* and *dominant*. Do you remember what that means?

If it is the fear of rejection that drives the proposer's behavior in the ultimatum game experiment, then it might be useful to elicit the beliefs concerning the responder's rejection behavior. Do you remember what to look out for?

All motives beyond payoff maximization must be viewed as a deviation in behavior from "rational self-interest" and it may be worth considering whether the differences in behavior that experimental design permits can be used to draw the most unambiguous conclusions possible about the underlying motives.

In all these considerations, one aspect that we have already discussed in great detail in ▶ Sect. 2.5.1 must always be taken into account – the experimenter demand effect. There is no general, specifically applicable rule on how to deal with it, apart from the ever-valid reminder that one must be aware of the fact that undesirable experimenter demand effects can occur. A sensible approach might be to consider what conclusions the subjects could draw from everything the experimenter does and whether among those conclusions there are also some that should not play a role in the experiment. We have also said a few things in ▶ Sect. 2.5.1 about how to avoid such undesirable effects.

2. Constraints under which decisions are to be made

The constraints under which decisions are to be made can take very different forms. Basically, everything that defines the decision-making situation in which the subjects are to be placed falls under this category. When designing these restrictions, a basic methodological decision is necessary. Should the experiment be designed in such a way that the decision in the laboratory runs as parallel as possible to real-world decisions, or is the aim to create an abstract decision-making situation that primarily serves to put a formal model to the test?

In terms of content, there are two important areas that can be designed in practically every experiment: first, the payoff conditions and, second, the information that the subjects receive. The main aspect of the payoff conditions is of course the payoff function, which constitutes the core of the experimental design (see ▶ Sect. 2.4.1). There are, however, other questions that need to be addressed. Should a show-up fee be paid (see ▶ Box 2.2)? How high should the initial endowment be? Will the subjects be given the money or do they first have to earn it in order to avoid the house money effect (which we discussed in ▶ Sect. 2.2.4)? With the payoff function and the initial endowment, the subjects' budget constraints are defined – and thus one of the most important constraints for economically relevant decisions. In addition to the monetary constraints that restrict our decisions in terms of income and prices, it is important to know what information we have when making decisions. It is an outstanding advantage of the experimental method that this aspect can also be fully controlled. This makes it possible to closely examine the impact of specific information on behavior. That is why it is crucial to very carefully consider what the subjects are to be informed about when determining the design. Closely connected to this is the question of whether and what form of communication between the subjects should be permitted. In ▶ Sect. 2.6 we discussed the factors that may play a role in this.

3. The manner of presentation (the frame)

In ▶ Sect. 2.5.3, we dealt in great detail with the question of what role the frame of an experiment can play. For this reason, we will only briefly go over the important issues that need to be considered in connection with the selection of the frame. First of all, every experimenter must realize that it really is necessary to make an active choice, because there is no such thing as an experiment without a frame. The second issue to be settled is whether one prefers as neutral a presentation of the decision problem as possible or whether the aim is to approximate a frame as it is actually found in the real world. Finally, the connection between the manner in which an experiment is presented and potential experimenter demand effects should be considered when designing this part of the experiment.

4. Experience and prior knowledge of the subjects of the experiment

People's prior knowledge or experience can systematically influence their behavior (see ▶ Sect. 2.3). If these factors are not controlled, there is a risk of having selection effects in the experiment and these should be avoided if possible. If in one treatment mainly economists participate and in another, mostly humanities students, this can lead to a difference that looks like an treatment effect, but which in reality can have other causes.

It is not only by way of academic subjects of study that such effects can arise. The laboratory experience of the subjects, for instance, may also play an important role. We will give an example. In a laboratory, an experiment is carried out in which the subjects are left in the dark as to how many parts there are in the experiment. The experimenter now adds another round or another run-through, unannounced. It may well be the case that this experience affects the expectations of the subjects of such an experiment the next time they participate in an experiment. They may then expect the same thing to happen again and this can certainly influence their behavior.

A similar effect may occur if subjects have already participated in a similar experiment before. This can easily happen because many experiments are conducted using the same basic design. The experience subjects gain in such experiments can influence their behavior in later repetitions. This is the reason why the software used to organize the recruitment of subjects and manage the pool of subjects opens up the possibility of maintaining an overview of the experiments in which a subject took part. Then, for example, when inviting people to a type X experiment, it is no problem to write to only those people who have never participated in a type X experiment before.

Now that the issue of how to control of the above items has been dealt with, some important elements of the design still need to be addressed. These have been discussed in detail elsewhere and will therefore only be briefly reiterated here. For one thing, a decision must be made as to whether a within-subject or between-subject design is to be used (see ▶ Sect. 2.7.3). Finally, it is necessary to determine whether the data provided by the experiment will conform to the statistical requirements that must be fulfilled for a meaningful analysis. To this end, the statistical procedures to be applied and the requirements these procedures place on the data must be determined. ▶ Chapter 4 of this book will deal with this topic in detail.

3.2.2 Instructions, Recruiting, Plan of Procedure und Pilot Experiment

Creating the Instructions

The subjects need to be informed about the course of the experiment and this is done with the help of instructions given to them. We have already made some basic considerations about what instructions should and should not contain (see ▶ Sect. 2.5.4). Here we are concerned with the more practical question of how instructions should be written.

Of course, there is no authoritative standard text, but in our experience it has proved useful to introduce the instructions by briefly informing the subjects in the experiment that they can earn money through their participation and whether it depends both on their own actions and those of other subjects in the experiment how much money they are paid in the end. It should also be emphasized that leaving the workstation and talking to other subjects during the experiment is prohibited. If the experiment involves communication between the subjects, this must of course be explained separately. How to get the experimenter's attention to ask questions, how long the experiment takes, whether there is a show-up fee and – if the experiment consists of several parts – how many parts the experiment has and how these parts are related are also typically explained in the instructions.

After this general information has been provided, it is time to describe the experimental design. It is important that this is done in such a way that every subject understands exactly what decision he has to make and what consequences this decision has for him and eventually for the other players. However, a caveat needs to be made in this connection. Particularly in experiments in which learning behavior is to be investigated, it is sometimes necessary *not* to tell the subjects everything that will happen. If they knew everything, there would be nothing left to learn. It must nonetheless be ensured that the subjects do not receive untrue information (see ▶ Sect. 2.3.1).

Instructions should be as simple as possible and not too long. The longer the text, the more likely the subjects will not read it to the end. When writing texts, scientists are used to assuming that their readers have a high level of expertise and intelligence, which is why it is sometimes difficult to explain simple things very precisely and clearly. But this is exactly what is necessary to avoid misunderstandings in relation to the experiment on the part of those subjects who – for whatever reason – do not pay close attention.

Instructions should explain the experiment comprehensively and, as already mentioned, under no circumstances should they contain untrue things. This does not mean, however, that instructions must contain everything that happens in the experiment. As already mentioned, in learning experiments, for example, it may make sense to leave the subjects in the dark about the true payoff mechanism. Sometimes it is also necessary to prevent the subjects from knowing how often the experiment will be repeated. This can be important, for instance, to avoid final-round effects. This brings us to a point where it can be difficult to decide whether we are still in the area of permissible omission or already in the area of forbidden deception. One example is public good experiments with a so-called restart (see Andreoni, 1988). At the beginning of the experiment, the subjects were only told about a normal public good experiment played over 10 rounds. Only after the 10 rounds had been completed were they informed that the game was to be played once more because "there was just enough time to play again". This is definitely a borderline case. On the one hand, the subjects were not told anything false. On the other hand, such an approach creates a reputation that in experiments a person can never be sure whether something else will follow. And this is not in the interest of experimental research.

Writing the Plan of Procedure

Once the instructions are written, it may prove useful to create a plan of procedure for the experiment. This is particularly the case when the different sessions and treatments are not always carried out by the same people. Assistants are frequently deliberately used to conduct an experiment in order to exclude possible experimenter demand effects that can emanate from a professor. As a consequence, it may well be that it is not always the same people who work in the laboratory. In this case, a plan of procedure is essential to ensure that all the experiments proceed in exactly the same way. This plan should describe as precisely as possible what is to happen during the experiment. This begins with the subjects entering the laboratory. Should they be admitted individually or as a group? What measures must be taken to maintain anonymity? How are the instructions distributed or read aloud? What is the procedure for responding to questions from the subjects? All these detailed questions will be dealt with in more detail in ▶ Sect. 3.3. It is vital that the plan of procedure describes all these details so that each and every person who conducts the experiment knows exactly what to do, how to do and when to do it, from the admission of the subjects to the final payment of the payoffs of the experiment. Creating a plan of procedure has another advantage: it facilitates the replication of the experiment. It is, of course, very helpful if such a plan exists and if it is made available to anyone who wants to repeat the experiment.

The Pilot Experiment

Once the plan of procedure has been drawn up and all the detailed issues described in it have been addressed, the experiment could in principle commence. But before doing so, it is often wise to run a pilot experiment. The purpose of such a pilot is to check whether everything runs exactly as imagined. An important point here, of course, is the software or the specific program that was written to conduct the experiment. Does it perform under realistic conditions – even if the users make mistakes while entering their data (as subjects sometimes do)? It is much more unpleasant to discover an error during the actual experiment than during a pilot experiment.

The way the pilot experiment is to be designed is entirely open. It is possible to either reduce the number of subjects in a session or invite the planned number. The former is cheaper, the latter is safer (for example, because it makes it easier to estimate the duration of the sessions). If the pilot experiment is to be used purely for testing the processes and the software, it can be run with people who know that it is a pilot experiment. These do not have to be recruited in the usual way. After all, it cannot be ruled out that something will go wrong and, in this case, the participants are prepared for it and the laboratory's reputation will at least not suffer. If, however, the aim is to gather valid data in the pilot experiment, there should be no deviation from the actual experiment when selecting the subjects, i.e. the same recruitment method and the same number of subjects must be used. Furthermore, the payoffs need to be real and equal to the payoffs of the planned experiment.

In addition to the software, the instructions should also be thoroughly checked in a pilot experiment. After the experiment, the subjects can be informed that they were involved in a pilot experiment and asked how easy it was for them to understand the instructions and how well they understood them. Experience has shown that subjects are very cooperative when asked where they see potential for improvement in the procedure of the experiment or in the formulation of the instructions.

After the completion of a pilot experiment and the evaluation of its results, the question arises as to how to deal with the data that was obtained. If the subjects were selected and paid off as they would be in the experiment, if everything ran smoothly and if no changes to the design or the way the experiment was carried out were necessary, there is nothing against integrating the data into the data set of the experiment. The pilot experiment therefore does not differ in any way from the other sessions in which the experiment is conducted. But what is to be done with the data if something goes wrong in the pilot experiment? As a rule, they disappear into oblivion, as does the entire pilot experiment, on account of being unusable for the experiment. Exceptions are, however, cases in which the selected parameterization proved to be unsuitable in the pilot experiment. Here it is in keeping with best scientific practice to report this, as this information is important for the interpretation of the robustness of the results. But it could also be useful for the pilot experiment to be published with the experiment, even if something was not successful. Reporting errors discovered during the pilot experiment provides others with an opportunity to learn from these errors. Making this possible may not always be easy (who likes to report on mistakes he made?!), but it might prove to be helpful for other experimenters.

Recruiting the Subjects

Before an experiment can be carried out, it is essential to confirm that suitable subjects are available. Recruitment is relatively easy if it is limited to students as subjects. In ► Box 3.1 we already pointed out that there are various software solutions for the administration of a pool of subjects, all of which are very user-friendly and make recruitment very easy. However, this presupposes that students register in the corresponding database, thus expressing their desire to participate in experiments. This can only be achieved if the laboratory carries out a certain amount of public relations work. This is not so difficult. Ideally, the university administration is cooperative and allows the laboratory, for example, to write to first-year students by email informing them of the laboratory, the possibilities of earning money and the registration procedure. Of course, the email should also contain a link to the corresponding portal. If there is no possibility to send electronic mail to the potential subjects, it is necessary to take the more difficult path and go through the lecture theaters to introduce the laboratory. Experience shows, however, that emails are much more successful because they make it extremely easy for the recipients to register in the database.

If the recruitment was successful, the laboratory possesses a pool of potential subjects for selecting those to be invited after the pilot experiment. The criteria used to do this can be very diverse, but it is crucial that they always take into account a principle that must be observed when inviting subjects: selection bias is to be avoided. For this purpose, it is necessary, for example, for the subjects to be *randomly* assigned to the different experimental treatments. The software used for the invitations is designed to do this, using a random selection procedure to choose the people to be invited for each treatment.

It is advisable always to invite a few people as substitutes, who only participate in the experiment if registered subjects do not show up. When inviting the subjects, it is important to inform them that they may act as a substitute and will therefore only be used if necessary. It is also important that the substitutes are paid for showing up, even if they are not used. Experience has shown that substitutes have no problem with their role if they receive monetary compensation for the costs they incurred to get to the laboratory. The number of substitutes needed and the frequency of subjects not showing up while not having an excuse depend largely on the reputation of the laboratory. If the laboratory has a reputation for making high payments to subjects and for responding to a subject's non-appearance, for example, by excluding the person from further experiments, then there will be relatively few no-shows. It may also be beneficial to make it clear to the subjects in the invitation that it is important that they actually appear because the research work involves taxpayers' money and can fail if the subjects simply do not show up.

In ► Sect. 2.3.2, we considered the question of whether students are suitable subjects for experiments. Most of the time this will be the case, but there might well be circumstances in which it is desirable to have either certain population groups in the laboratory or a representative cross-section of these groups. In such a case, the recruitment process is of course completely different from that with students - most of all, it is much more difficult.

Question
Can you state the advantages and disadvantages of recruiting only students for an experiment?

The biggest challenge is to avoid selection processes when recruiting non-students. This is very difficult because ultimately there is very little control over who actually participates and whether a selection has taken place in the decision to participate. Recruitment becomes slightly easier if a specific group is to be addressed, for example pupils or kindergarten children. In this case, cooperative schools or kindergartens must be found to help with recruitment. It becomes more difficult if non-student adults are wanted, especially if the selection is to be representative. There are several ways to get hold of such subjects. Ideally, organizations or providers that offer a representative sample for conducting experiments (unfortunately, usually for a fee) can be contacted, or the experiment can be conducted within the framework of a representative survey (see, for example, Dohmen et al. 2011). Otherwise it is possible to try using the participant pool of online work markets such as MTurk (see, for instance, Horton et al. 2011), an advertisement in social networks, a newspaper advertisement or, if a cooperative journalist is found, an article in the newspaper (see, for example, Bosch-Domenech et al. 2002). Random selection from the phone book is also very complex. Since unsolicited phone calls are either not allowed or not popular, it is advisable to write a letter beforehand announcing a phone call and explaining the issue. An example providing an illustration of how difficult it is to avoid selection processes is the fact that many people no longer allow phone book entries with their addresses, while those who still do may well systematically differ from the average. Whatever process is used for recruiting, one should be prepared for the fact that a large number of contacts are needed before the required number of subjects can be acquired. This is probably the reason why so few experiments with non-student adults are found in the literature.

3.3 Conducting an Experiment

3.3.1 Access to the Laboratory, Instructions, Unusual Incidents

Access to the Laboratory

Once the pilot experiment has been evaluated, all the necessary design adjustments have been made and enough subjects have registered for the experiment, the actual experiment can proceed. The first step, of course, is to get the subjects into the laboratory. To do this, a list of names of the registered subjects must first be created. The question of how the subjects actually get into the laboratory depends largely on the specific experiment. The issue to decide here is how to manage the required level of anonymity between the subjects. If it is essential that the subjects have no opportunity to identify themselves, then it makes little sense to invite them all to the laboratory together. In such cases, a somewhat more complex procedure is required. Staggering the arrival time in the laboratory is not recommended since, first, this can easily go wrong and, second, it can lead to the first subjects having to wait a long time until everyone is there and the experiment can commence. It is easier to specify an appropriate number of meeting points in the institution and to ask each subject to go to one of these meeting points. There they are fetched individually by the experimenters and led to a workstation in the laboratory, where they are not visible to the other participants. Using walkie-talkies has proven to be very helpful in this process, allowing those who pick up the subjects to communicate with each other. In this way, it is easy to prevent chance encounters between the subjects on their way to the laboratory.

If the anonymity of the subjects is not an important aspect of the experiment, the complicated process of fetching the subjects can be dispensed with and they can simply be sent to a location near the laboratory. This can be a separate room or a corridor. Once everyone is gathered, the substitutes find out whether they can participate or go home after receiving their compensation. Two tasks then follow. First, the names of the subjects are checked so that after the experiment the names of those who took part in the experiment and of those who may have been absent without an excuse can be entered in the subject database. The second task is to assign the subjects to the various roles. In most experiments, there are different roles: buyers or sellers, proposers or receivers and so on. Although it is not uncommon for there to be only one role, for example in the provision of public goods, the experiment is still run in several groups, so the groups have to be made up. It makes good sense to combine the two tasks. When the names are checked, the subjects draw "lots" that randomly assign them to a role or group. A well-organized laboratory holds suitable objects, such as table tennis balls, wooden balls or the like, that can be used as lots. Drawing lots for roles and group memberships ensures that the assignment is randomized, which is extremely important in order to avoid selection effects. At the same time, identification numbers can be drawn with the lots. Obviously, this has to be done in such a way that the experimenter cannot see the identification number. When making decisions in the experiment, the subjects can then enter their number instead of their name. This increases the anonymity of the decisions.

Instructions for the Subjects

There is no set rule as to how the instructions are to be communicated to the subjects. However, we recommend first handing them out in writing, printed on a sheet of paper (not online), and then, if possible, reading them out loud. Reading the text aloud almost always has the effect that the subjects simultaneously read the text on their sheets, thus ensuring that they have read it to the end. If the instructions are not read aloud, this effect is lost and the experimenter can only assume that everyone actually has read everything to the end.

Whether it is possible or not to read the instructions out depends on the conditions of the experiment. For example, if the subjects are in soundproof booths that they should not leave because they are not permitted to identify each other, reading aloud is pointless, unless the booths are equipped with loudspeakers so that communication is possible. Similarly, if two or more treatments are conducted simultaneously in the laboratory, reading the (different) instructions out does not make any sense. If all subjects are in the same laboratory room and no special arrangements to ensure anonymity have to be made and if all subjects participate in the same treatment, there is no reason why they should not be called together as a group and the instructions read out. However, reading out instructions should be as homogeneous as possible across sessions and treatments (i.e. ideally the same experimenter should be involved). Once all subjects have read the instructions, they should have the opportunity to ask questions. In ▶ Sect. 2.5.4, we explained in detail why it is better not to have these questions asked publicly, but privately, i.e. in a conversation between the subject and the experimenter. In the same section, we also described in detail how comprehension tests should be designed in order to avoid unwanted influences on the subjects.

Unusual Incidents

When a lot of experiments are carried out in a laboratory, at some point isolated incidents occur that are normally considered to be rather unlikely to happen. Even if they are very rare occurrences, some kind of emergency plan that can be implemented in such a case should be kept in mind. One of the emergencies that can occur is the total crash of the computer system. Normally, such a system cannot be restored in a few moments, and it is particularly difficult to estimate how long a repair will take and whether a resumption of the session is possible or not. There is usually no alternative but to stop the experiment and send the subjects home. In this situation, no secret should be made of the cause of the disruption. The subjects are more inclined to appreciate that the experiment cannot go on if they know why. But at such moments it is important to remember the reputation of the laboratory, which will benefit if the subjects are at least compensated for the costs they have incurred by participating in the abandoned experiment, since they will not receive any payment for the experiment as such. Although the laboratory does not receive any useful data in return for the compensation payments to the subjects, it does preserve the laboratory's reputation that it is worth participating in experiments.

Another type of unlikely but possible incident is that subjects drop out during the session, for such reasons as they feel sick, the symptoms of an influenza infection are developing or they cannot cope with the confinement of the booth in which they are sitting for long enough. In such a case, of course, the health and well-being of the subject have the highest priority. It is therefore essential to resist the temptation to persuade the man or woman concerned to continue to participate in the experiment. Participation in an experiment must be strictly voluntary. In order not to get into such a situation in the first place, however, it can be helpful to ask all subjects before starting the experiment whether they see themselves as being able to spend the time needed in the laboratory.

If illness nevertheless occurs, it naturally depends on the type of experiment how it continues. Basically, the other participants should not suffer any disadvantage if someone drops out. Let us look at an example. Suppose an experiment is played in groups of four people. One person drops out. The three remaining subjects must then be adequately remunerated. If no decisions have been made, a lump sum can be paid to cover the opportunity costs of participation. If some decisions have already been taken, they can if necessary be extrapolated for the entire experiment and paid out accordingly. Of course, the data generated by the group with the absent subject cannot be taken into account in the evaluation of the experiment.

3.3.2 Organizing the Payments to the Subjects

Once all the decisions have been made and the experiment is over, it is time for the payments to be made to the subjects. Before this can happen, there is occasionally a problem that we would like to discuss briefly. The behavior of the subjects can vary greatly, and this may also manifest itself in the fact that the individual participants in the experiment solve the decision problems at very different speeds. This in turn may mean that individual subjects finish the experiment much earlier than others. What is the best way to handle this?

If the payment does not depend on the speed at which decisions are made, but only on the decisions themselves, then the earlier one leaves the laboratory, the higher the hourly rate of pay. This creates strong incentives to make decisions as quickly as possible. However, this is not in the interest of the experimenter, because speed can easily be at the expense of care. Subjects should think carefully and very precisely about their decisions and not hastily. Therefore, it should not pay off to be faster than the other subjects in the experiment. This is an important reason why payment should not be made until *all* the subjects are finished. The disadvantage of this rule is that it may happen that individual subjects have to wait idly for a relatively long time before payment is made. This effect can be alleviated by applying the principle "We will only continue once everyone is finished" throughout the experiment. For example, if several rounds are played, the next round should only be played after the last round has been completed.[7] Nevertheless, waiting times in the laboratory can occur, and the subjects may find this unpleasant. Experienced subjects therefore take something to read with them to the laboratory in order to make good use of the waiting time. However, this can lead to very different opportunity costs of time for the subjects, so it is not an ideal solution. However, the use of own electronic media should be strictly prohibited. These should not be used, in particular, because they could allow the subjects to communicate with each other or with third parties during the experiment.

There is another compelling reason for not making payment until everyone is finished. If somebody were to be paid off while the experiment is in progress, it would inevitably lead to those who are not yet finished being disturbed and having the feeling that they have to hurry, because others can already leave. This should be avoided at all costs. The subjects do not need to know how quickly the other subjects perform their tasks and restlessness in the laboratory is inherently not good for an experiment.

Once all the subjects have completed the experiment, payment can be made. How it is organized depends on whether the experiment is double blind or not. We will start with the payment in a non-double-blind design.

Ideally, payment should not be made in the same room as the experiment. If this cannot be avoided, it should at least be ensured that the anonymity of the payment is otherwise secured. An important reason for this is that the information about the payment of the others could lend itself to making a social comparison. For one thing, a competitive element could be brought into the experiment that is not desired. Experiments are not generally about entering the subjects in a contest. However, subjects could come up with the idea of organizing one by trying to be better than the others. If they do not know what the others earn, this contest cannot take place.

Payment should therefore be arranged in such a way that each subject is paid out individually, without the others being able to observe how much each person receives.

7 Note that this feature is included in, e.g., the software zTree.

This is easiest if payment is made in another room or in the corridor in front of the laboratory. Of course, every subject must sign for receipt of the money. In order to keep the payments secret from the other subjects in the experiment, an individual receipt could be presented to each subject, and this must be signed. The number of receipts that the laboratory then has to manage, however, quickly becomes very large. Therefore, a procedure is recommended in which all the subjects and their payoffs can be entered in a list on one sheet. For this purpose, a template of strong cardboard or plastic that is much larger than a normal sheet of paper is made. A slot that allows only one row of the list to be visible is cut into this template, while all the other rows are hidden. When a subject comes to the pay station from the laboratory, the list is presented to her, she only sees her name and acknowledges receipt of the money without having a chance to see what the others have earned. The laboratory then only needs to manage one sheet of paper per session to prove to the administration that the experiment funds have actually been paid out.

Payment in a Double-Blind Design

In a sense, the moment of payment is the critical point in a double-blind treatment. How can we ensure that a subject receives his money, acknowledges receipt of this money and that he still can be sure that no one can see how high his payoff was? There are several answers to this question. Let us start with the receipt. A solution to this problem requires that the university administration plays along and places a certain amount of trust in the experimenters responsible. After the experiment, a list is created of the total amounts earned by the subjects of the experiment. If a certain amount was earned more often, it must be listed accordingly. The subjects then acknowledge with their signatures that they have received one of the amounts in the list without revealing the amount. This is a simple and pragmatic solution to the problem, which can be easily implemented if the administration has the confidence that the experimenters will not cheat and use false information to fill up the laboratory's coffee-kitty.

In order for it to be possible to hand over the money to the subjects without being able to recognize their identity, double-blind designs use identification numbers on cards that the subjects draw randomly and in a concealed manner at the beginning of the experiment. During the experiment, they identify themselves with this number so that it can be observed what action the subject with the number XY took, for example. At the end, the payoff for XY can also be calculated. One way of organizing payment without the subject being seen is to have a laboratory staff member in a separate room sitting behind some form of screen or partition. The subject passes the card with the identification number over the screen; the employee determines the payment for this number and also passes the money in an envelope over the screen. In this way, XY obtains his or her money without those behind the screen knowing who XY is.

If this procedure is still not secure enough, the following can also be applied. Once the payoffs have been calculated, the money is placed in padded envelopes bearing the identification number. The envelopes are placed in a cardboard box that is deposited in the corridor in front of the laboratory. Next to it is another box containing empty envelopes of the same type, as well as the pen with which the money envelopes were labeled. One after the other, the subjects are asked to go into the corridor, take an empty envelope and mark it with their identification number. They insert the card with their number in the envelope. Then they take the envelope labeled with their number and containing their money out of the cardboard box and replace it with the envelope containing their card. This ensures that no one, neither subsequent subjects nor experimenters, can determine which subject had which identification number. This procedure is cumbersome, but it is foolproof, so to speak, and therefore particularly suitable when it is important to assure the subjects of the experiment very credibly that their actions cannot be observed.

References

Andreoni, J. (1988). Why free ride?: Strategies and learning in public goods experiments. *Journal of Public Economics, 37*, 291–304.

Bosch-Domenech, A., Montalvo, J. G., Nagel, R., & Satorra, A. (2002). One, two,(three), infinity,...: Newspaper and lab beauty-contest experiments. *American Economic Review, 92*(5), 1687–1701.

Dohmen, T., Falk, A., Huffman, D., Sunde, U., Schupp, J., & Wagner, G. (2011). Individual risk attitudes: Measurement, determinants, and behavioral consequences. *Journal of the European Economic Association, 9*(3), 522–550.

Fischbacher, U. (2007). z-Tree: Zurich toolbox for ready-made economic experiments. *Experimental Economics, 10*, 171–178.

Greiner, B. (2015). Subject pool recruitment procedures: organizing experiments with ORSEE. *Journal of the Economic Science Association, 1*(1), 114–125.

Horton, J. J., Rand, D. G., & Zeckhauser, R. J. (2011). The online laboratory: Conducting experiments in a real labor market. *Experimental Economics, 14*(3), 399–425.

The Experiment from a Statistical Perspective

J. Weimann, J. Brosig-Koch, *Methods in Experimental Economics*, Springer Texts in Business and Economics, https://doi.org/10.1007/978-3-319-93363-4_4

To call in the statistician after the experiment is done may be no more than asking him to perform a postmortem examination … he may be able to say what the experiment died of.
R.A. Fisher, Indian Statistical Congress, Sankhya, 1938

Overview
The statistical analysis of the data obtained in an experiment is an elementary part of an experimental investigation. It makes it possible both to interpret the results of an experiment in an appropriate way and to support the experimental examination of the research question. It also allows the experimental setup to be improved before the actual experiment commences. Our main objective is to develop a broad guide to the use of statistical methods that systematizes and presents the content of the most important classes of methods and identifies the most important prerequisites for their application.

4.1 Introduction

If a research question is to be answered experimentally and with the aid of statistical methods, the experiment must be designed in such a way that it answers this question as well as possible. "As well as possible" in this chapter means that the choice of method has been made in such a way that the formal method of analysis is appropriate to the statistical nature of the data generated so that they are compatible.

The interplay between the research question, the design of the experiment with the resulting raw data and the statistical data analysis can be compared (at least to some extent) with cooking a dish. As with cooking, in an experiment well-structured preliminary planning is of fundamental importance. In cooking, this means *before* we go to the shop and buy ingredients and *before* we commit ourselves to a particular cook who will later transform these ingredients into a culinary delight, the dish, the ingredients and the cook must be precisely matched with each other in order to ultimately achieve the desired success. In addition to knowing which recipes have proven to be particularly good, we must also know the special qualities of each of the available cooks in order to get the "best" out of the ingredients for the dish.

The same applies to conducting an experiment. *Before* we send subjects to the laboratory to generate raw data for us and *before* we commit ourselves to some statistical method of analysis, the research question (the dish), design of the experiment (the recipe), the resulting raw data (the ingredients) and the statistical analysis (the cook) must be precisely matched with each other. A bad recipe leads to a bad dish – even the world's best chef can hardly change that. And a poorly designed experiment leads to a weak scientific result – even the most sophisticated method of analysis cannot change that. On the other hand, a particularly well-qualified cook can still get "that certain something" out of a good recipe and a well-founded analytical method can derive an even more significant scientific insight from a well-designed experiment.

In experimental practice, there is unfortunately a very slight risk that the individual components of a study are not, or only insufficiently, well matched. Aspiring scientists in particular are often under great pressure to publish, which creates the desire to carry out the experiment as quickly as possible and to deal with the statistical matters only once the experimental data are available. At this stage of the process, it is not uncommon to discover that essential experimental treatments are missing, that important variables have been recorded in an unsuitable way or not at all, or that only a small part of the data can be evaluated meaningfully with a suitable statistical method. Yet it is then often too late and parts of the experiment have to be rectified or perhaps carried out again. Not only is this noticed by journal referees, but at the end of the day it also costs more money and time than if a little more of both had been invested in a detailed and, most importantly, "thought-through" planning of the experiment before it was carried out.

From a statistical point of view, the course of an experimental study should be divided into a *design phase* and an *execution phase*. The *design phase*, which is to be carried out first, consists of the following tasks and typical issues:

- *Operationalizing the research question*:
  - What are the central constructs for which data must be collected during the experiment in order to answer the research question?
  - Can these constructs be *measured* as variables?
  - How should these variables be measured?
  - Which of them is the dependent variable?
  - Which of them are independent variables?
- *Structuring the statistical design:*
  - Which variables are to be manipulated in which way by the experimenter (choice of treatments)?
  - Which variables can I control and how can an undesired variation of the dependent variable be minimized?
  - What is the observational unit and what is the experimental unit?
  - How should a sample of subjects be selected?
  - How many subjects do I need to show correctly that a certain effect "exists" with a given probability?
  - What groups of subjects should be formed and what method should be used to form these groups?
  - Will variables be measured on several levels (e.g. within-subject and between-subject)?
  - How frequently and when should each subject's variable be measured?
  - Which are the qualitative variables and which are the quantitative variables?
- *Translating the research question into a statistical hypothesis or a statistical model:*
  - What formal relationship could exist between the observed variation of the dependent variable and the variation of the independent variables?
  - Which are the fixed-effect variables and which are the random-effect variables?
- *Choosing suitable statistical methods of analysis:*
  - What is the purpose of my statistical analysis:
    - To provide a descriptive presentation of the data and the treatment effects?
    - To make a statistical conclusion concerning the population from which the sample is drawn (inference)?
    - To make a prediction based on an estimated model?

- What are the main statistical characteristics of the experimental design or the resulting data (answers from previous questions)?
- What analytical methods can be used in view of the main statistical characteristics?

We have addressed many of these issues in the first three chapters of this book. Here, in the last chapter, we will focus on the statistical analysis. As soon as the theoretical design phase has been completed, it is the turn of the practical execution phase. This includes actually conducting the experiment (see ► Chap. 3 of the book):

- *Computer-assisted processing of the data*
  - Are there missing values?
  - Multiple measurements: long format vs. wide format;
  - Conversion of the data into the format of the statistics software;
  - Are there outliers?
  - Are there subjects who have obviously made arbitrary decisions?
  - What are short yet understandable variable names?
  - Creating new variables from (a combination of) already collected variables (e.g. group averages);
  - Creating a list of variables with descriptions.
- *Computer-assisted analysis of the data*
  - Describing the data using key indicators;
  - Graphical representation of the data;
  - Fitting the statistical model to the data by estimating the model parameters;
  - Model diagnostics;
  - Making inferences;
  - Predictions using the estimated model.
- *Conclusions*
  - Can the treatment effects be verified statistically?
  - Can the model explain the observed data well?
  - Are further experimental treatments necessary?

Important

Every experiment should always be actively designed in a goal-oriented manner before it is actually performed. This is the only way to produce data from which meaningful and valid conclusions can be drawn. No method of analysis, however sophisticated, can compensate for qualitative shortcomings in the experimental design.

There are entire textbooks available that deal with many of the tasks and issues mentioned above. For this reason, the statistical part of this introductory textbook is very much limited in its scope.

First of all, we can only present a small fraction of the methods available. Which method we have and have not included was significantly influenced by the following factors: practical relevance (do the assumptions allow a trouble-free application in practice?), complexity (is the method sufficiently easy to understand?) and popularity (which method is used particularly frequently in renowned experimental economic journals?). One consequence of this is that a particular procedure, which is used in accordance with this guide, does not necessarily have to be the best possible one. Generally, there are several valid statistical methods for one and the same research question and it may be appropriate to use methods different from those discussed here. Our classification is therefore not a rigid and binding selection rule, but is only a first general guideline. The very selective nature of this statistical guide also means that the structure of a traditional statistics textbook is thrown overboard. In particular, jumping to very different sections of the statistics will be unavoidable.

Second, we will only deal with any method presented here to the extent necessary for its fundamental nature, and therefore its area of application, to become clear. We will try to dispense with formal presentations as far as possible. This of course means that it will not be possible to show how to derive a method or how exactly the theory behind it is to be understood. For these purposes, we refer the interested reader to the relevant statistical literature.

4.2 Operationalizing the Research Question

4.2.1 Construct Validity

Operationalizing the research question means translating the basic idea of the experiment, which was initially formulated verbally, into a non-verbal form that is compatible with statistical methods. Essentially, the aim is to find measurable variables for the constructs of the research question being investigated that best capture the construct. Experiments that possess this property are called *construct valid* (Leonhart 2008). At first glance, establishing construct validity appears quite simple in some research questions. If we want to investigate altruism, for example, it makes sense to measure the amounts allocated in a dictator game. The amount given could then be a measure of the degree of the behavioral construct "altruism". But as is so often the case, the devil is in the detail. Where, for example, is the dividing line between self-interest and altruism? Can we still call someone who has 100 euros and gives away 1 cent altruistic? And are we really measuring unconditional altruism or does the willingness to give something away depend on other things (such as the experimenter demand effect)?

It is obvious at this point that construct validity can only exist if the construct itself is *unambiguously* defined. This problem becomes even more pronounced for other behavioral constructs. For example, it is scarcely possible to examine "reciprocity" in general in a way that is construct valid. A distinction is made between direct reciprocity (B helps A because A helped B), upstream reciprocity (B helps C because A helped B) and downstream reciprocity (C helps A because A helped B before) (Nowak and Siegmund 2005). There are also views of reciprocity based on higher order intentions and expectations (e.g. Rabin 1993; Dufwenberg and Kirchsteiger 2004; Falk and Fischbacher 2006) and those based on types (e.g. Levine 1998) or emotional status (e.g. Cox et al. 2007). In order to examine reciprocity in a construct-valid way, it must be decided in advance what kind of reciprocity is involved.

Important
When translating the research idea into measurable variables, attention must be paid to construct validity. This requires a clear definition of the construct in advance. This, in turn, often requires extensive research of the theoretical literature.

4.2.2 Types of Variables

In order to test a research idea experimentally, it is necessary to generate different types of variables. For example, suppose that the research hypothesis is that "the amounts offered in the ultimatum game are lower if the first mover is playing against a computer instead of a human being (and he or she knows this)". In this case, the *dependent* variable is the amount offered by the first mover. This can also be called the *outcome variable* because it is observed by the experimenter and is the result of a personal decision of each individual subject. An *independent* variable is expected by the experimenter to have an influence on the dependent variable, but not vice versa. In accordance with our research hypothesis, we expect the binary variable "computer opponent" (yes/no) to have an impact on the amounts offered. In a controlled experiment, the values of these independent variables are set systematically rather than simply being observed by the experimenter. In the above example, the experimenter measures the dependent variable "amounts offered" once under the value "yes" and once under the value "no" so that a comparison of both conditions is possible and the research hypothesis can be tested. In this case, the independent variable is also called a *treatment variable*, because its values represent the "treatments" or comparison conditions of the experiment under which the dependent variable is observed. The values of the treatment variables are called (treatment) *conditions.*[1]

Some further points need to be considered if the study is to draw a *causal* conclusion about the dependent and independent variables (and this is the main purpose of controlled experiments). If we observe a difference in the amounts offered once under each of the conditions "computer opponent – yes" and "computer opponent – no", we must be able to rule out that this difference was caused by other influences. In the worst case, the variable "computer opponent" actually has no influence whatsoever on the "amount offered", with the difference arising only because we measured the amount offered under the condition "computer opponent – yes" on Monday morning at 7 o'clock and the amount offered under the condition "computer opponent – no" on Friday afternoon at 4 o'clock. If the time of day does indeed have a causal influence on the amount offered, but this is not explicitly part of our research hypothesis or our study, this variable is an example of what is called a *confounding variable*. Confounding variables blur the causality between dependent and independent variables because they have a "hidden" influence on the dependent variable that is not explicitly part of the experiment. The great

1 Treatment variables are usually *factor variables*, i.e. variables that can only take a limited number of values. The values of a factor variable are generally referred to as *levels*.

advantage of a laboratory experiment over a field study, for example, is that various treatments can be carried out in a relatively strict *ceteris paribus* environment (Latin for "with all other conditions remaining the same"). In our example, we could immediately exclude the possible influence of the "time of day" variable by performing both treatments on the same day of the week at the same time. If the time of day variable is included in the experiment and kept constant over all treatment conditions, we have "controlled for time of day". This means that the variable "time of day" cannot be responsible for the variation in the amount offered – because it remained unchanged. In this way, the original confounding variable "time of day" becomes the *control variable* "time of day".

Unfortunately, there are also confounding variables that cannot be controlled for. These are mainly such factors that make up the individual personality of a subject. Examples are intelligence quotient, income of parents, allergies, education, political sentiments, spatial ability, physical fitness and many more. Of course, not all possible uncontrollable variables are relevant to our own experiment, since many have no connection whatsoever to our dependent variable. Nonetheless, we would be well advised to carefully consider what, on the one hand, has a high probability of influencing our dependent variable and, on the other hand, can vary from subject to subject while at the same time remaining beyond our control. When we talk about *uncontrolled variables* in the following, we will always assume that they have an impact on our dependent variable. We will distinguish between those whose value is *measurable* for each subject, and those whose value cannot be measured. "Income of parents", for instance, is quite easy to measure, whereas "political orientation" is much more difficult.

4.2.3 Control, Randomization and Sample Size

Regardless of whether or not an uncontrolled confounding variable is measurable, its impact on our dependent variable should be removed from the experiment as far as possible; otherwise a clear causal conclusion with respect to our treatment variable is no longer possible. The advantage of the confounding variable being measurable is that the subjects can be arranged according to the value of this variable. *Blocking* exploits this fact. Instead of directly setting the confounding variable at a constant value for all the subjects, we divide the subjects into blocks so that within a block the confounding variable takes on a constant value and is thus controlled for. For example, if we know that the subjects' gender influences our dependent variable, but we have no interest in explicitly modeling it, we simply divide the subjects into men and women and conduct the treatments for each of the blocks. Gender is then a *block variable*, which cannot influence the dependent variable in any of the blocks.[2]

It is much more difficult to eliminate the impact of confounding variables that are not controllable *and* not measurable. A 100% control of such variables is hardly possible since many of them are not only not measurable, but also unknown and their influence is therefore "hidden". Nevertheless, there is a simple statistical trick that can mitigate their impact. The basic idea is to form two groups of subjects across which the possible

2 We will discuss a specific block design later in ▶ Sect. 4.3.2.

confounding factors are distributed as evenly as possible. This is done by *randomly* assigning each subject to one of the groups (*randomization*). In the process, it should be ensured that the groups consist of a *sufficiently large* number of *independent* subjects. If there were only one single subject in each of the two comparison groups, the probability would be very high that the two subjects would by chance differ greatly in terms of an unmeasurable confounding variable.

It is similarly problematic if there are several subjects in each comparison group but they do not make independent decisions within their respective group (twins, close relatives, friends, etc.). In this case, the whole group would have to be interpreted as one "big" subject, causing a problem similar to that of a single subject. Diligently carrying out the randomization when creating the groups, however, will lead to the individual differences of the subjects being balanced out on average across the two groups. For example, if we have to assume that a subjects "political orientation" variable can have an impact on the giving behavior in the dictator game, the amount given by particularly conservative subjects in both groups will approximately balance each other out in a randomization. Since the same applies to particularly social democratic subjects, the effect of "political orientation" is balanced out not only across the groups, but also within the groups. In this sense, we have created two *homogeneous* and thus comparable groups. If we now administer the treatment to one of the groups (e.g. "play against a computer") and not to the other ("do not play against a computer"), in the best case a difference in the amount given can only be attributed to the treatment variable "computer opponent (yes/no)". Possible unobserved confounding variables have now been controlled for as far as possible.

Important

In a laboratory experiment, the central variable is the dependent variable. Changes in this variable are due to the influence of explanatory variables and various confounding factors. If the observed change in the dependent variable is to be attributed to a change in the explanatory variable induced by the experimenter, the three most important concepts to be considered are:

1. **Control** (all the unwanted influences that can be kept constant should be kept constant);
2. **Randomization** (create comparison groups that are homogeneous on average by leaving it to chance which subject is placed into which group);
3. **Sample size (or replication)** (ensure a sufficient number of independent observations in a treatment, i.e. sufficiently large groups of subjects who do not systematically exhibit the same behavior).

4.2.4 Scales of Measurement

Another important approach to classifying experimental variables is the level or scale of measurement. For many methods of statistical analysis, the format we use to measure makes a difference since the different scales of measurement contain different information, e.g. measuring age in the format "12", "18", "50", "80" or in the format "very young", "young", "old" and "very old".

The scale with the lowest information content or the lowest scale is the *nominal scale* (also called categorical scale). The term "nominal" comes from Latin and means "belonging to the name". In fact, the value of a variable measured on a nominal scale has no other function than to assign a unique name. Imagine, for example, the local city office (with only one clerk) at a certain time. Now we happen to ask two people sitting there for their social security number. The answers, or realizations, are nothing other than a numerical designation for the person interviewed and do not allow any further conclusions apart from a simple distinction of that person from another person (F12345 is a different person from M12345) or the unambiguous assignment of a person to a *category* (F12345 is a woman and M12345 a man). Categorically scaled data can easily be used to determine the absolute and relative frequencies of a category, which in turn form the data basis for suitable statistical methods of this class. Other examples of category variables are "religion" (Protestant/Catholic/Muslim/...), "accept offer" (yes/no) or "hair color" (black/brown/...).

A little more information is provided by a variable measured on the *ordinal scale* (also termed rank scale), whose values always represent an ordering or sequence. In our local city office example, we could again randomly ask two people, but this time for the processing number drawn at the entrance. Assuming the two respondents have the numbers 110 and 90, then we not only have information about an identifier, as with a nominal scale, but in addition we know that person 90 will come before person 110 in terms of time. We also know that person 100's request will be processed at some point between numbers 90 and 110. What is not known with an ordinal measurement of time is whether exactly as much processing time passes between the numbers 90 and 100 as between the numbers 100 and 110. It is the size relation of the numbers to each other alone that determines the order or rank of the realizations, whereas the differences in value of the numbers are meaningless.

As soon as differences in value of the measured quantity play a role, a *metric scale* (also named cardinal scale) is used. Identical differences on this scale always correspond to identical differences in the measured attribute. An example of a metrically measured attribute is age in years. Cardinal scales are often further differentiated. If the scale has an absolute zero point, then we speak of *ratio scales*. For example, the variables number of dental fillings, minutes of a day spent watching television, or the net wage of a student on vacation have absolute zero points, because a zero value is defined and makes sense. In these cases, it is always possible to make a meaningful ratio statement, such as "Peter has twice as many fillings as Anne". If this zero point is missing or it has been set at some arbitrary lower limit, the metric scale is an *interval scale*. For example, the attribute body weight does not have a natural zero point, because people with a body weight of zero do not exist at all. Furthermore, it is unknown beforehand what the smallest possible realization will be. If we were to stipulate that no person will weigh less than 50 kg or weigh more than 150 kg, then 50 kg would be the defined zero point. We would then have an interval scale that ranges from 0 kg to 100 kg and it is clear that a realization of 40 kg does not represent twice the weight of a realization of 20 kg on this scale.

4.2.5 Random Variables and Their Distribution

In the statistical modeling of the relationship between variables, the dependent variable is interpreted as a *random variable*. In an ultimatum game in which 10 single non-divisible plastic coins are to be allocated, it is clear from the outset that the number of retained coins will be an integer between 0 and 10 and that the decision space is limited by these same numbers. But the experimenter does not know in advance which number will actually be selected. From the point of view of the experimenter, a subject is therefore not much more than a random generator producing numbers between 0 and 10 according to a certain probability distribution. The elementary events of this random process are the realizable numbers. However, there are also random processes in which the realizations are not numerical in nature. Imagine two tosses of a single coin. If H = head and T = tail, the set of possible elementary events is the sample space $\Omega = \{HH, HT, TH, TT\}$. It is not easy to calculate using these realizations, so each of the events is assigned precisely one number according to a given practical rule. For example, the experiment could be a competition in which one euro is awarded each time a head is flipped. A reasonable rule would then be HH → €2, HT → €1, TH → €1, TT → €0.

Such an assignment can also be found in the ultimatum game above. For example, if we want to know how much money a subject keeps for himself, the relevant assignment rule would be: 0 coins → €0, 1 coin → €0.50, 2 coins → €1.00,..., 10 coins → €5.00. A (one-dimensional) *random variable X* is a specified mapping that assigns exactly one real number to each possible outcome of a random experiment and that real number is called the realization, x.

Which values of a random variable are most likely and which are less likely is determined by their *distribution*. The so-called *density function* of a discrete random variable indicates the probability with which a certain value occurs. The outcomes of rolling a dice, for example, are evenly distributed, each with a respective probability of 1/6. In the ultimatum game, on the other hand, we can reasonably expect that not all the amounts given will occur with equal probability. However, the exact distribution of the amounts is unknown. In the case of a continuous random variable, such as the time it took the subject in the ultimatum game to make his decision, the probability of an individual value cannot be specified. If an infinite number of values exist, the probability of a single value must be infinitely close to zero. For this reason, with continuous variables, it is only possible to indicate specific probabilities for ranges of values, with the total area below the density function always being 1. The cumulative (continuous) *distribution function* is, mathematically speaking, the integral of the continuous density function. The value of the function at a point x thus indicates the probability with which the random variable assumes a value less than or equal to x.

The inverse function of the distribution function is called the *quantile function*. It reverses the role of input and output of the distribution function, so to speak. In concrete terms, this means that it provides us with the quantile, i.e. the possible range of x values, within which a value with a specified probability of occurrence may be found.

Most statistical distributions have certain parameters which, depending on the value they have been set to, determine the shape of the density function. The three most important parameters are *expected value*, *variance* and *degree of freedom*. The expected value is the average of all the values drawn, if we (theoretically) draw a random sample infinitely often under the given distribution. For example, since there is an equal probability of rolling each number on a (normal) dice, the expected value is $1/6 \cdot (1 + 2 + 3 + 4 + 5 + 6) = 3.5$. The expected value of a distribution is a location parameter that provides information about where the theoretical mean value is located on the number line. The *variance* is the mean square deviation of all the realizations of the expected value and thus represents information about the dispersion of the random variable. The greater the variance, the wider and flatter the density function.

The mother of all distributions is the normal distribution. Its parameters are the expected value μ and variance σ^2. The probability density is bell-shaped and symmetrical around μ, where it has the highest density function value. Every normally distributed random variable can be converted by a simple transformation into a *standard normal distribution* with $\mu = 0$ (center of the x-axis) and variance $\sigma^2 = 1$.

Other important distributions are not parameterized directly using expected value and variance, but indirectly using what is termed *degrees of freedom*, which influence the expected value and/or variance. The (Student's) t-distribution, for example, has such degrees of freedom, with the shape of its density function more and more closely approximating that of the density function of the standard normal distribution with increasing degrees of freedom.

In addition to the uniform, normal and t-distribution, there are of course a large number of other statistical distributions, each based on its own specific random variables. However, since only the shape and the parameterization differ, but not the basic nature, we will not go into the different types further. For those who are looking for a good overview and want to calculate the function values of density, distribution and quantile functions quickly and easily at the same time, we recommend the tool PQRS described in the following box, which will often be used in this part of the book.

Box 4.1 Working with Statistical Distributions

The manual calculation of the function values of a density, distribution or quantile function of a probability distribution can be very complex. It is also cumbersome (and rather old-fashioned) to look them up in tables that list the function values for some parameter constellations of the distribution. So why not leave the calculating work to the computer? A suitable program is the free tool *PQRS* (Probabilities, Quantiles and Random Samples) available at ▶ http://www.pyqrs.eu/home/. It contains the formulas for both the density and the distribution function of all the major probability distributions. We use version 3.4 (not the Python version). After starting the program, the interface shown in ◘ Fig. 4.1 appears.

The default setting in PQRS is a normal distribution with expected value 0 and variance 1 (standard normal distribution). The density function of this distribution is displayed in the pdf tab. The total area under this function represents the probability 1 or 100%. In ◘ Fig. 4.1, this is divided into two equal parts with a probability of 50% each and the separation is at 0, so we can conclude that the probability of drawing any negative number from the standard normal distribution is the same as drawing any positive number, namely 50%. With the two arrows below the graphic we can change the quantile, or the x-value, and thus the size of the two areas relative to each other. For example, if we want to know the probability of drawing a number less than or equal to −2, we enter −2 in the number field between the arrows, confirm with the Enter key and get the graph shown in ◘ Fig. 4.2.

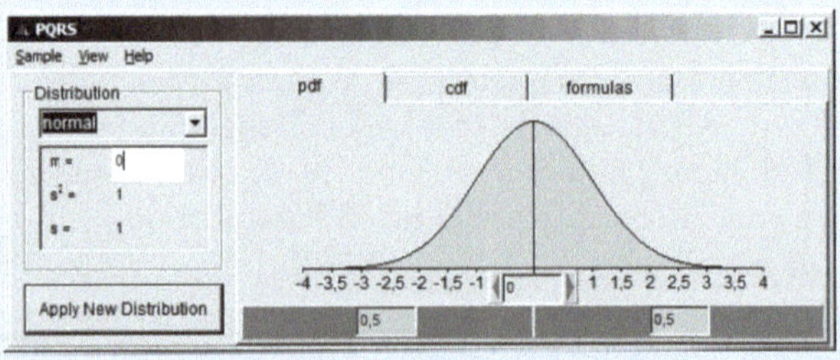

◘ **Fig. 4.1** Normal distribution

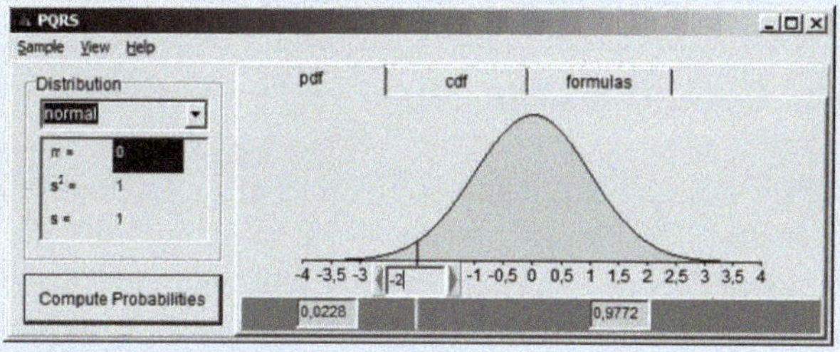

■ **Fig. 4.2** Normal distribution quantile −2

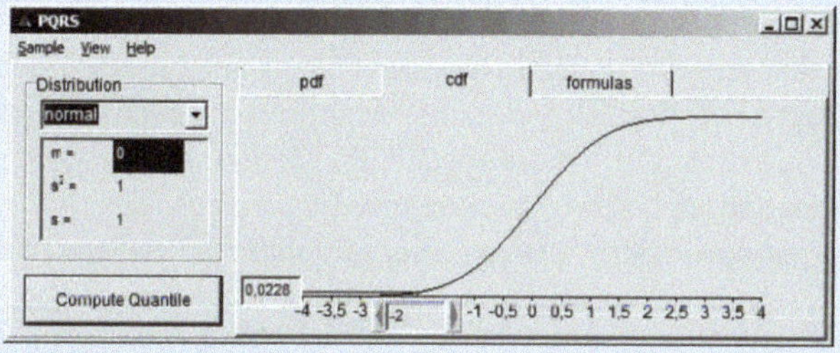

■ **Fig. 4.3** Distribution function of the normal distribution

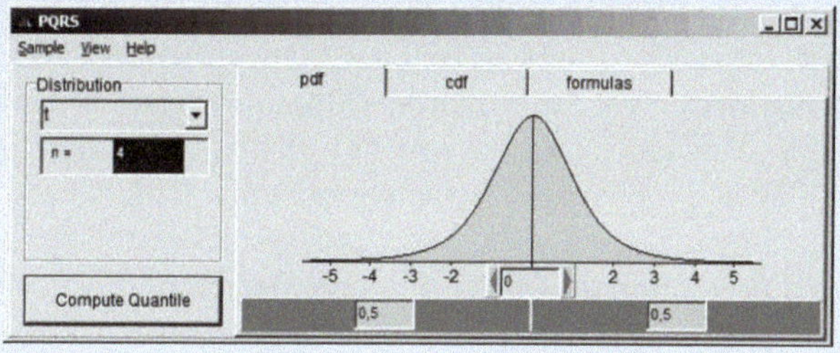

■ **Fig. 4.4** *t*-distribution

The left number field represents the sought-after probability of 2.28% and the right number field represents the probability of drawing a number greater than −2, 97.72%.

We obtain the corresponding distribution function by clicking on the cdf tab (cumulative distribution function; see ■ Fig. 4.3).

Here we see two number fields, one of which we can freely specify, while the other is calculated by the PQRS. If we enter a value in the lower field, the value of the left number field is calculated, corresponding to the calculation of a distribution function value. This shows a probability for a given quantile. If we reverse the procedure and enter a value in the left field, it would be the same as calculating a quantile function value. Then we get a quantile at a given probability. This procedure is needed later to calculate the critical values for testing hypotheses.

The option of drawing a random sample from the selected distribution is also practical. To do this, we select "Sample" from the top menu and then "Draw random sample". After entering the desired sample size *n* (preset to 10) and the number of decimal places (preset to 3) we get 10 random numbers (which of course change after each new run).

If we now select another distribution under "Distribution", such as the *t*-distribution, the parameters of this distribution automatically appear in the field below. If we enter 4 as the degree of freedom here, we obtain, for example, the *t*-distribution shown in ■ Fig. 4.4, with which we can carry out all the steps discussed above in the same way.

4.3 Creating the Statistical Design

4.3.1 Compiling the Observation Units

Selecting a certain number of subjects from a total population is referred to as *sampling* in statistics. The specific design of the sampling not only influences the costs of the experiment, but also the type and strength of the statistical statement to be made at the end of the experiment.

First, it is necessary to determine *what* the sample should be drawn from. This set of subjects is called the *population*. For most experiments that take place at universities, the largest possible population is the number of all enrolled students. Sometimes people are also recruited outside the university for laboratory experiments and added to the subject pool. Basically, the smaller and more specific the population from which subjects are drawn, the more specific the statistical statement that can be made about this population. Accordingly, statements based on small specific populations cannot be generalized particularly well to larger populations (see "External Validity", ▶ Sect. 1.5).

Second, some thought needs to be given to the *sample size*, i.e. the question "*How many* subjects do I draw from the specified population?" Unfortunately, in experimental practice this question is often answered *solely* on the basis of the budget, true to the motto: "We simply take as many subjects as we can pay for, regardless of whether this number is large or small enough". Of course, the budget is a binding constraint. In the neurosciences, for instance, laboratory times are extremely expensive, so that sample sizes are (often have to be) in the single-digit range. However, such small samples are problematic, especially from the point of view of inferential statistics. The probability that a statistical hypothesis test correctly identifies an actual effect as present (this is called the power of a test) decreases drastically with smaller samples. In other words, even if in reality there is a relatively strong and scientifically relevant effect in the population, it will at best be recognizable as a "random artifact" and not as a statistically significant effect. On the other hand, there is also a "too large" in terms of sample size, since having samples that are too large can make statistical hypothesis tests too sensitive. This means that even the smallest, possibly scientifically insignificant effects become *statistically* significant.[3] It is thus already clear that statistical significance should not be confused with scientific significance. Depending on the sample size, both can be completely different. This is because statistical significance is strongly influenced by the sample size, whereas the true effect to be detected in a population is not.

Third, there are several ways to obtain a sample. There are two basic types of sampling: one in which the probability of a subject being drawn can be specified, and one in which this is not the case. The best-known representative of the first variant is the *random sample*. From a population of size N, n subjects are *randomly* selected. The probability of each subject being drawn is then $1/N$. What is important here is that every subject has the same chance of being drawn. If this is not the case, the sample is *not*

3 We will discuss statistical significance and the relationship between sample size, effect size and power in more detail in ▶ Sect. 4.5.

representative of the population and is referred to as a *biased sample*. It is important to ensure that it is actually only chance that determines whether a member of the population is drawn or not. This means, for instance, that the experimenters must not make the selection personally.

A disadvantage of random sampling is that we have to draw a rather large sample in order for it to be representative, i.e. to reflect the structure of the population sufficiently well. Consider, for example, a population of 1000 students of economics. Of these, 100 students come from a poor home ("P") and receive state financial assistance and 900 students have rich parents ("R"), who finance their studies. If we were to draw an extremely small sample of size 2, there are only three results to distinguish: (P, P), (P, R) or (R, P) and (R, R). All three results have different probabilities of occurrence, but none of the samples can even begin to reflect the frequency structure of the population (10% P and 90% R). If the parental income plays a role in the context to be examined, then it is clear that this cannot be examined with a sample of size 2.

The main error here is that the sample is too small and unrepresentative. If we draw 100 people instead of only 2, a realistic result would be for example "8 times P and 92 times R", which provides a much better representation of the true structure of the population. If the sample were as large as the population itself, we would have the exact frequency structure of 100 times P and 900 times R. Of course, that size of the sample does not makes any sense.

A slight variation of the random sample is *systematic random sampling*. Here we would select, for instance, every tenth person from a randomly ordered list of subjects. Thus, each person has a selection probability of 1/10 *before* the list is created and it is still guaranteed that the selection is not based on preference. This procedure naturally presupposes that the list or the population itself is not subject to any order.

If it is clear that a (sufficiently large) random sample is not affordable and a representative sample is still required, then *stratified sampling* is a good possibility. The population is first divided into subpopulations (strata), with the subjects within each subpopulation having at least one common characteristic that distinguishes them from the subjects of the other subpopulations. A random sample is then drawn from each stratum. Each of these samples must make up the same proportion of the total of all samples as each stratum in the total population. For example, our population has 1000 subjects, of which 200 are male and rich (20%), 400 are male and poor (40%), 100 are female and rich (10%) and 300 are female and poor (30%). Our study requires that even with relatively small samples each of these four strata is represented in relation to the above proportions. For a sample size of 10, we would draw a random sample of size 2 from the stratum "male and rich", a random sample of size 4 from "male and poor", a random sample of size 1 from "female and rich" and one of size 3 from "female and poor". Overall, our sample has the same frequency structure as the population. Stratified sampling only works, of course, if the most important characteristics of all the subjects of the entire population are known, so that suitable strata can be formed. In practice, it is therefore particularly important to use a subject pool in which the most important characteristics of the subjects (gender, course of studies, experimental experience, etc.) are comprehensively documented and well maintained.

Non-probability sampling always leads to biased samples since not every subject has the same probability of being selected. In what is termed a convenience sample only those people who can easily be reached by the experimenter (such as students in a lecture) are invited. Samples in which participation in the experiment is on the subject's own voluntary initiative (voluntary response sample) are also biased since there is a self-selection within the population of those who have the greatest interest in the experiment.

Important
Larger, more general populations enhance the external validity of the experiment. At the same time, it is advantageous to know the main characteristics of all the elements of the population, which means it must not be "too large". The way in which the sample is selected influences the character of the sample and thus also that of the experimental data. As far as sample size is concerned, it is also possible for this to be "too large" or "too small".

4.3.2 How Do Experimental Treatments Differ?

It is possible to classify experimental treatments according to the *number of factor variables* and their *type* as well as the *number of possible values*. In a *single factorial design*, only a single variable is changed. If this is a binary variable with just two values, or levels, we speak of a *1 × 2 factorial design*. A distinction is made between *quantitative* or numerical factor variables (e.g. 10 km/h and 20 km/h) and *qualitative* or categorical variables (e.g. "slow" and "fast"). 1 × 2 factorial designs can be evaluated particularly easily since only the mean values of the dependent variables are usually compared under these two treatment conditions. Ideally, this difference is due to the treatment itself and is therefore called the (simple) *treatment effect*. The quantitative difference between the two values is called the size of the treatment effect or the (unstandardized) *effect size*. If, on the other hand, the factor variable has more than two levels, the treatment is called *multilevel factorial design*. In this case, the mean values of the dependent variable can be compared pairwise for every two levels or simultaneously for all levels. We will discuss each method of evaluation in more detail later.

A design with two factors is considerably more complex than a single factorial design. A diagram is a good way to get a better understanding. For example, if we want to experimentally investigate how the factors "games against the computer" (Comp: no/yes or 0/1) and "the experimenter knows who I am" (Anon: no/yes or 0/1) affect the giving behavior in a dictator game, then this hypothetical 2 × 2 factorial design could be represented as a cube plot, as shown in Fig. 4.5.[4]

The numbers in bold represent the (hypothetical) average amounts given for an endowment of 100 laboratory dollars. Each of these figures was determined in a separate experiment with the same number of subjects. The effect of Comp without anonymity is $\Delta C_1 = 12 - 32 = -20$. The fact that a subject is playing against a computer seems to reduce his willingness to give if he knows that the experimenter knows him. Even under anonymity this effect of Comp is negative, $\Delta C_2 = 5 - 21 = -16$. These effects are called the *simple effects* of the factor Comp under non-anonymity and under

4 A real cube is formed when three factors with two levels each are considered.

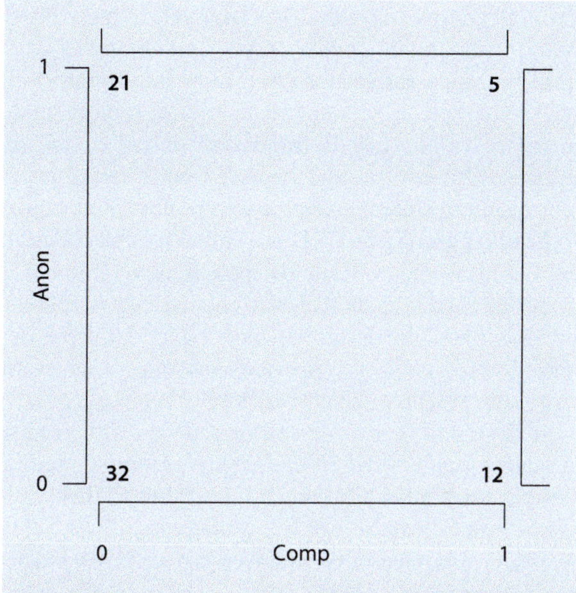

■ **Fig. 4.5** Cube plot of an example 2 × 2 factor design

anonymity. The *main effect* ΔC of Comp is the average of both simple effects, i.e. $\Delta C = (\Delta C_1 + \Delta C_2)/2 = -18$. In the same way, the simple effects and the main effect of Anon can be calculated: $\Delta A_1 = -11$ and $\Delta A_2 = -7$, resulting in $\Delta A = -9$.

Note that in the experiment Anon = 1 and Comp = 1, two factors were changed simultaneously, whereas this was not the case in the basic experiment Anon = 0 and Comp = 0. This means that the 5 lab dollars given cannot be compared directly with the 32 lab dollars given, since it is not clear to which factor the difference can be attributed. It is nevertheless important to carry out the treatment Anon = 1 and Comp = 1 because it provides *two* additional estimators: one for the effect of Comp with Anon = 1 and one for the effect of Anon with Comp = 1, which means that twice the amount of usable information is obtained with only one further treatment. This allows us to determine, in particular, whether the effect of Anon depends on the level of Comp, or whether the effect of Comp depends on the level of Anon. If either is the case, it is referred to as an *interaction* between the factors Comp and Anon. In our example, we see that the effect of Comp without anonymity is four times higher than with anonymity. Likewise, the effect of Anon without a computer is four times higher than with a computer. There is therefore an interaction, slight though it may be.

Everything we said about the 2 × 2 design also applies to multifactorial designs with three factors (cube design) and more (hypercube design). If k is the number of factors with two values each, then the number of treatments to be carried out in a full factor design is 2^k. Since the number of treatments and thus the costs grow exponentially in k, the practical relevance of full factor designs (at least in economics) decreases rapidly with the number of factors.

Another way to classify experimental designs is based on how subjects are assigned to each treatment combination. The simplest design is the *completely randomized design* (CRD). As the name suggests, the assignment of the subjects is completely random across all groups. In general, it is advisable to try to divide the subjects into equally large groups or treatment combinations. Uneven group sizes do not represent an insoluble problem from a statistical point of view, but usually complicate the analysis unnecessarily. For example, if we want to divide 100 subjects in a 2 × 2 design, we could put the numbers 1–4 in a bag 25 times each, mix them well and have all 100 people draw a number.[5] This ensures that we have 4 equal, *unrelated* groups. With a 1 × 2 design, the control and treatment groups would be unrelated. Since each subject can only be in one *or* the other group, the CRD is a between-subject design. Some advantages of the CRD are that it can theoretically be applied to any number of treatments, it allows unequal group sizes and it lends itself to comparatively simple statistical methods of analysis. On the other hand, the design must not be used if there are some doubts about the process of randomization. This is the case as soon as many treatments are tested in small groups and/or the available subjects are highly heterogeneous.

If the subjects are very heterogeneous with regard to a *measurable* confounding variable, a *randomized block design* is recommended (also see ▶ Sect. 4.2.2). Here the subjects are divided into "blocks" according to their various characteristics. The characteristic property of these blocks is that the similarity of two subjects within a block must be greater than that of two subjects between two blocks. In the simplest case, the confounding variable, or block variable, on the basis of which the blocks have been created, has only two values. For example, if we know that the gender of a subject influences giving behavior in the dictator game, but at the same time we have no scientific interest in this effect, then the influence of gender on the variation of the dependent variable is undesirable and should be eliminated. The sample is divided into the two blocks, "women" and "men", and the experiment with all the treatments is carried out separately in each block (*complete* randomized block design). It is important to assign the subjects to the control and treatment groups in the same way within each block. In other words, within-block randomization must not differ across different blocks. The overall effect of the treatment is obtained by combining the effects of each block. Since gender remains unchanged in each of the blocks, it is guaranteed that each treatment involves the same number of men and women. The effect on the giving behavior (e.g. comp) that is actually of interest is adjusted for the influence of "gender" and can therefore be estimated more effectively than without block formation. If comp did not, in reality, have an actual effect and we were to use a non-blocked, randomized design, it is possible that we would still obtain a difference between the treatment and the control groups, with this being caused by a random unequal distribution of gender across the control and treatment groups.

The principle of forming blocks can also be extended to block variables with many levels. In *matched-pairs design* a block consists of two subjects who are similar with regard to a particular level of the block variable. For example, in the ultimatum game we would like to measure the effect of Anon on the amount offered and block "parental income" because we suspect that this could be a confounding variable. For simplicity's sake, only 6 subjects with incomes of (in thousand euros) 41, 38, 69, 68, 35 and 39 are available. We then have the three pairs (35,38), (39,41) and (68,69) in a matched-pairs

5 Of course, the same procedure can be simulated faster on a computer.

Fig. 4.6 Cube plot of an example 2 × 2 factorial design

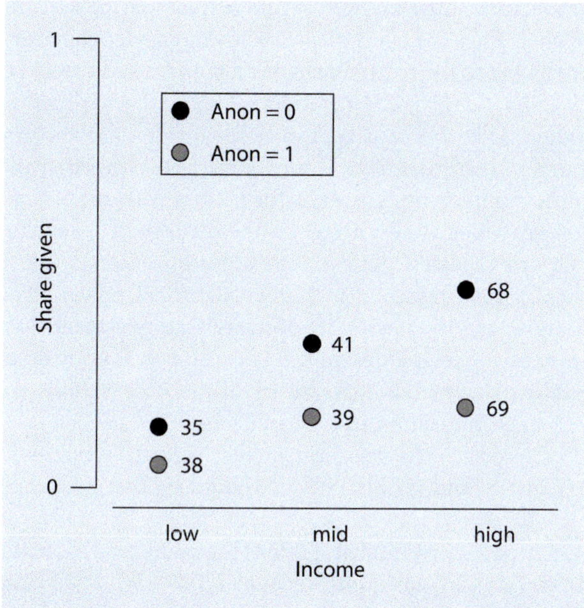

design, with a random decision subsequently being made in each block as to which subject enters the treatment group (Anon = 1) and which one enters the control group (Anon = 0). The outcome variable of both subjects is then measured in each of the blocks. A hypothetical result is shown in Fig. 4.6.[6]

We see that the treatment (Anon = 1) leads to a lower offer in all the blocks. If income were the only variable to be controlled for, and since in each block the confounding variable was more or less constant, we could conclude that Anon exerts a causal influence on the amount offered. However, if we would like to find out *how large* the treatment effect is, a problem with blocking quickly becomes apparent: treatment variables and block variables can interact. In other words, it may not be possible to measure the treatment effect independently of the level of the block variable. In our case, we see that the larger the block variable, the greater the effect of Anon (distance between the black and gray dots in each pair). Rich students therefore react more to the introduction of anonymity than poor students. Therefore, in a statistical model based on a block design, block effects and interaction effects must always be explicitly modeled as well.

In the *repeated measures design*, each subject undergoes several measurements, either in one and the same treatment at different times (*longitudinal design*) or in different treatments, naturally also at different times (*cross-over design*). The sequence of treatments a subject goes through is again randomized. In each case, multiple measurements generate a within-subject structure with several observations for each subject. The main statistical problem with multiple measurements is the interdependence of the observations. In a 1 × 2 factorial design with multiple measurements, we get a control

6 These data are only for instructional purposes and are not intended to be realistic.

group (measured at level 1) and a treatment group (measured at level 2), which are *related*. Thus, the effect measured using the dependent variable can no longer be clearly attributed to the treatment, since it could just as easily be a time or sequence effect (e.g. learning, familiarization, fatigue). Counterbalancing the order (balancing) often comes to our aid in this case, i.e. two homogeneous groups are formed and one group is measured in the order level 1, then level 2 and the other in the order level 2, then level 1. More complex designs exist for factors with more than two levels, such as the *Latin square*, which is also based on the concept of balancing (Leonhart 2008). The advantages of repeated measurements are lower costs due to fewer subjects, lower error spread, thus resulting in higher statistical power than comparable between-subject designs, and the possibility of measuring treatments over time (dynamics). The disadvantages of such a design are that it involves considerably more complex methods of analysis due to the dependency of the observations and weaker causalities owing to sequence, time and carry-over effects.

Of course, the selection of experimental designs presented here is by no means exhaustive and is only intended to give a first impression of how variable the specific structural set-up of an experiment can be. For more comprehensive and detailed presentations, we recommend one of the many existing textbooks that focus on experimental design. These include Box et al. (2005), Wu and Hamada (2009) and Morris (2010).

4.4 Statistical Tests

In everyday life, we all too often find ourselves drawing completely unscientific invalid conclusions, such as "A friend of mine was once robbed in City A and so it is a criminal city" or "A seatbelt isn't necessary. After all, I've never had an accident". Even without formal analysis, we can be fairly certain these conclusions generalize far too much, since they are based on only one observation. But how can concrete statements be made about the quality of a conclusion? How certain can an experimenter be that an observed effect is not completely random? In such situations, tools from inferential statistics come to our assistance. The focus is on what is known as the statistical *hypothesis testing*. This can be used to check how consistent a general statement about the characteristics of a population is with the observed laboratory data or with the sample.

4.4.1 Formulating Testable Hypotheses

The starting point of a hypothesis test is what is known as the *research hypothesis*. It usually postulates the content of the research question, i.e. a difference or an effect with regard to a scientifically interesting characteristic of the population under consideration. In the ultimatum game, for example, the following assertions could be made:

1. *RH1*: Northern Germans do not offer exactly half of the amount they have available on average, but either more or less than half.
2. *RH2*: Northern German men offer less on average than northern German women do.

The first research hypothesis postulates that the average amount offered by all northern Germans (unknown population parameter) differs from a specific, predetermined value of this population parameter of 50%. For verification using statistical hypothesis testing, it suffices in principle to take a single sample, i.e. a subset of all northern Germans, to determine the mean amount offered and to check whether this sample value differs sufficiently significantly from 50%. We will discuss later what "sufficiently significantly" means in this context.

The second hypothesis postulates an effect between two different populations. This effect could be tested by taking a sample from each population (control and treatment) and comparing the respective means of the amounts offered.

In order for such verbal research hypotheses to be tested using standardized, statistical methods, they must first be brought into an equally standardized, non-verbal form. The main problem here is to adequately capture the verbal content of the research hypothesis using a single quantitative population parameter. In our example, this would be the unknown, average share of the amount to be split that all Northern Germans offer to the other player in the ultimatum game, i.e. the population mean μ. Thus RH1 could be translated into the statistical hypothesis $\mu \neq 0.5$ and RH2 into the statistical hypothesis $\mu_m < \mu_f$ or $\mu_m - \mu_f < 0$. A characteristic of such a statistical formulation of the research hypothesis is that there is no equality sign in it; an equality sign would be synonymous with the statement that there is no difference or no effect.

There are basically two possible approaches to testing the research hypothesis:

1. Approach A: We assume that the research hypothesis is true and try to find evidence *in favor of* the research hypothesis.
2. Approach B: We assume that the opposite of the research hypothesis is true and try to find evidence *against* the opposite of the research hypothesis.

Basically, what we assume to be *true* is formulated as the *null hypothesis* H_0 and the opposite or complement of this as the alternative hypothesis H_1, which for example RH1 would mean:

- *Approach A:*
  1. H_0: $\mu \neq 0.5$ (research hypothesis assumed to be true)
  2. H_1: $\mu = 0.5$
     (Goal: accept H_0)
- *Approach B:*
  1. H_0: $\mu = 0.5$ (hypothesis assumed to be true)
     H_1: $\mu \neq 0.5$ (research hypothesis)
     (Goal: reject H_0)

Since approach A aims to test the research hypothesis directly, it could at first be considered preferable to the indirect approach B. But as we can easily see, approach A contains an inequality sign in the null hypothesis. This ambiguous formulation therefore allows any number of true values for $\mu \neq 0.5$. A true μ value, for example, cannot be both 0.6 and 0.4. The null hypothesis must therefore always include the unambiguous case of equality, leaving only the "more complicated", indirect approach B as an option. This principle of statistical testing is comparable to the presumption of innocence in a court case. The initial or null hypothesis is: "The defendant is innocent." Instead of showing directly that a defendant is guilty, more or less strong evidence that is not consistent with the innocence of the defendant is presented by the prosecutor. If this evidence is strong enough, the assumption of innocence is no longer valid and the defendant is found guilty. If, however, it is not possible to produce sufficiently strong evidence against the assumption of innocence, the defendant is not found guilty because his previously assumed innocence could not be called into question beyond reasonable doubt. The null hypothesis is assumed to be true until the data collected are sufficiently strong against it and it must be rejected. As soon as this is the case, the alternative hypothesis is indirectly accepted. However, if the data cannot refute the null hypothesis, it must still be assumed that it is true and the research hypothesis is not accepted. Since only the null hypothesis is tested in a hypothesis test and evidence is sought against it, a null hypothesis can only be rejected or not rejected but, strictly speaking, not accepted.

In contrast to the two-tailed (two-sided or non-directional) research hypothesis RH1, the research hypothesis RH2 postulates a one-tailed (one-sided or directional) hypothesis, because it suggests the postulated effect is in one direction More precisely, the hypothesis $\mu_m - \mu_f < 0$ is a left-tailed hypothesis, because it is postulated that the difference $\mu_m - \mu_f$ is to the left of zero, i.e. in the negative range. If the sign were reversed, this would correspond to a right-tailed formulation. With one-tailed hypotheses, the case of equality must also be included in the null hypothesis, so that for RH2 we have:

$$\mathrm{H}_0\colon \mu_m - \mu_f \geq 0 \text{ versus } \mathrm{H}_1\colon \mu_m - \mu_f < 0.$$

Alternatively, the simplified formulation

$$\mathrm{H}_0\colon \mu_m - \mu_f = 0 \text{ versus } \mathrm{H}_1\colon \mu_m - \mu_f < 0$$

is often used. The simplification is permissible because whenever $\mu_m - \mu_f = 0$ is (not) rejected, $\mu_m - \mu_f \geq 0$ will also (not) be rejected, i.e. "equal" implies "greater than or equal to".

There is no generally accepted answer to the question of whether there should be a one-tailed or two-tailed formulation of the hypotheses as there is still no consensus in statistics on the circumstances in which one method is clearly superior to the other (Sheskin 2000). Only if one direction of testing can be ruled out from the start, owing to a rejection of the null hypothesis in this direction simply being far too unlikely, is a directional hypothesis preferable. For example, we might be interested in comparing the average body weight of men and women. Before making this comparison it could reasonably be expected that, if there is a statistically significant difference between the average weights, it must be in favor of men, because it is not possible for the true average weight of women to be more than that of men. So the actual hypothesis is preceded by a conjecture about the true situation and, depending on whether it can be agreed upon or not, this more or less suggests a one-tailed hypothesis. If μ_m is the average body weight of men and μ_f that of women, then we would test $\mathrm{H}_0\colon \mu_m - \mu_f = 0$ versus $\mathrm{H}_1\colon \mu_m - \mu_f > 0$. If our initial conjecture is in fact true, we obtain a more statistically significant difference using this formulation than if we had used the two-tailed hypothesis formulation, $\mathrm{H}_0\colon \mu_m - \mu_f = 0$ versus $\mathrm{H}_1\colon \mu_m - \mu_f \neq 0$. The rejection area to the right of zero is only half as large in the two-tailed formulation as it is in the one-tailed formulation. In other words, the probability of correctly rejecting a null hypothesis that is actually false is greater when using a one-tailed test than when using a two-tailed test.

Let us return to the ultimatum game with the research hypothesis H_1: $\mu \neq 0.5$. We randomly select ten pairs of A and B players and have the players in each pair play the ultimatum game against each other. The share that A offers his opponent B is recorded in all 10 games and the arithmetic mean is 0.21. In view of this result, it could be assumed that the null hypothesis would have to be rejected because the number 0.21 is "quite far away" from the number 0.5. But how sure can we be about rejecting the null hypothesis? What does "quite far" mean in this context? And if the null hypothesis can be rejected, from which value on would the null hypothesis no longer be rejected? These questions will now be answered.

4.4.2 How Inferential Statistics Works

The basic idea behind the solution to the above decision problem is actually very simple: we check whether the ten amounts offered could be expected to occur as they did if the null hypothesis were true. If the null hypothesis in our ultimatum game is true, then the realized offers must come from a distribution with an expected value of $\mu = 0.5$. A specific value for the spread of this distribution must also exist. It is either known from the outset, which is rarely the case in practice, or it needs to be estimated in advance on the basis of the sample data obtained. Let us assume for the moment that the amounts offered in our ultimatum game are normally distributed with an expected value of $\mu = 0.5$ und a variance of $\sigma^2 = 0.01$. For example, a sample that would seem to be consistent with the null hypothesis is (0.50, 0.77, 0.38, 0.59, 0.25, 0.46, 0.68, 0.71, 0.51, 0.41). The mean value is $\bar{x} = 0.526$ and all the values would be spread roughly around the value 0.5. In this particular case, we can even be certain that this sample comes from this distribution, because the data were obtained from PQRS using the normal distribution with an expected value set at $\mu = 0.5$ and a variance of $\sigma^2 = 0.01$. In an experiment, however, the data are of course not computer-generated, but are the result of decisions made by the subjects. This means that even with an apparently "correct" observed sample, such as the one above, we can never say with certainty that this sample actually originated from the distribution we assumed. Worse still, with these ten numbers alone, we cannot even make a probability statement in relation to this assumption. This is only possible once we have found a random variable that provides the best possible estimate of the distribution parameter being tested in the hypothesis (here the expected value μ) on the basis of the obtained random sample data. With regard to the hypothesis H_0: $\mu = 0.5$ versus H_1: $\mu \neq 0.5$, an obvious candidate for such a *test statistic* (or test value) is, of course, the sample mean $\bar{x}$. In fact, it can be shown that $\bar{x}$ is an unbiased estimator of the population mean μ, i.e. with an infinite number of repetitions it at least on average matches the value μ. Mathematically, the test statistic is a sampling function that converts a vector of numbers (our sample) into a single real number. Statistically, the test statistic is always a random variable that takes on a new value for each new random sample. This means that a new sample, and thus a new mean $\bar{x}$, would result if we have all ten subjects play the ultimatum game again. The distribution the test statistic $\bar{x}$ follows for each new sample is called the *sampling distribution* or *null distribution* if we assume that the null hypothesis is true and the sample therefore does in fact stem from a distribution with an expected value of $\mu = 0.5$. In order to make concrete probability statements, we need specific values of the density and distribution functions. The determination of these values is explained in the following box.

Box 4.2 Performing a Hypothesis Test Using PQRS

After starting the PQRS program presented in ▶ Box 4.1, we select the desired distribution under the "Distribution" option. Our ultimatum game null distribution is a normal distribution. For simplicity, we first assume that the parameters of this distribution are known, using the values $\mu = 0.5$ and $\sigma^2 = 0.01$. We therefore change the default values to the correct values and then click on "Apply New Distribution". The image shown in ◘ Fig. 4.7 then appears.

A *hypothesis test* addresses the question of whether it is "likely" that the realized value of the test statistic originates from the null distribution or whether the value is so uncommon that it is simply "too unlikely" that it does so. What "likely" and "too unlikely" mean in this context must, of course, be specified beforehand. For example, we could specify that a random variable that assumes a *critical value* less than 0.304 or greater than 0.696 is "too uncommon" to originate from the above null distribution with $\mu = 0.5$ and $\sigma^2 = 0.01$, because the probability of this is only 5%. If, for instance, a value of 0.8 is obtained and we conclude that the random variable is not drawn from the null distribution, there is still the possibility that it does indeed originate from the null distribution and we were only unlucky to observe an improbable but nevertheless possible occurrence of 0.8. In this case, our conclusion is wrong and the probability of making such an error is 5%. The specification of this error probability (or the critical values associated with it) is in principle arbitrary, but a *significance level* of 5% for statistical conclusions has become established as a common standard.

In order to make a probability statement in relation to the test statistic, we must of course know its null distribution. Even if we know that the amounts offered by each subject are normally distributed with an expected value $\mu = 0.5$ and a variance $\sigma^2 = 0.01$, this does not mean that the mean of these data also exactly follow this distribution. It can be shown that the null distribution of the mean of normally distributed random variables is again a normal distribution with the same expected value (here $\mu' = \mu = 0.5$), but with the modified spread $\sigma' = \sigma/\sqrt{n} = 0.1/\sqrt{10} = 0.0316$. The spread σ' of the test statistic is called the *standard error*, SE. When using the sample mean $\bar{x}$ as an estimator of the true population mean μ, the standard error naturally depends on the sample size, since a mean based on a large sample is a more accurate estimator of the population mean than a mean based, for example, on only two observations. The critical values of the null distribution with $\mu' = 0.5$ and $\sigma' = 0.0316$ at a significance level of 5% are 0.438 and 0.562. We obtain these values by distributing the probability mass at the significance level of 5% over the two ends of the normal distribution, i.e. 2.5% at each end. To do this, we enter the value 0.025 in the red number field and read the quantile in the grey number field between the two arrows (see ◘ Fig. 4.8).

We obtain the upper critical value of our two-tailed hypothesis test by entering 0.025 in the blue number field. Our concrete sample yielded the value $\bar{x} = 0.526$, i.e. it is statistically reasonable to interpret it as "typical" or "not overly extreme" because it lies within the acceptance

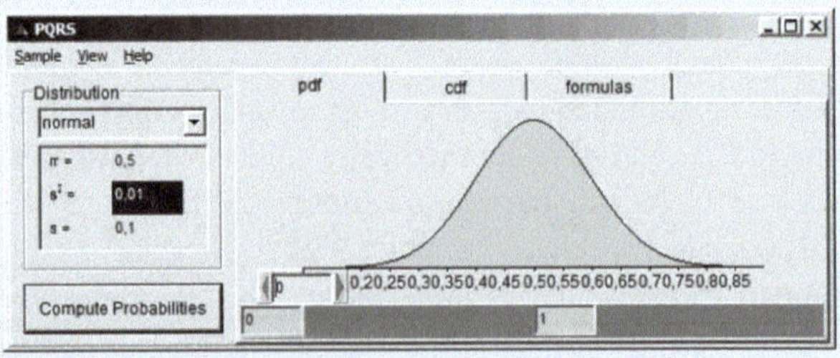

◘ **Fig. 4.7** Normal distribution

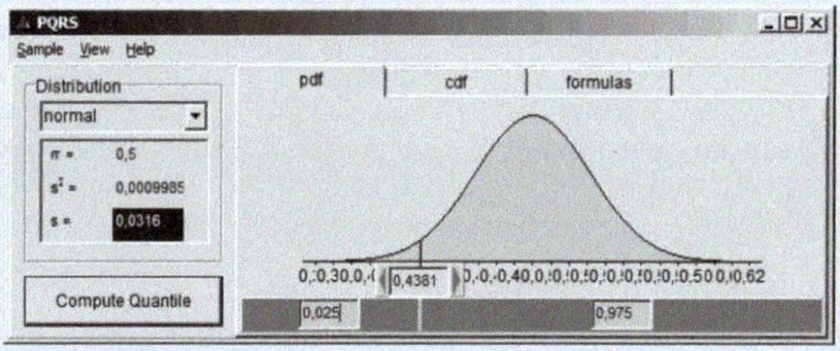

Fig. 4.8 Critical value 2.5%

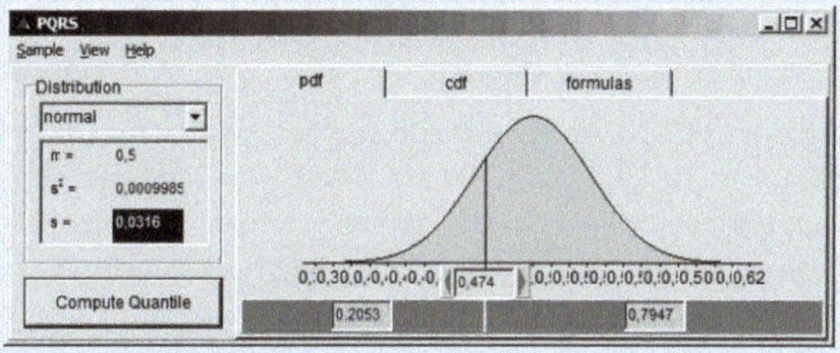

Fig. 4.9 Critical value

region between 0.438 and 0.562. We would consider any value that is less than 0.438 or greater than 0.562 to be too unlikely to be statistically consistent with the null hypothesis $\mu = 0.5$, as the probability of such an event is only 5%. The hypothesis that our sample comes from a distribution whose expected value $\mu = 0.5$ cannot therefore be statistically rejected.

An equivalent method to arrive at the (same) test decision is the comparison of the *p-value* with the significance level. The *p*-value is the probability of the realized value of the test statistic or a more extreme value occurring. For a two-tailed test H_0: $\mu = 0.5$ versus H_1: $\mu \neq 0.5$, this would be the probability of observing a value of $\bar{x}$ that is greater than 0.526 or less than 0.5–0.026 = 0.474. We enter all the relevant values in PQRS and obtain the image shown in Fig. 4.9.

This shows that the probability of the event $\bar{x} \leq 0.474$ is just 0.2053. Because of the symmetry of the normal distribution around μ this corresponds precisely to a *p*-value of $2 * \text{P}(\bar{x} \leq 0.474) = 2 * 0.205 = 0.41$. Since this probability mass is greater than the significance level of 0.05 (i.e. the probability mass in the rejection region), the null hypothesis is not rejected. Unfortunately, the procedure described above is only practical if you have a computer with PQRS or another statistics program at your disposal. In order to determine the critical values or the probability masses of the null distribution, a different table would be needed for every possible value of μ and σ^2, i.e. theoretically infinitely many. This is why the random variables of any desired normal distribution are often standardized. If we know that $\bar{x}$ is normally distributed with an expected value of μ' and a standard deviation of $\sigma' = \sigma / \sqrt{n}$, then $z = (\bar{x} - \mu')/\sigma'$ is *standard normally distributed* with $\mu = 0$ and $\sigma = 1$. With this transformation, only one table is needed for any parameters μ and σ^2 of the normal distribution, i.e. a table of the standard normal distribution. In our case, the standardized test statistic $z = (\bar{x} - \mu')/\sigma' = (0.526 - 0.5)/(0.1/\sqrt{10}) = 0.82$ and the question equivalent to the non-standardized procedure is: is the value $z = 0.82$ sufficiently different from $\mu = 0$ that we can conclude with an error probability of 5% that the sample does not come from a distribution with $\mu = 0.5$? The critical values of a standard normally distributed variable at a significance level of 5% are approximately −1.96 and 1.96. Since $z = 0.82$ lies within this acceptance region, we again do not reject the null hypothesis. We once again calculate the *p*-value, which is $2 * P(z \leq -0.82) = 2 * 0.205 = 0.41$. We already know this number from the previous paragraph, which confirms that both test methods (one with the untransformed test statistic and the other with the transformed test statistic) are equivalent.

If the null hypothesis is not rejected, as in our case, then there is no statistically significant difference between the realized value of the test statistic $\bar{x} = 0.526$ and the value of the population parameter $\mu = 0.5$ formulated in the null hypothesis. We then say: "The null hypothesis cannot be rejected" or "The value $\bar{x} = 0.526$ is not statistically significantly different from 0.5". However, if the *p*-value is below the significance level, the difference is statistically significant. The degree of significance is usually marked with one to three stars, similar to the distinction given to chefs. It is often set as follows: one star (*) for $p < 0.100$, two stars (**) for $p < 0.050$ and three stars (***) for $p < 0.010$.

4.4.3 Possible Errors and Power of a Test

As clear as our test decision with the described procedure may be (reject H_0 or not), it must always be kept in mind that no statistical test can determine whether a hypothesis is *actually* true or false. Even if the test statistic of the sample is in the critical region and we come to the conclusion that the null hypothesis should be rejected, it can still be true. The larger we choose the critical area or the significance level, the more likely this so-called *Type I error* is. Let us imagine for a moment that the null hypothesis is in fact true. A good test statistic will then provide values that are on average far from the critical range. However, if we now increase the critical area, it will become increasingly likely that the test statistic of the sample will still fall into the critical area.

Now let us imagine that the null hypothesis is in fact false. In this situation, we would be making an error by not rejecting the null even though it is false (*Type II error*). For this, too, there is a probability, β, which can be represented as an area under a density function. However, this density function is different from the previous situation because the null distribution only applies if H_0 is actually true. The distribution that applies if H_1 is true is therefore called the H_1 distribution. ▪ Figure 4.10 compares the two distributions and identifies possible rejection and non-rejection areas for a one-tailed test.

Let us assume the test statistic of our sample lies to the left of the critical value. We are then in the non-rejection region of the H_0 distribution and will not reject the null hypothesis. If the null hypothesis is actually true, there is a high probability of $1 - \alpha = 95\%$ that we are making the correct decision. But in the case of the null hypothesis actually being false (i.e. the H_0 distribution does *not* apply), our decision

▪ **Fig. 4.10** H_0 distribution versus H_1 distribution and their rejection and non-rejection regions for a one-tailed test

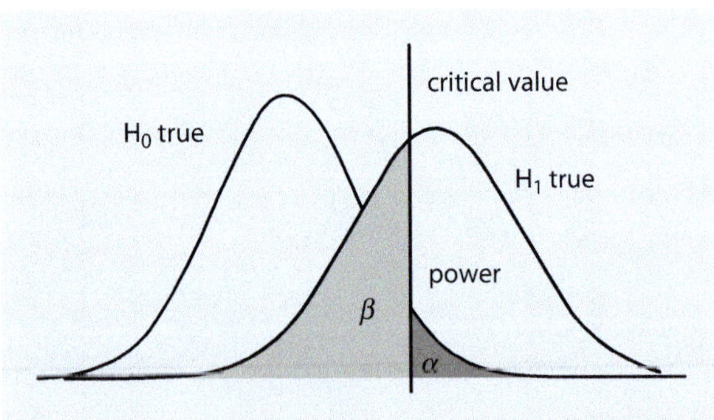

means we are not rejecting a false hypothesis. The probability β for this Type II error can be read from the H_1 distribution, as this applies if H_1 is true (or H_0 is false). The area β then represents the acceptance region of a false null hypothesis.

Now let us assume that the test statistic lies to the right of the critical value. We then reject the null hypothesis. If H_0 is in fact true, we are actually rejecting a correct hypothesis with our decision. The probability of this Type I error is α. However, if the null hypothesis is actually false, the H_1 distribution on the right applies and we have correctly rejected a false hypothesis. The probability of this event, $1 - \beta$, corresponds to the area of the rejection region of the H_1 distribution. $1 - \beta$ is the probability of correctly rejecting a false null hypothesis. At the same time, it is the probability that our test correctly shows the research hypothesis H_1 to be true. In this context, we speak of the *sensitivity* or power of a test, because $1 - \beta$ says something about how suitable the entire procedure (drawing of the sample, application of the test statistic) is for identifying the presence of an effect (which is formulated in the research hypothesis) that actually exists. As we can see, the smaller β is, the greater the power. A summary of these cases is shown in ◘ Table 4.1.

The interdependence of these two types of error can immediately be seen in ◘ Fig. 4.10. The greater the probability of a Type I error α, the less likely the Type II error becomes, and vice versa. It therefore does not make any sense to set one of the error probabilities as small as desired, because the other error then becomes increasingly probable. Ultimately, a balance must be struck in which the consequences of both types of error should be taken into account when making the decision.

Important

If the research question is formulated in the form of a statistical hypothesis, hypothesis tests can be used to draw statistical conclusions regarding this hypothesis. There is, however, always a certain probability of errors. No hypothesis test in inferential statistics can determine whether the hypothesis is actually true or false.

It is clear from what has been said so far that the sensitivity or power of an experiment is of great importance. In the following section, we introduce the basic idea behind determining the statistical power.

◘ **Table 4.1** Summary of error probabilities

	Truth	
	H_0 true	H_0 not true
Rejection H_0	**Type I error**	*correct*
	(Prob. α)	
Non-rejection H_0	*correct*	**Type II error**
	(Prob. $1 - \alpha$)	(Prob. $1 - \beta$)

4

4.5 Power Analysis

4.5.1 Basics

The aim of the following discussion is to present how power analysis works. This can best be achieved if we start with very simplistic assumptions that are consequently far removed from practice. In order to determine the statistical power of real experiments, it is generally necessary to deviate from these simplifying assumptions. This in turn means that the analysis no longer resembles a "standard procedure", which can simply be processed step by step. Indeed, power analysis very quickly becomes very complex and time-consuming, requiring advanced statistical concepts, which is why there are textbooks on this topic with all the related problems and approaches to finding possible solutions (Cohen 1988; Ellis 2010; Murphy et al. 2014).

The simplest case of a power analysis is probably the following. We draw a single sample from a normal distribution with a known variance σ^2 and calculate its mean. This value is used to check whether or not the hypothesis "The population mean μ takes on a certain value, μ_0" can be accepted. We could, for example, again imagine the percentage amounts given by a dictator in the dictator game. These would be normally distributed with a population variance $\sigma^2 = 0.25$, or $\sigma = 0.5$. The sample size would be $n = 25$ and the initial hypothesis would be that on average 40% of the available amount is not given, i.e. we test H_0: $\mu = 0.4$ versus H_1: $\mu \neq 0.4$.

Our test statistic is the mean $\bar{x}$ of the sample of size n = 25. If the population from which this sample is drawn is normally distributed with μ and σ^2, then the mean $\bar{x}$ is also normally distributed with the same expected value μ, but the changed variance $\sigma^2/n = 0.25/25 = 0.01$. The sample size n in the denominator of the formula takes into account the fact that the larger the sample from which the mean is formed, the more accurately a sample mean corresponds to the true population mean.

If the null hypothesis is true, i.e. μ actually assumes the value 0.4, then the sample distribution (null distribution) looks like the one shown in ◘ Fig. 4.11.

The rejection regions, each with a probability mass of 2.5%, are bounded by the critical values 0.204 at the left end and 0.596 at the right end of the distribution.

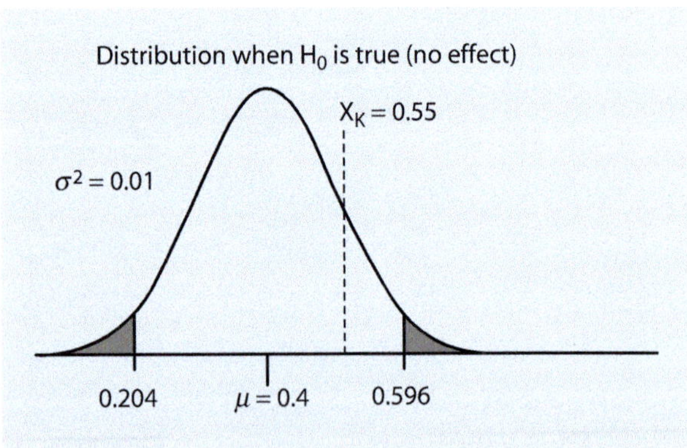

◘ **Fig. 4.11** Sample distribution in the example

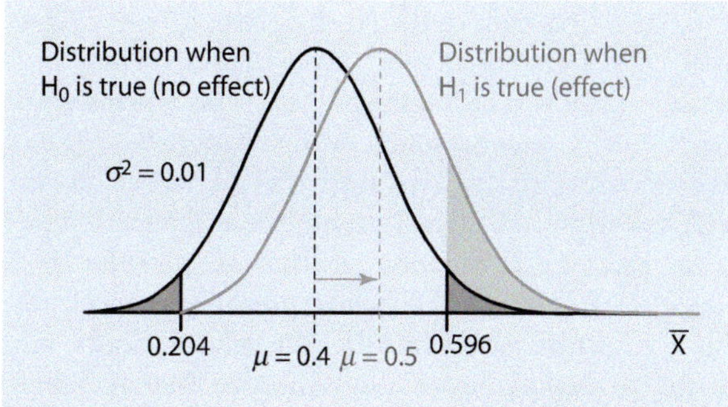

■ **Fig. 4.12** Sample distribution in the example with false null hypothesis

Now let us assume that the null hypothesis is false. In a two-tailed research hypothesis, this means that any deviation of the true μ upward or downward from the value $\mu_0 = 0.4$ leads to a false null hypothesis. For example, the true population mean could be $\mu = 0.5$ (with the same variance). The null hypothesis then contains an error of "size" +0.1 and the center of the correct sample distribution is exactly 0.1 further to the right than assumed in the null distribution (see ■ Fig. 4.12).

Since we always assume that the null hypothesis is true in the hypothesis test, our decision is based on the null distribution with its critical values and we therefore again reject the null hypothesis if the test statistic, i.e. the sample mean, is less than 0.204 or greater than 0.596. But *if* we are left of 0.204 or right of 0.596, and *if* the H_1 distribution is actually the "correct" sample distribution, we are no longer making an error (Type I) by rejecting H_0, but are in fact making the *right* decision, since the null hypothesis is actually false this time. The overall probability that the test statistic is less than 0.204 or greater than 0.596 can be determined using the "correct" (light gray) sample distribution. It is the sum of the area below the density function on the left of 0.204 and on the right of 0.596. With the help of PQRS, we determine that the left probability mass is 0.0015 and the right one is 0.1685.

So if the null hypothesis is false because the true population mean $\mu = 0.5$ instead of $\mu = 0.4$, the probability that we correctly reject the false null hypothesis is $0.0015 + 0.1685 = 0.17 = 17\%$. The same probability results if we make the same error in the other direction, i.e. the true μ is 0.3 instead of 0.4. The probability 17% is the sensitivity or statistical power of the two-tailed hypothesis test presented here. Of course, it should always be borne in mind that this particular number only applies to the specific deviations from the true value of 0.1 or − 0.1. Strictly speaking, therefore, we have not determined the power under the hypotheses $H_0: \mu = 0.4$ versus $H_1: \mu \neq 0.4$, but under a much more concrete alternative hypothesis, namely $H_0: \mu = 0.4$ versus $H_1: \mu = 0.5$ *or* $\mu = 0.3$. Basically, we can only calculate the power as described above if we have already specifically postulated the size of the error we would make in the alternative hypothesis on the incorrect assumption that the null hypothesis is true. What is notable about this is that the probability of still making a correct test decision progressively decreases, the "less false" the null hypothesis is; or, to put it more simply, the test becomes less and less sensitive, the less H_0 and H_1 differ.

To illustrate this, we imagine, for example, that the true population mean is $\mu = 0.401$ and the null hypothesis is still $H_0: \mu = 0.4$. Then the H_1 distribution is only slightly shifted by 0.001 to the right of the H_0 distribution. This means $\alpha \approx 1 - \beta$, i.e. the significance level of 5% in the case a valid null hypothesis is only slightly less than the power $1 - \beta$ in the case of an invalid null hypothesis (the probability of correctly rejecting a false null hypothesis). If we now tested the null hypothesis H_0: $\mu = 0.4$ on the assumption that it is *true*, we would make a Type II error in $\beta \approx 95\%$ of all cases and not reject the (very likely) false null hypothesis. The larger the error postulated by the null hypothesis, the further to the right the H_1 distribution is away from the H_0 distribution and the greater the power $1 - \beta$. Conversely, the more precisely we specify the region in the research hypothesis H_1 in which we assume the true value to be, the greater the probability that we will accept the research hypothesis if the true value actually lies within the region defined by H_1. This can be achieved particularly if not only the absolute deviation of the true value μ from the postulated value μ_0 is specified, but also the direction of the deviation. For example, the one-tailed test of H_0: $\mu = 0.4$ versus H_1: $\mu = 0.5$ *ceteris paribus* would be more statistically powerful than the two-tailed test H_0: $\mu = 0.4$ versus H_1: $\mu = 0.5$ *or* $\mu = 0.3$. The one-tailed case yields the sample distributions shown in Fig. 4.13.

The left rejection region disappears and the right rejection region now represents a probability mass of $\alpha = 5\%$ (instead of 2.5% earlier). The resulting power is the probability mass to the right of the critical value 0.565 under the orange sample distribution. With the help of PQRS, we find that this probability is $1 - \beta = 0.2595 = 25.95\,\% > 17\,\%$. The difference in power between the two-tailed and one-tailed tests is thus *ceteris paribus* almost 9%.

Important

Hypothesis tests do not *per se* have a certain power or sensitivity. In particular, the probability of correctly confirming a true research hypothesis depends on how precisely the research hypothesis is formulated or how much information about the effect to be investigated is known in advance.

It is very important to realize that the experimenter can influence the power of the experiment. As the following remarks will make clear, this influence is by no means limited to the way the experimenter formulates the research hypothesis.

Traditionally, experiments contain a control or baseline treatment and a treatment in which the variable to be tested is intentionally changed by the experimenter. The research hypothesis postulates that the treatment has an effect on the observed variable and the

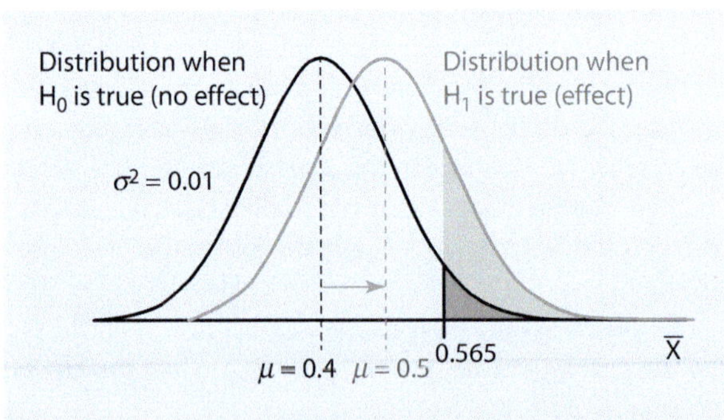

Fig. 4.13 Sample distribution for the one-tailed test in the example

null hypothesis postulates that the treatment has no effect. For example, a treatment in the dictator game experiment could be to have the subjects communicate before the dictator announces his allocation. If the dictator game is now played by two independent groups, one without communication ("*A*" as in anonymous) and one with communication ("*C*" as in communication), then it is possible to investigate whether the two samples come from a population with the same expected value $\mu_A = \mu_C$ or whether the treatment has caused the population mean to change, with the result that $\mu_A \neq \mu_C$. The difference between the parameters in question, $\Delta = \mu_C - \mu_A$, which arises due to the treatment, measures the strength of the effect of communication on the average amounts allocated. In general, the *effect size* is a measure of the difference between an expected parameter under the null hypothesis and this expected parameter under the alternative hypothesis.

First, we consider the case in which the null hypothesis $\mu_A = \mu_C$ or H_0: $\Delta = 0$ is true and communication has no impact on the population mean of the allocations. The sample distributions of both means are then identical since the populations also have the same distribution.

Now we assume that the research hypothesis is valid, i.e. communication has an impact on the average amounts given of $\Delta < 0$ or $\Delta > 0$. In this case, we have two *different* populations: average dictator allocations without communication and average dictator allocations with communication. In the case of $\Delta > 0$, the corresponding sample distributions are as shown in ▪ Fig. 4.14.

If we draw a sample from the population "without communication", then the average amounts given follow the left sample distribution. This distribution is identical to the original null distribution. If we draw a sample from the population "with communication", the average amounts given follow the right distribution, whose expected value is $\mu_C > \mu_A$. The horizontal distance between the two peaks corresponds to the (unstandardized) effect size of communication.

Since we are only interested in whether there is a difference between the two expected values and not in the extent of this difference, we can use the distribution of the difference Δ under the null and alternative hypotheses, $\Delta = 0$ versus $\Delta > 0$, as a basis, instead of the distributions of $\overline{x}$. This is displayed in ▪ Fig. 4.15.

The dark distribution is the distribution we draw from when the null hypothesis is true, and the light distribution is the distribution we draw from when there is an actual

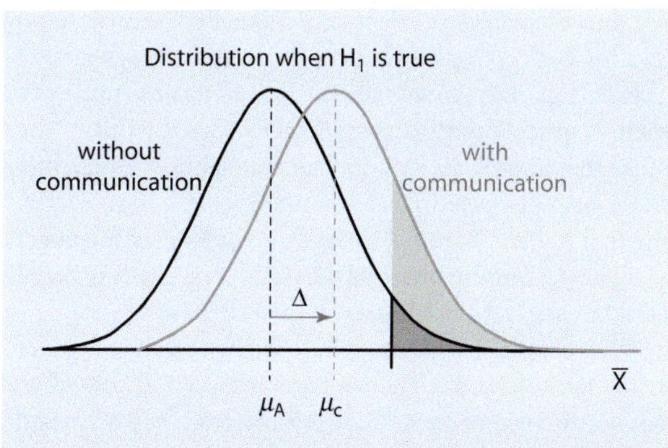

▪ **Fig. 4.14** Sample distribution with true research hypothesis in the example

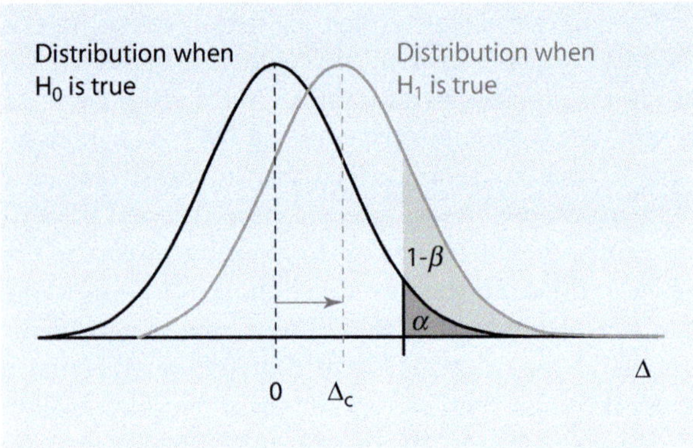

Fig. 4.15 Distribution of the difference Δ under the null and alternative hypotheses

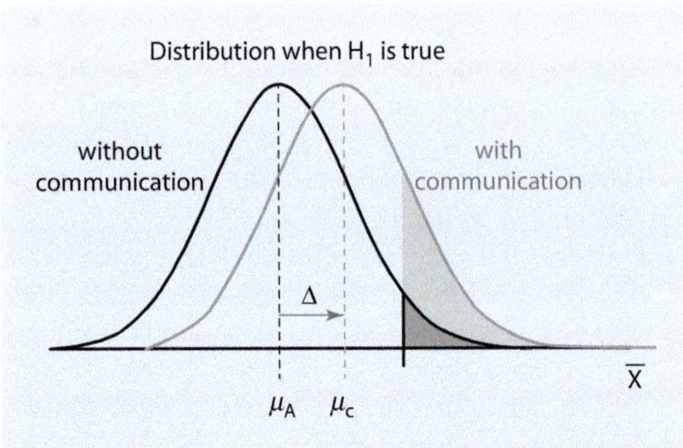

Fig. 4.16 Increased power due to larger effect size

effect size of Δ_c. The probability that our hypothesis test will show the existence of any effect (because we correctly reject the null hypothesis) is the power $1-\beta$, given that the true effect size is Δ_c. As can easily be seen from Fig. 4.16, the probability of correctly identifying an existing effect is greater if the true effect size is larger.

This means that the power of a two-sample test and the effect size always go hand in hand. There is little point in qualitatively talking about the power of a two-sample study without reporting an effect size. Take, for example, a power of 90%. In 90% of all cases, an actual effect of communication on the average amount allocated is then also reported as existing. In the first instance, that in itself sounds good. However, this quality assessment is quickly put into perspective if we start from the (admittedly hypothetical) situation that almost all dictators allocate almost everything if there is communication, i.e. there is an extremely large treatment effect. In this light, it would be almost disappointing that the test does not show an actually existing effect as being present in at least 99% of all cases. On the other hand, if the true effect is very small, a power of as little as 60% could possibly be used as a quality criterion.

The problem with the whole thing is that the actual effect is unknown. We do not know by how much the average amounts allocated would increase within the entire population in the case of communication. If we knew this, we would not need hypotheses or inferential statistics. Therefore, there are only two possible ways of incorporating power into the statistical evaluation of an experiment. The first is to estimate the true effect size based on the difference of the observed sample means, $\overline{x}_C - \overline{x}_A$. This absolute difference, however, depends on the unit of measurement. A difference in the allocated amounts of 1000 cents would be large relative to a difference of 10 euros, although in both cases the same amount of money is involved. For this reason, standardized effect size measures are often used, although they are not without criticism (e.g. Lenth 2001). The most common of all effect sizes in a two-sample test that compares the means is Cohen's *d* (Cohen 1988), which simply divides the difference in the sample means by the population standard deviation assuming $\sigma_C = \sigma_A = \sigma$:

$$d = \frac{\overline{x}_C - \overline{x}_A}{\sigma}$$

For example, if $\overline{x}_C = 0.55$, $\overline{x}_A = 0.4$ and $\sigma^2 = 0.025$, then $d = 0.9487$. Cohen himself proposed a "T-shirt" rating of small ($d < 0.2$), medium ($0.2 < d < 0.5$) and large ($d > 0.8$).

The second, much less controversial, way to deal with the unknown true effect size is to leave it unknown from the outset and instead, prior to performing the experiment, specify a separate effect size to be used for the power analysis. This is then no longer a formal statistical matter, but rather a subjective issue that must be answered from the perspective of the respective scientist. The smallest possible effect size that still holds *practical significance* in the respective discipline is usually used. We could, for example, discuss with other scientists how large the effect of communication on the amounts given by dictators would at least have to be in order to achieve any scientific significance at all. Alternatively, it is also possible to look for further studies that have already estimated effect sizes in a similar context, which we could then adopt.

In our dictator game experiment, for instance, we could specify $\Delta_c = 0.1$ and then ask ourselves what the probability is of a particular hypothesis test showing a practically relevant treatment effect of 0.1. If it turns out that this probability is only 30%, for example, we should consider how to redesign the study in such a way that the power is increased. This brings us to the next quantity that influences the power and this is also affected by the experimental design.

The *sample size n* can be freely set by the experimenter within certain limits. Frequently, too little attention is paid to this freedom in the design when planning the experiment. It is particularly important because the size of a random sample has a decisive influence on the probability of Type I and II errors, and thus also on the power of the experiment. As already discussed, the larger the sample, the better it represents the population from which the sample is drawn. It follows directly from this that, in the attempt to determine the parameter of the population that is assumed as true in the null hypothesis (e.g. μ), the test statistic (e.g. the mean of the sample) will get increasingly closer to this parameter. In other words, the larger the samples, the less the value of the test statistic will deviate from the true population parameter. If there is less spread in the test statistic, its distribution is narrower and higher, i.e. probability mass is removed at the ends and added near the expected value (distribution mean). Graphically, this means that, all other things being equal, the error probabilities α and β decrease, since they are represented by a smaller area under the density function.

The more subjects we send to the laboratory at the same time, the less likely we are to report a non-existent effect (Type I error) or an existing effect as non-existent

(Type II error). This also makes it clear that increasing the size of the sample increases the power of the test without necessarily increasing the probability of a Type I error. From a statistical point of view, the sample should therefore always be as large as possible. From a practical, experimental point of view, however, there are a number of reasons for at least an upper limit to the sample size. For instance, the financial resources, the availability of subjects and the capacity of laboratories are generally limited.

Furthermore, extremely large samples lead to a "very high" sensitivity of the test, meaning that any actual treatment effect, no matter how small, which may be completely insignificant from a practical point of view, becomes statistically significant if the sample is sufficiently large. This is also called the "Fallacy Of Classical Inference" and may be interpreted as a disadvantage. On the other hand, it is entirely possible to classify effects as statistically significant but economically unimportant.

In summary, when it comes to samples, size is indeed important, and it can certainly be a case of "too little" or "too large". Samples that are too small (underpowered studies) lead to a large dispersion of the test statistic values around the true value and thus to large standard errors or broad sample distributions. The probability that the test statistic happens to land sufficiently close to the true value to correctly reject the null hypothesis therefore becomes smaller and smaller. For this reason, even large, practically significant effects tend to remain statistically insignificant and thus undiscovered in smaller random samples. Conducting a low-power study is needless and wastes resources since such a study is unable to reliably identify actual effects.

Furthermore, samples that are too large (overpowered studies) also waste scarce resources. If we invite, say, 100 subjects and can prove with 80% probability that an effect of size 0.001 exists, but the minimum practically significant effect is 0.1, and this could also have been proven with 50 subjects and the same power, then we can save the monetary and non-monetary costs (payments, time, stress, etc.) of 50 subjects. Leonhart (2008, p. 82) summarizes very aptly:

> *"The optimal sample size is large enough to ensure that an effect relevant to practice can be statistically verified. On the other hand, it is small enough to make only practically significant effects statistically significant."*

The optimal sample size is therefore a point to consider carefully when designing an experiment. But how do we determine the *optimal* sample size?

4.5.2 BEAN and the Optimal Sample Size

To begin with, the optimal sample size of an experiment as such cannot be calculated, since such a calculation still contains too many degrees of freedom, even with complete information on population parameters (especially variance). This becomes clear if the four interacting factors of a power analysis are taken into account once again. The abbreviation BEAN is often used as a mnemonic:

1. **Beta** β (probability of a Type II error or power)
2. **Effect size** (true influence of the treatment on the measured variable)
3. **Alpha** α (probability of a Type I error)
4. **N** (Number of subjects or sample size)

In themselves, none of these factors is a consequence of any "natural law" and clearly defined in advance. Rather, they can all be set more or less quite freely. The determination of the optimal sample size is therefore not simply a static calculation carried out in the same way for each experiment, but largely involves "coordinating" subjective factors (e.g. "What probability of a Type I error am I prepared to accept?") or exogenous facts (e.g. "How large is my laboratory?").

Uniform standards have already developed for alpha and beta in most sciences. The willingness to falsely not reject a null hypothesis is very small and is almost always quantified by an error probability (significance level) of only 5%. A Type II error is usually considered to be less serious. Therefore, beta (if at all) is set to four times the alpha level (20%). The resulting power of 80% is also increasingly being seen as a prerequisite by third-party sponsors when financing experimental studies, especially if the consequences of a Type II error are particularly drastic and/or the amount of funding is particularly high. This figure, originally recommended by the American Psychological Association (APA), now seems to be establishing itself as the norm in other disciplines as well.

If two of the four factors are given, the fourth remaining factor can be calculated by defining a third factor. In other words, the BEAN of a hypothesis test has three degrees of freedom. For example, if both error probabilities are given, only two mutually inverse questions can be considered:

1. If exactly n test subjects can be financed, what is the minimum effect size that can be correctly identified as being "present" with a probability of 80%?
2. If the smallest possible, practically significant effect is Δ, how many subjects are needed to correctly identify it as being "present" with a probability of 80%?

Let us start with the simplest case, the example from the previous section. There, a single sample of group size $n = 25$ was drawn. Now we will test H_0: $\mu = 0.4$ versus H_1: $\mu = 0.5$ (right-tailed). The population variance of the normally distributed allocations is again $\sigma^2 = 0.25$, or $\sigma = 0.5$, with an unknown population mean μ. We determined the power by calculating the probability density of the H_1 sample distribution in the rejection region of the H_0 sample distribution. The critical value of the null distribution was $c = 0.565$. If $F\left(x_0;\mu_1;\sigma/\sqrt{n}\right) = p\left(x \le x_0\right)$ denotes the cumulative distribution function of the sample distribution, which specifies the probability of the event $x \le x_0$ for a given quantile x_0, then the following applies to the power in our example

$$1-\beta = 1-F\left(c;\ \mu_1;\ \frac{\sigma}{\sqrt{n}}\right) = 1-F\left(0.5645;\ 0.5;\ \frac{0.5}{\sqrt{25}}\right) = 0.2595 = 25.95\%.$$

This yields the unique relation

$$\beta = F\left(c;\ \mu_1;\ \frac{\sigma}{\sqrt{n}}\right).$$

Theoretically, the probability of the Type II error can be calculated with this formula if the H_1 distribution function F is known. Even if they are not visible at first glance, the equation contains all our BEAN components: **B**eta (β) and the **N**umber of subjects (n) can be located directly in the formula. **A**lpha (α) is implicit in the critical value c since there is a unique relationship between the two factors for the given parameters μ_0 and σ_0 of the null distribution. The **E**ffect size $\mu_1 - \mu_0$ is calculated using the value of μ_1 given by the alternative hypothesis and the value of μ_0 implicit in the null distribution or c.

The practical problem with this equation is that it cannot easily be solved using the arguments of F, such as n. The distribution function F cannot be converted into an elementary primitive function and so numerical methods have to be used, which is why it is best to leave it to the computer to do the arithmetic. A suitable program is needed for this and we present a selection of different alternatives in the following box.

Box 4.3 Software for a Power Analysis

Most programs developed specifically for power analysis have a fairly clear trade-off between ease of use/performance and the cost of using the program. Generally there is a free application for almost every power analysis. One problem, however, is that one program supports a certain subset of analyses and another supports another, so that several programs may be required in order to obtain the most complete range of applications possible. This problem does not (theoretically) exist for programs whose source code is also freely available (open source) because here users can put together their "own" power program, provided they have the relevant programming and statistical know-how.

Among the open source solutions is a rather spartan, but nevertheless powerful and easy to use program DSTPLAN (▶ biostatistics.mdanderson.org/SoftwareDownload/ProductDownloadFiles/DSTPLAN_V4.5v.zip). It was developed by biostatisticians at the University of Texas and covers the most common power applications. In principle, the program works according to the BEAN procedure, which we have just explained. Any five of six values, which essentially represent BEAN elements, can be entered and the program calculates the sixth value.

A more modern and more powerful open source solution is the modular program package R (▶ www.r-project.com). Modular means that the basic configuration is limited to the most basic statistical functions and everything else is integrated into this basic configuration in the form of modules or packages, as required. For a power analysis, for example, the "pwr" package is recommended. This needs to be installed on top of the basic configuration of R to maintain its functionality. The package then provides several commands for power analysis. In addition to the rather artificial case of normally distributed variables with known population variance, the "pwr" package naturally also contains commands for more practical tests such as *t*-tests or *F*-tests. If the range of available applications is not sufficient, the package (or any other) can easily be extended using the user's own commands.

The tool G*Power (▶ www.gpower.hhu.de) is very popular among the free closed-source solutions because – like the PQRS - it provides a compact and clear graphical user interface. It is also nice to be able to graphically display the density functions of the respective H_0 and H_1 distribution. The program contains the most important power analyses from a practical point of view. Purely didactic examples, such as our work example with the known population variance, cannot be reproduced. Since the source code is not public, the functionality of this program cannot be extended or changed directly by the user.

A free R-based online solution for several basic tests can be found at ▶ www.powerandsamplesize.com. In addition to detailed explanations, a graphical user interface and graphical representations, R-codes with which the selected procedure can be implemented locally in R are also given.

For those who do not want to compromise on the functionality and ease of use of power analysis software, there are powerful but also very expensive commercial software packages available, such as NCSS Pass 14 (▶ www.ncss.com) or nQuery Advisor + nTerim (▶ www.statsols.com). In our opinion, StudySize (▶ www.studysize.com) offers one of the best compromises in price and functionality of all commercial programs. For less than $100, StudySize also offers Monte Carlo simulations and power analyses for nonparametric tests under a graphical user interface. A 30-day trial version is also available free of charge.

4.5.3 Power Analysis and the "Hard Truth" of its Results

One of the main applications of a power analysis is the *a-priori* determination of the necessary sample size. In the simplest case, we simply change our work example and ask ourselves how many subjects we need to get the default value of power of 80% instead of 25%. In the StudySize program, we select "Normal Distribution" from the "Distribution Parameters" menu and confirm "New Calculation" with the "Continue" button. Then we fill in the fields as shown in ◘ Fig. 4.17 and click on the "Sample Size" button.

This shows us that we would have to invite 155 subjects to get 80% power if our standard effect size is 0.2! Now, assuming we can finance $n = 40$ subjects, what minimum effect size can be correctly identified as "present" with 80% probability? In this case, we leave d unspecified and specify $n = 40$ instead, using the above program to calculate that $d = 0.3931$.

The value $d = 0.3931$ corresponds to a (non-standardized mean difference) of $d\sigma = 0.1965$. This is the smallest possible effect that we can correctly identify as present with a probability of 80% if we can test "only" 40 subjects at the same time.

Let us pretend for a moment that the above power analysis is the result of a real study. This result would then be very sobering. To be able to correctly identify an absolute effect of 0.1 in 80% of all cases as present, we need 155 subjects. This specification may well blow our budget. How could such a situation be handled?

One point we need to consider is more cost-effective ways of increasing the power of a statistical study, in addition to increasing the size of the sample, under otherwise identical conditions. One possibility would be to take measures that lead to a reduction of the standard error. The less the mean of a sample is spread around the true population mean, the lower the risk of an random extreme value that could be wrongly interpreted

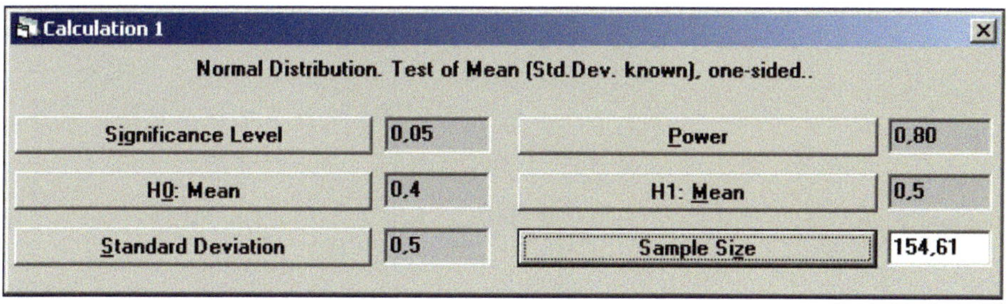

◘ **Fig. 4.17** StudySize power analysis

as an effect and that, in reality, was only caused by the standard error. The confounding variables already discussed are a significant factor influencing the standard error. Individual differences in the subjects (preferences, intelligence, receptiveness, motivation, etc.) can "blur" the true treatment effect because they randomly influence the measured variable without having been explicitly considered. From a statistical point of view, it would therefore be desirable to keep the pool of subjects as homogeneous as possible, for example by only recruiting students of economics in the 3rd semester. However, the more specific the group of individuals from which the subjects are selected, the more specific the possible effects that are identified become. These may then apply to that specific group of subjects, but not necessarily to others.

Factors that might lead to arbitrary behavior by some subjects also increase the standard error. Comprehension problems or signs of fatigue on the part of the subjects should already have been avoided in the planning phase. Against this background, it is especially important to formulate the instructions clearly and unambiguously. The subjects have to know exactly what they are doing so that they do not randomly generate an effect that does not actually exist, thus unnecessarily reducing the sensitivity of a statistical inference.

Parametric tests are also more powerful than their nonparametric counterparts. Again, the intuition behind this statement is quite simple. Nonparametric tests normally use only the ranks of the observations and not the observation itself. The transition from a metric variable to an ordinal one inevitably leads to a loss of information concerning the effect under investigation. This information could have been used in a test to improve the statistical power. For example, suppose we had measured a dependent variable in the control and treatment groups in the following extreme form (see ■ Table 4.2).

Without having to try a test, it should be clear that the treatment has a very large quantitative effect. Specifically, the sample mean has increased from 3 to 25. A parametric t-test for equality of population means gives a p-value of 0.0002 and thus the null hypothesis "no effect" is clearly rejected at a significance level of 5%. The nonparametric counterpart to this test would be a Mann-Whitney-U test. This uses only the ranks of the values as a basis for information and checks the probability that the three lowest ranks are in one group and the three highest ranks in the other group. In our example with a sample size of $n = 3$, we get the ranks 1, 3, 2 in the control group and the ranks 5, 6, 4 in the treatment group. This is already the least random distribution of ranks across the two groups that is possible, as the three lowest ranks are included in the control group and the three highest ranks are included in the treatment group. This means that there is no other distribution that delivers an even smaller p-value.

■ **Table 4.2** Values of the dependent variables in the control and treatment groups (example)

Control group	Treatment group
2.5	24.8
3.7	28.1
2.8	22.1

However, this smallest possible p-value is still 10% at $n = 3$, meaning that the null hypothesis at a significance level of 5% can never be rejected. As a result, the probability of correctly rejecting a false null hypothesis (the power of the test) is zero. In order for this test to be able to detect an existing effect, we would have to accept a significance level of more than 10%.

Last but not least, the specific design of the experiment has an influence on the statistical power. For example, a repeated measures design, in which several consecutive measurements are recorded for one and the same person, is more powerful than one in which each person is measured only once. So if only a certain number of subjects can be organized and we have already exhausted all other possibilities of increasing the power, each subject could be faced with the same decision-making problem several times, thus increasing the behavioral information collected per subject. Similarly, under otherwise equal conditions, we have more power in a within-subject design with paired observations per subject than in a between-subject design with two independent groups.

Important

The goal of the power analysis of an experiment is to coordinate the four mutually influential BEAN parameters (beta, effect size, alpha, *N*) in such a way that the probability of correctly identifying a truly existing effect with this experiment is sufficiently high. None of these parameters is a "natural constant", fixed at some particular value. The spread of the test statistic is particularly important. Keeping this as small as possible is a creative process that affects alpha and beta as well as the effect size (as far as it is standardized).

4.5.4 Misapplications and Misunderstandings in Power Analyses

Experimental economists are chiefly concerned with the discovery and quantification of real behavioral effects. As a rule, the investigation is preceded by the assumption that this effect actually exists. For example, if one wants to investigate to what extent communication influences the willingness to cooperate in a prisoner's dilemma, the initial assumption is that the treatment with communication actually causes a behavioral effect, otherwise an explicit investigation would simply incur costs, and little knowledge would be gained. In other words, the primary interest of an experimental economist is to show the *existence* of an effect, rather than its non-existence.

Closely related to this is the aforementioned probability $1\text{-}\beta$ that a test variable correctly rejects a false null hypothesis, i.e. it detects a real existing effect (H_0 is in fact false and H_1 is actually true). Therefore, if an experimental economist is testing a hypothesis because he suspects that an effect exists, this probability is more important than the probability $1\text{-}\alpha$ of correctly stating (we do not reject H_0) that an effect does not exist (H_0 is in fact true).

Against this background, it is reasonable to assume that in every statistical analysis of an experiment, it is standard practice to provide information on the power of the test and the effect size. However, this is not the case. Instead, the empirical significance level (p-value) is almost automatically nearly always compared with the theoretical significance level (α). Whenever the first value is greater than the second, the data collected is considered as being not sufficiently consistent with the existence of an effect and the null hypothesis is not rejected. If no effect *really* did exist, this decision would almost always be the right one, because at a significance level of 5%, this probability would be 95%. The point is, however, that *we do not know whether there is actually no effect*. One could just as well exist, which in practice is even assumed in advance. And under this assumption, nothing is known about the likelihood of making the right decision when an effect is detected. Without any *a priori* information about the power of a test, it seems premature to reject the research hypothesis simply because the null hypothesis is not rejected. In a low power test, say 40%, the probability of making a Type II error is 60%. This means that in 60% of all cases an effect that actually exists would be declared non-existent. On the other hand, it also seems too hasty to conclude that an impressive effect was discovered simply because the null hypothesis was rejected. The probability that we will correctly show a real effect by rejecting the null hypothesis is only 40%. A coin toss would be more reliable than this test in detecting an effect. The significance or non-significance should at least be combined with rough indications of the power of the test used.

We now know that a power analysis is necessary to generate useful and informative statistical inferences. Unfortunately, however, power analysis is similar to many other important things in our everyday lives, such as nuclear fission, the Internet or a kitchen knife: they must be used *correctly* in order to generate the desired benefit. If used incorrectly, the adverse consequences can be severe in some cases. In the following, we will briefly discuss the most common errors in the application of power analyses.

First of all, the term power *analysis* itself is often responsible for misunderstandings and the misapplication of the process of determining the power. This is not an analysis in the sense that existing data are analyzed. Rather, power is used to try to compare and evaluate different possible experimental scenarios (Do I invite 10 or 20 students?, Do I use a within-subject or between-subject design?, Will I use parametric or nonparametric tests?, etc.). In a power analysis, the factors influencing the power of an experiment are carefully harmonized with each other, without reference to a specific data set.[7] Power analysis is therefore to be understood as a design tool that can be applied *before* the experiment, not as an analytical tool that evaluates existing data retrospectively.

In practice, it is, alas, observed time and again that retrospectively calculated power is used, or rather "abused", as an explanation for the results observed experimentally. Let us assume that an experiment provides data showing an effect that is not statistically significant, i.e. we are not in a position to reject the null hypothesis. Now we take a measure of the size of this observed effect and calculate (using the sample size and significance level) a value for the power. The false argument, which can be observed again and again in this context, is: "Since the probability of correctly rejecting a false null hypothesis is high, but we have *not* rejected it, the null hypothesis must very probably be true." The point is, however, that it is not in fact possible to have high power if the null hypothesis has already been rejected. Power is the probability of correctly rejecting a false null hypothesis. If we have *not* already rejected it, there is no longer any likelihood

7 We will discuss an exception in the course of this section.

of correctly rejecting it. The error in the design of retrospective power is that we cannot make any statements of probability about events that have already been observed. Imagine, for example, rolling a dice. *Before* we roll the dice, we can say, "There is a 1/6 chance that we *will* roll a 4." This is equivalent to the statement, "We throw the dice (theoretically) infinitely often, then we will get a 4 in 1/6 of all cases." Now suppose we actually roll the dice and observe a 4. Then the statement, "With a probability of 1/6 we have rolled a 4," is simply nonsense. The fact is we *have* rolled a 4 – no more and no less. Similarly, it makes no sense to make a probability statement about experiment data already observed – and power *is* a probability.

The second fallacy in retrospectively calculating power is to assume that this calculated value provides information that goes beyond that provided by the *p*-value. Hoenig and Heisey (2001) show that there is a *clear* inverse relationship between the *p*-value and the retrospective power of any hypothesis test. The higher the *p*-value, the smaller the retrospective power, and vice versa. The non-significance of a study is therefore *always* accompanied by a low retrospective power and it makes no sense to explain or "excuse" non-significance with a low retrospective power.[8] Lenth (2000) notes very aptly in relation to this:

» *If my car made it to the top of the hill, then it is powerful enough to climb that hill; if it didn't, then it obviously isn't powerful enough. Retrospective power is an obvious answer to a rather uninteresting question.*

The following discussion thread, published in a neighboring discipline, is one example that shows that this fact is obviously not common knowledge: ▶ http://core.ecu.edu/psyc/wuenschk/stathelp/Power-Retrospective.htm. Apparently, an editor of a scientific journal is calling for *ex post* power analyses of the existing results in order to re-evaluate the significance or non-significance from a different perspective. Of course, the question of the reliability of the result to be published is entirely justified – after all, one could have made an error by rejecting the null hypothesis. And information on how likely this error is would, of course, be revealing in this context. The point is, however, that retrospective power is not this probability. Retrospective power is not the probability of our correctly rejecting or having correctly rejected a false null hypothesis in a study. Rather, it is the probability that we will correctly detect an existing effect in a *future* study if it is assumed that the *true* variance and *true* effect size of the population exactly correspond to the *observed* values of the previous study. In this sense, retrospective power of one study can be used for the next, possibly better designed, study, but not as a measure of the quality of a study that has already been conducted.

Just like retrospective power, the *p*-value represents a single realization of a random variable. Each time the experiment is repeated, a different value will result. For this reason, no information about the reliability or accuracy of the result can be derived from a single *p*-value. In most cases, however, a small *p*-value is interpreted as a reliable signal for the existence or non-existence of an effect, true to the motto: the smaller the *p*-value, the "better" or the more reliable it is. What is usually completely ignored, however, is the question of how much this value is scattered when the experiment is repeated.

---

8 It can also be shown that the retrospective power is about 50% if the *p*-value is equal to the significance level (Lenth 2007).

If the dispersion of p over several samples were very small and close to the determined p-value, this would not be too bad. However, Cummings (2013) impressively shows by means of a simulation that even in a standard experimental setting (see the following footnote), almost any p-value between 0 and 1 can be realized with a similarly high probability in multiple replicated experiments. One p-value alone would thus give an extremely unreliable signal about the (non-)existence of an effect, and the question of whether an effect can ultimately be shown to be significant or not would largely be a matter of luck.[9]

A valid method to learn a little more about the reliability of a result is the calculation of confidence intervals, because they combine the information of a point estimator with the information about the accuracy of this point estimator.

Important

Power analysis is a tool for designing the experiment *before* it is carried out. One main application is planning sample size. Retrospectively calculated power cannot be used to explain the results obtained. Furthermore, it does not provide any information about the reliability or "confidence" of the experimental result.

4.6 Choosing Statistical Tests

4.6.1 What Should be Taken into Consideration?

The "right" choice of methods for the analysis of experimental data always lies between two extremes. One extreme is a completely arbitrary decision to use a particular method of analysis, which is then applied entirely without reflection. The other extreme is the assumption that there is only one method of analysis that is perfectly suitable for each experiment. Both approaches are, of course, equally wrong. On the one hand, it is certainly possible and necessary to limit the number of methods that can be used. All experimental data have certain characteristics that rule out certain statistical analyses while allowing others to be performed. On the other hand, an experiment is never so specific that only one optimal method of analysis can be used. We can therefore say goodbye to the idea of a clear guideline providing an exclusive type of statistical analysis for each type of experiment, as well as to the idea that a free choice exists.

The basic approach for choosing suitable methods of statistical analysis first of all involves matching the formal requirements for the application of a method with the given characteristics of the data. All methods of inferential statistics, correlation analysis and regression analysis are based on certain assumptions. Violating these assumptions has serious consequences to varying degrees, ranging from "The analysis leads to completely false results and in no way describes the real relationship examined," to, for example, "The analysis leads to inaccuracies which must be taken into account when interpreting the results."

9 Cummings calls this effect "Dance of the p-values" and demonstrates it on YouTube (e.g. ▶ www.youtube.com/watch?v=5OL1RqHrZQ8)

In this section, the basic classification criteria of popular statistical methods are briefly introduced so that it is possible to broadly organize the data obtained using a range of methods acceptable for those data. The main objective of this section is therefore to avoid the most serious errors in choosing a method of statistical analysis. It is particularly important to see these considerations as part of the experimental design, which takes place *before* the actual experiment. Once the data are available and it is only then noticed that no suitable procedure to analyze them exists, it is usually too late for corrections. It is therefore not the experiment alone that determines the subsequent statistical method. During the design phase of the experiment, the experimenter should already be considering all the possible methods that are to be applied after conducting the experiment. Of course, here too there can also be "too much of a good thing". Too much importance should not be attributed to the influence of a statistical method on the design of the experiment either. There is little point in first looking for an elegant or particularly "in" method of analysis and only then pondering which scientific question could be investigated with it. In this sense, the statistical analysis is *always* subordinate to the experiment's research question and not vice versa. Furthermore, it is not necessary to apply a method simply because it is permitted from a formal point of view. As a matter of principle, statistical data analysis should always be based on expert knowledge, and a statistical method should only be used if the results can provide a real insight into the research question being investigated experimentally. An *ad-hoc* application of a method "for the sake of the method only" should be avoided, since the statistical analysis then often misses the point of the original question.

4.6.2 Classifying Test Methods

Statistical hypothesis tests can be categorized using several criteria. One of the most basic distinguishing features of statistical hypothesis tests is the number of groups or samples the test is comparing. If only one group is being examined, it is possible, for example, to test whether its mean is consistent with a certain population parameter that is assumed to be true. In this way, a comparison is made between the specific sample and a postulated true value of the population using one-sample tests. If, on the other hand, two groups are to be compared, for example in a classical control and treatment group comparison, it is assumed that the samples were taken from two separate populations. In this case, other tests must be used. Other tests have been developed for comparisons between more than two groups.

As soon as several groups are to be compared, the choice of a suitable test depends on whether the groups are statistically independent of each other (unrelated or unpaired) or not (related or paired). This question is largely answered by the experimental design used. Testing individual subjects in a number of experimental conditions or groups unavoidably leads to related samples. By their very nature, two successive decisions of the same person cannot be independent of each other. It does not matter whether the person makes the decisions one after the other in two different treatments (cross-over design) or in one and the same treatment (longitudinal design). If, on the other hand, each subject is a decision-maker only once, it can be assumed under conditions of full anonymity and no feedback that the decision of one person does not influence the decision of another person.

The third criterion influencing the choice of the statistical methods is the question as to which assumptions about the probability distribution of the variables apply. Two broad classes of methods are available, parametric and nonparametric, depending on the answer. Parametric methods only provide meaningful results if specific assumptions about the form (e.g. normal distribution) and the parameters (e.g. mean, variance, degrees of freedom) of the distribution apply. Sometimes finding this out is quite straightforward but in most other cases at least some uncertainty remains. As long as the sample is very large (about 100 or more), this uncertainty hardly plays a role due to the central limit theorem. Even if the true distribution is not normally distributed and a parametric test requires the normal distribution, this test will still provide reliable results for large samples. For this reason, it is said that parametric tests are robust (against deviations in the distribution) for large samples. For small samples, however, it is highly advisable to be sure that the assumptions concerning the distribution of a test are correct. Even small deviations from the assumed distribution can make a test result completely unusable.

Nonparametric (also distribution-free) methods are an alternative to this. They do not depend on the form and the parameters of the distribution of the population from which the sample was taken. However, this does not mean, of course, that nonparametric procedures do not require any assumptions. The assumptions are only less restrictive than in the parametric case. Nevertheless, the question as to why distribution-dependent methods still exist at all is justified. To put it briefly, the advantage of not being dependent on distributional assumptions is directly associated with a disadvantage. Most nonparametric methods use ordinal (rank) data. Unfortunately, when metrically scaled variables are converted into ranked data, information from the original sample is inevitably lost. This loss of information means that distribution-free tests can less reliably detect actual group differences as statistically significant than comparable tests that use distribution assumptions. The probability of identifying an effect as significant when it actually exists (the power) is never greater with distribution-free methods than with the parametric ones. Unfortunately, the smaller the sample, the more serious this disadvantage becomes. And so small samples mean that distribution-free methods are not better *per se*. Especially in cases where the data are scaled metrically and it is very certain that they stem from a normal distribution, it is the parametric procedures that are usually the lesser of two evils, even with small samples. A low robustness in this situation is probably less of a disadvantage than a low power. However, if there is uncertainty about the distribution and/or the original data are already ordinal, there is a good argument for using nonparametric methods. Both classes thus have their *raison d'être*.

As long as large samples are involved, there is no need to worry too much about which class is the better choice. A parametric test then has only a slightly higher power than its nonparametric counterpart, but the latter may be somewhat easier to perform. The parametric variant is robust to different distributions and the nonparametric variant is independent of these anyway. Does that solve the problem of choosing the right method? Alas, it does not. It is unfortunate that it is precisely in experimental economics that the samples are rather small. Neuroeconomic studies using magnetic resonance imaging are substantially more expensive, and therefore, even sample sizes of 10 are considered large there. The quality of a statistical conclusion in such situations can significantly depend on whether parametric or nonparametric methods are used. The fact that most economic experiments are analyzed using nonparametric methods is less due to the small samples than to the fact that the data are by nature not normally distributed and ordinal (Davis and Holt 1993).

As an alternative to the parametric/nonparametric distinction, the scales of measurement of the data to be examined can also be used. Tests that analyze metric data are often classified as parametric tests and tests that evaluate nominal or ordinal data are often classified as nonparametric (Sheskin 2000). In the following, we will focus on the scales of measurement.

Important

For the initial choice of a statistical hypothesis test, the following criteria at least must be considered:

1. **One or more groups?**
2. **Related or unrelated groups?**
3. **Parametric or nonparametric data or scales of measurement of the data?**

4.6.3 How Do I Choose a Specific Test?

With the help of the three classification criteria presented in the previous section, it is relatively easy to create a rough selection scheme that can be used to group frequently used hypothesis tests.

An example is shown in ▫ Table 4.3. Each row distinguishes between the different scales of measurement, with the first row containing parametric tests and the last two rows containing nonparametric tests. In the columns, we distinguish between the analysis of a single sample or two samples and the question of whether the latter are statistically independent or not.

Although the classification scheme in ▫ Table 4.3 is quite general, the tests listed represent, of course, only a very small selection. The number of available tests is simply too large to be presented in this book. First, there are a number of other tests that would

▫ **Table 4.3** A simple classification of test methods. The word "test" was omitted from every name for space reasons

Design			
	1-sample	*2-sample*	
Scale		*independent/between-subject*	*dependent/within-subject*
metric	z, t	t	t
ordinal	Kolmogorov	Wilcoxon rank-sum, Mann-Whitney U	Wilcoxon signed-ranks
nominal/ categorical	binomial, multinomial	Fisher's exact X^2 $(2 \times k)$	McNemar

fit into the scheme in Table 4.3, but we cannot present them for space reasons. Second, the classification features of Table 4.3 could be further extended. For example, we do not compare more than two samples. There are entire books available that only deal with particular subsets of all the tests, their theoretical background and specific characteristics. We therefore refer readers who need as comprehensive a guide as possible to what we consider to be the most helpful works.

Sheskin (2000) is probably the most comprehensive guide to statistical hypothesis testing. In well over 1000 pages, parametric and nonparametric test methods are presented in detail, with a distinction being made between related and unrelated groups as well as between single and multi-sample tests. For each test, it describes (i) which hypothesis is tested, (ii) what the main requirements and assumptions are, (iii) what an example might look like, (iv) how the particular calculations of this test are performed and (v) how the test results need to be interpreted. This tome can be highly recommended as a detailed reference work for experimental scientists who attach great importance to hypothesis tests.

The guide by Kanji (2006) is also very useful. 100 tests are classified into parametric and distribution-free tests as well as one-sample, two-sample and multi-sample tests and discussed very concisely on one or two pages each. This guide is therefore similar in structure to that of Sheskin (2000), but concentrates only on the essentials in the description of the tests and in the examples. This book of just under 250 pages is highly recommended for a "quick reference".

In addition, there are some excellent textbooks dealing specifically with nonparametric tests and also have the structure of a classified guide. Siegel and Castellan (1988) has established itself as a classic and is still indispensible in experimental economic practice today. A quick overview in the form of a table presenting all the tests and referring to the respective chapters can be found on the inside back cover of the book. Conover (1999) also focuses on nonparametric testing. Although there is no tabular guide for the selection of tests, the individual methods and the steps in their calculation are explained in greater detail and in a very clear manner.

In the following, we will present the tests listed in Table 4.3 while focusing on four aspects. First of all, we will discuss which type of research questions the test is suitable for and how the hypothesis to be tested is formulated. Next, we will briefly discuss the specific prerequisites that must be met in order to be able to use the test. In the third step, we will present the respective test statistic and its distribution resulting from the validity of the null hypothesis (null distribution). This is the necessary prerequisite for carrying out the test. The final step is usually an example that serves as the basis to replicate running the test in a preferred statistics program.

4.6.4 The *z*-Test und *t*-Test for One Sample

The *z*-test for one sample examines whether the mean $\bar{x}$ of a random sample is sufficiently consistent with a given population mean μ_0 that is assumed true. If the difference between $\bar{x}$ and μ_0 is significant, the data do not support the hypothesis that the sample was drawn from a population with a mean $\mu = \mu_0$. Accordingly, the null hypothesis is H_0: $\mu = \mu_0$ and the alternative hypotheses are H_1: $\mu \neq \mu_0$ or H_1: $\mu < \mu_0$ or H_1: $\mu > \mu_0$.

Since this is a parametric method, an important prerequisite is that the sample was taken from a normally distributed population with a known variance σ^2. The sample size of a z-test should comprise at least 30 observations.

The test statistic (z-value) is the standardized mean $\overline{x}$ of the sample and is calculated according to the formula

$$z = \frac{\overline{x} - \mu_0}{SE}.$$

This random variable is standard normally distributed with a mean of 0 and variance of 1 or, more succinctly expressed, $z \sim N(0;1)$. The standard deviation of the test statistic is

$$SE = \sqrt{\sigma^2 / n},$$

where σ denotes the standard deviation of the random variables being investigated in the population and n denotes the sample size.

In most cases, the distribution parameter σ^2 is unknown and must therefore be estimated in advance. In this case, the test statistic is

$$t = \frac{\overline{x} - \mu_0}{\widehat{SE}} \sim t_{n-1},$$

and the test is the corresponding t-test. The estimated values are marked with a cap. Unlike the null distribution of the z-test, the null distribution of the t-test is different for different sample sizes, as it depends on the degrees of freedom n-1. The estimated standard error is

$$\widehat{SE} = \sqrt{\hat{\sigma}^2 / n}$$

with the estimated population variance

$$\hat{\sigma}^2 = \frac{1}{n-1} S_{xx},$$

where $S_{xx} = \Sigma(x_i - \overline{x})^2$ denotes the sum of all squared deviations of x_i from its mean (also referred to as the variation of x).

Example

The scores of a nationwide math test are normally distributed with a mean of $\mu = 78$ points and a standard deviation of $\sigma = 12$ points. The teacher of a particular school wants to test whether his newly introduced method of teaching math has a positive significant influence on the point score students achieve. His research hypothesis is therefore $H_1: \mu > 78$.

The 36 students in his course obtained an average score of $\overline{x} = 82$ from the values 94, 68, 81, 82, 78, 94, 91, 89, 97, 92, 76, 74, 74, 92, 98, 70, 55, 56, 83, 65, 83, 91, 76, 79, 79, 86, 82, 93, 86, 82, 62, 93, 95, 100, 67, 89. The test statistic is then $z = (82-78)/(12/\sqrt{36}) = 2$.

If we do not know the true variance of the population, we calculate the variation of the sample $S_{xx} = \sum (x_i - 82)^2 = 4892$. From this, we determine $\hat{\sigma}^2 = \frac{1}{n-1} S_{xx} = \frac{1}{36-1} 4892 = 139.77$ and thus $\hat{\sigma} = \sqrt{139.77} = 11.82$. This is quite a good estimate, because the true value was $\sigma = 12$ points. The estimated standard error is then $\widehat{SE} = \frac{11.82}{\sqrt{36}} = 1.97$ and the test statistic is $t = (82-78)/1.97 = 2.03$. If the degrees of freedom $v = n - 1 = 35$ and the value of the test statistic $t = 2.03$ are entered in the t distribution of PQRS, ◘ Fig. 4.18 results.

The p-value is 2.5% < 5% and we reject H_0: $\mu = 78$ at a significance level of 5%.

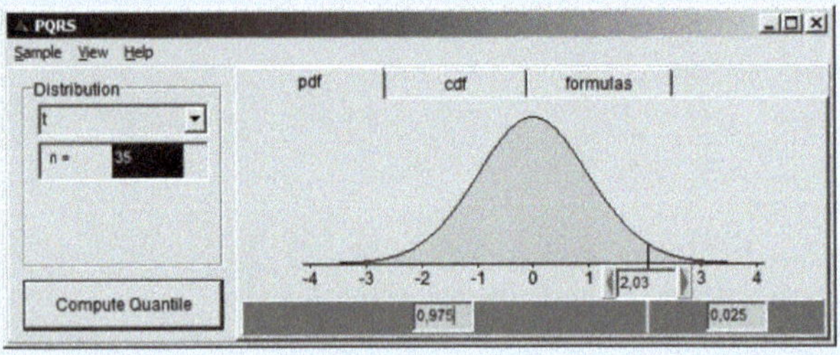

◘ **Fig. 4.18** Performing a one-tailed t-test with one sample

4.6.5 *t*-Test for Two Independent Samples (Between-Subject Comparison)

In order to compare two samples, we need to modify the one-sample t-test. First, we assume that no one is represented in both samples at the same time and that the realizations of one sample are not in any way influenced by those of the other sample. The test will determine whether the means $\overline{x_1}$ and $\overline{x_2}$ of these two independently drawn samples differ so much that it can be concluded that a significant difference between the population means exists. If the difference between $\overline{x_1}$ and $\overline{x_2}$ is significant, the data do *not* support the hypothesis that the samples were taken from populations with the same mean, $\mu_1 = \mu_2$. Therefore, the null hypothesis is H_0: $\mu_1 - \mu_2 = \mu_0$, generally with "no difference", i.e. $\mu_0 = 0$, being tested. The alternative hypotheses are then H_1: $\mu_1 - \mu_2 \neq \mu_0$ or H_1: $\mu_1 - \mu_2 < \mu_0$ or H_1: $\mu_1 - \mu_2 > \mu_0$.

Since we are still in the realm of parametric methods, it is necessary to assume that the each sample was randomly selected from its own normally distributed population. The two populations have the same, albeit unknown, variance σ^2, but it is not necessary for the samples to be of equal size. It is crucially important that the subjects are randomly assigned to the different treatments in a between-subject design. Only a successful randomization can ensure that selection effects can be avoided (see ▶ Sect. 4.2.3). The standard errors are estimated as in the t-test above.

The test statistic is t-distributed with $v = n_1 + n_2 - 2$ degrees of freedom, where n_1 and n_2 are the respective sizes of the samples, and is the standardized difference between the two sample averages

$$t = \frac{\overline{x_1} - \overline{x_2} - \mu_0}{\widehat{se}} \sim t_{(n_1+n_2-2)}.$$

The standard error is calculated from a weighted mean of the sample variances as follows

$$\widehat{SE} = \sqrt{S_1^2 + S_2^2} = \sqrt{\hat{\sigma}_1^2\left(\frac{1}{n_1} + \frac{1}{n_2}\right) + \hat{\sigma}_2^2\left(\frac{1}{n_1} + \frac{1}{n_2}\right)}$$

with

$$\hat{\sigma}_1^2 = \frac{1}{\mathrm{n}_1 + \mathrm{n}_2 - 2}\mathrm{S}_{x_1 x_1}$$

$$\hat{\sigma}_2^2 = \frac{1}{n_1 + n_2 - 2}S_{x_2 x_2}.$$

4.6.6 *t*-Test for Two Dependent Samples (Within-Subject Comparison)

A further modification of the *t*-test is required if the realizations of one sample are not independent of those of the other sample. This is always the case in a within-subject design of an experiment, since one subject makes decisions in two different treatments or samples. Therefore, there are pairs of measured values in which the decision of the same subject is found in both treatments. The null hypothesis is H_0: $\mu_1 - \mu_2 = \mu_0$, with $\mu_0 = 0$ usually being tested, and the alternative hypotheses are H_1: $\mu_1 - \mu_2 \neq \mu_0$ or H_1: $\mu_1 - \mu_2 < \mu_0$ or H_1: $\mu_1 - \mu_2 > \mu_0$.

Once again, the samples are randomly drawn from each of the normally distributed populations of unknown but equal variance σ^2. The test statistic is the same as in the two-sample case using independent samples. The standard error is calculated from a weighted mean of the sample variances, corrected by the degree of correlation between the two samples

$$\widehat{SE} = \sqrt{S_1^2 + S_2^2 - 2\rho S_1 S_2},$$

where ρ represents the (Bravais-Pearson) correlation coefficient between the two samples. It is calculated using

$$\rho = \frac{S_{x_1 x_2}}{\sqrt{S_{x_1 x_1} S_{x_2 x_2}}}.$$

The following also applies

$$S_1^2 = \hat{\sigma}_1^2\left(\frac{1}{n_1} + \frac{1}{n_2}\right)$$

$$S_2^2 = \hat{\sigma}_2^2 \left(\frac{1}{n_1} + \frac{1}{n_2} \right)$$

and

$$\hat{\sigma}_1^2 = \frac{1}{n_1 + n_2 - 2} S_{x_1 x_1}$$

4

$$\hat{\sigma}_2^2 = \frac{1}{n_1 + n_2 - 2} S_{x_2 x_2}.$$

4.6.7 Kolmogorov Test

The Kolmogorov test is one of what is termed goodness-of-fit tests. These tests examine whether the distribution of the values of a sample are those that would be expected based on a specific, pre-defined distribution. This means that this test provides statistical evidence as to whether or not the assumption of a particular distribution is fulfilled. For this purpose, the empirical distribution function F_x of the sample, i.e. the proportion of observed x-values that are smaller than or equal to a specific x-value (for *all* real x-values), is compared with the pre-defined or presumed distribution function F_0. The test statistic D measures the degree of agreement and is the maximum distance between F_x and F_0.

The null hypothesis postulates concordance between the theoretical and the empirical distributions, and the alternative hypothesis states that the sample does not originate from the theoretical distribution. For this reason, a two-tailed hypothesis, which allows a deviation in both directions, is usually used in practice. In contrast to most other tests, with the Kolmogorov test we do *not* want the null hypothesis to be rejected, since we usually expect the assumption concerning a particular distribution to be confirmed (e.g. normal distribution). The more dissimilar the data are to the reference distribution, the higher the probability that the null hypothesis will be rejected.

Prerequisites and Special Features

Technically speaking, the Kolmogorov test requires a continuous random variable (with at least ordinal values). It can be shown that only with this condition is the distribution of the test statistic D *independent* of the actual form of the true distribution from which the sample was drawn, thus making the Kolmogorov test a truly distribution-free method if this condition is met.

The alternative for discrete data is a χ^2 goodness-of-fit test, which in turn requires a relatively large sample size to generate valid test decisions. If the Kolmogorov test is nevertheless to be applied to discrete data, it is necessary either to accept significantly more conservative test decisions or to use certain modifications of this test that specifically address discrete data (Conover 1972).

The Test Statistic and the Null Distribution

The test statistic is the maximum difference in value between the two cumulative *distribution functions*, that of the reference distribution F_0 and that of the empirical distribution F_x

$$D = max\left|F_x - F_0\right| \sim F_D.$$

The test statistic follows a unique but not common null distribution F_D that does not depend on F_x provided that F_x is continuous. The critical values of D are tabulated (Massey 1951) up to a sample size of n = 35. They can be calculated for larger samples at a significance level of 5% according to the formula

$$D_{crit} = \frac{1.3581}{\sqrt{n}}.$$

This test (including the critical values) is already included in almost every statistics program, thus rendering the use of tables unnecessary here as well.

4.6.8 The Wilcoxon Rank-Sum Test and the Mann-Whitney *U* Test

The Wilcoxon rank-sum test is a popular alternative to the *t*-test when it does not appear realistic to assume a normal distribution and/or the data are not scaled metrically. Like the *t*-test, it compares the equality of the "central points" of two independent samples. As the name of the test indicates, the Wilcoxon rank-sum test is based on ordinal rank data. Arithmetic means no longer exist for these data and we generally speak of "central tendencies" to compare groups.

The data consists of one observation each of the n_1 and n_2 random variables $(x_1, \ldots, x_{n1})$ and $(y_1, \ldots, y_{n2})$. Stochastic independence exists both within each sample $(x_1, \ldots, x_{n1})$ and $(y_1, \ldots, y_{n2})$ and between the variables x_i and y_i. All the random variables are continuous and measured on at least an ordinal scale.

First, the $n = n_1 + n_2$ observations across both samples are assigned ranks. The smallest of the n observations receives the smallest rank, the largest observation the largest. Ideally, n different ranks can be assigned for n observations. Then the sums of the ranks of each sample are calculated separately to obtain the rank totals R_1 and R_2. The test statistic used for the Wilcoxon rank-sum test is the smaller of the two rank sums obtained from the samples, i.e. $R^* = \min\{R_1, R_2\}$. The *p*-value, for the calculation of which the null distribution is required, is then the probability of obtaining this rank sum R^* or a more extreme one (in the sense of the research hypothesis H_1).

To determine the null distribution, we first calculate the total number of possible variations V which could be used to distribute the n ranks between the two samples of sizes n_1 and n_2. It is possible to show that

$$V = \frac{(n_1 + n_2)!}{n_1!n_2!}$$

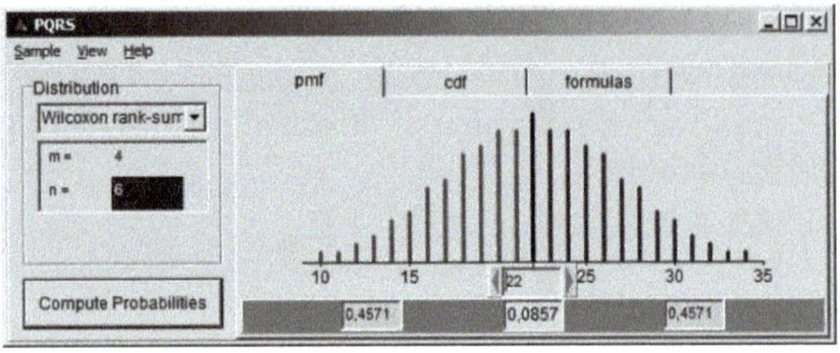

Fig. 4.19 Null distribution of a Wilcoxon rank-sum test with two independent samples

4

holds. The null distribution then consists of the V possible rank combinations in the sample that has the smaller rank sum R^*. To obtain the p-value, the rank sums of all the combinations that are less than or equal to (for the left-tailed test) or greater than or equal to (for the right-tailed test) R^* are counted and this number is divided by V.

In PQRS, the null distribution is obtained by selecting "Wilcoxon rank-sum" under "Distribution" and entering the size of the first sample (e.g. $m = 4$) in the "m" field and the size of the second sample (e.g. $n = 6$) in the "n" field. After clicking on "Apply New Distribution" the image shown in Fig. 4.19 is displayed.

By moving the slider on the abscissa (here set to the center, 22), we can again specify any quantiles and read their corresponding probabilities in the line below. In this case, for example, we see that the probability of obtaining a value of the test statistic R^* that is less than 22 is no more than 0.4571 = 45.71%. The probability of obtaining exactly 22 is 0.0857 = 8.57%. Thus, the probability of obtaining a value less than or equal to 22 is the sum of these two probabilities, i.e. 0.4571 + 0.0857 = 0.5428 = 54.28%. With large samples and limited measurement accuracy, it is not uncommon to obtain equal values (ties) within a group or between both groups. The number of ties then corresponds to the number of ranks that should actually be assigned but cannot because it is not possible to distinguish one value from another. In practice, all these values are then assigned the mid-rank, based on the average of the previous and the following rank. For this purpose, it makes sense to first write the observations from both samples in a single table arranged according to rank. An example is given in Table 4.4.

The value 9.5 occurs a total of four times. The mid-rank in this case is (20 + 15)/2 = (16 + 17 + 18 + 19)/4 = 17.5.

If n_1 or n_2 are greater than 20, the null distributions of both variants can be approximated by a normal distribution with

$$\mu = \frac{n_1(N+1)}{2}$$

and

$$\sigma^2 = \frac{n_1 n_2 (N+1)}{12}.$$

The test statistic R^* can then be transformed into standard normally distributed z-values in the usual way.

An alternative method that always leads to the same test result as the Wilcoxon rank-sum test is the Mann-Whitney U test. For each rank in one group, we determine

Table 4.4 Procedure for ties

Group 1	Group 2	Rank
…	…	…
8.7		15
	9.5	(16) – > 17.5
9.5		(17) – > 17.5
	9.5	(18) – > 17.5
	9.5	(19) – > 17.5
	10.2	20
…	…	…

how many smaller ranks there are in the other group (the so-called rank scores). The sum of these scores is then used in the same way as the rank sum in the Wilcoxon test, i.e. the test statistic S^* is the smaller of the two score sums. When approximating the null distribution of S^* with a normal distribution, only the expected value has to be adjusted:

$$\mu = \frac{n_1 n_2}{2}.$$

The variance is the same as for the Wilcoxon R^* test statistic.

Example

The aim is to test whether students of economics (ECON) give significantly less in the dictator game than students of humanities (HUM). For this purpose, there are $n_1 = 4$ observations in group ECON and $n_2 = 5$ observations in group HUM (see Table 4.5).

The test statistic is the rank sum $R^* = R_1 = 12$, since this number is the smaller of the two rank sums. Furthermore, for the total number of variations V with which $N = 9$ ranks can be distributed between two samples of sizes 4 and 5:

$$V = \frac{(4+5)!}{4!5!} = 126.$$

The possible rank totals for the sample ECON are arranged according to size:

1. $1+2+3+4=10$
2. $1+2+3+5=11$
3. $1+2+3+6=12$
4. $1+2+4+5=12$
5. $1+2+4+6=13$

… …

126. $6+7+8+9=30$

Table 4.5 Example data for the Wilcoxon rank-sum test

ECON		HUM	
Allocation	Rank	Allocation	Rank
13	4	25	9
14	5	15	6
8	2	9	3
6	1	22	8
		19	7
Sum	$R_1 = 12$		$R_2 = 33$

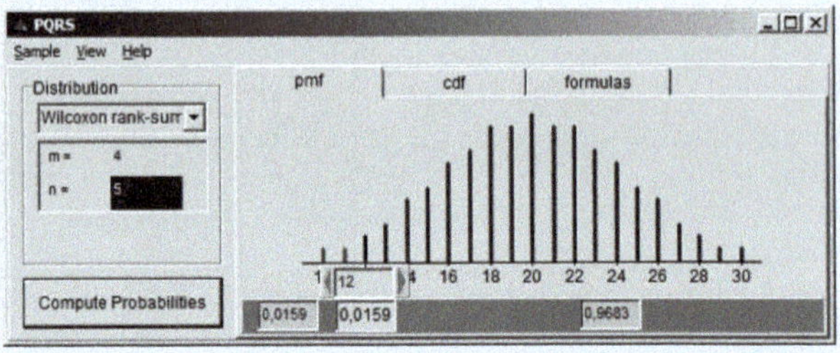

Fig. 4.20 Performing a Wilcoxon rank-sum test

We see that the rank sum is less than or equal to $R^* = R_1 = 12$ in only 4 out of 126 possible cases. Accordingly, the *p*-value is $4/126 \approx 0.0317 = 3.17\%$ and the null hypothesis would be rejected at a significance level of 5%. In PQRS, this value is obtained by determining the null distribution for $n_1 = 4$ and $n_2 = 5$ (*m* and *n* in PQRS) and selecting quantile 12.

As can be seen from Fig. 4.20, the probability of obtaining an R^*-value of 12 or less is the sum of 0.015873 (probability of $R^* < 12$) and 0.015873 (probability of $R^* = 12$), i.e. again approximately $0.0317 = 3.17\%$.

In the Mann-Whitney *U* test, the scores are as shown in Table 4.6.

$S^* = S_1 = 2$ and $V = 126$. The ordered score totals in the ECON group are as follows.

1. $0+0+0+0=0$ (Ranks 1, 2, 3, 4)
2. $0+0+0+1=1$ (Ranks 1, 2, 3, 5)
3. $0+0+0+2=2$ (Ranks 1, 2, 3, 6)
4. $0+0+1+1=2$ (Ranks 1, 2, 4, 5)
5. $0+0+0+3=3$ (Ranks 1, 2, 3, 7)

… …

126. $5+5+5+5=20$.

Again, the value of the test statistic (here, $S^* = 2$) is not exceeded in only 4 out of 126 cases, i.e. the *p*-value is analogous to that in the Wilcoxon-rank sum test, $4/126 = 0.032 = 3.2\%$. The corresponding null distribution in PQRS is displayed in Fig. 4.21.

As we can see, the only difference is an adjusted quantile ($S^* = 2$ instead of $R^* = 12$); however, the corresponding *p*-value in the Mann-Whitney U test is the same as that in the Wilcoxon rank-sum test.

Table 4.6 Scores in the Mann-Whitney *U* test

ECON			HUM		
Allocation	Rank	Score	Allocation	Rank	Score
13	4	1	25	9	4
14	5	1	15	6	4
8	2	0	9	3	2
6	1	0	22	8	4
			19	7	4
Sum		$S_1 = 2$			$S_2 = 18$

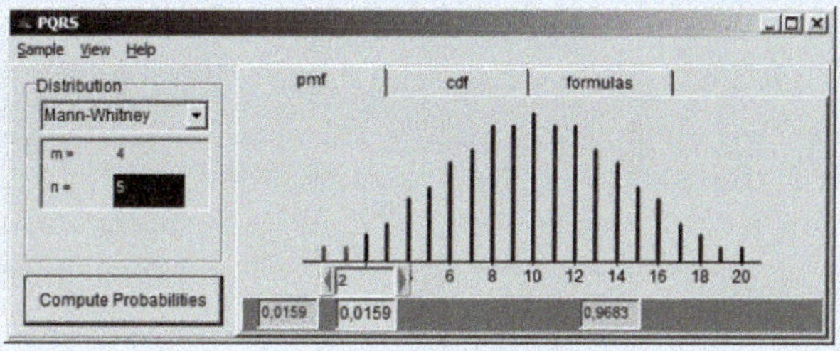

Fig. 4.21 Performing a Mann-Whitney *U* test

4.6.9 Wilcoxon Signed-Rank Test (Two Dependent Samples)

Just as the Wilcoxon rank-sum test can be seen as a nonparametric counterpart to the *t*-test with two independent samples, the Wilcoxon signed-rank test can be used as a nonparametric alternative to the *t*-test with two dependent samples. It is one of the standard tests for ordinal data in a within-subject or matched-pairs design.

The Wilcoxon signed-rank test is based on the differences in the values of the two samples. Although the direction of the difference is taken into account, the size of the difference is only included in the test statistic in the form of an ordinal ranking.

The hypotheses are the same as those in the Wilcoxon rank-sum test. If the null hypothesis is valid, it is assumed that the differences originate from a population that is symmetrically distributed around the median of 0. This means that prior to the actual sample realization, the probability of ranks 1, 2, 3 etc. is as high as the probability of ranks −1, −2, −3 etc., i.e. 0.5. Drawing the sample (or carrying out a treatment) can also be imagined as flipping n coins numbered from 1 to n, with the positive number on one side of the coin and the negative number on the other side. This property of the equal probabilities of the signs is important in the derivation of the null distribution.

The data consist of one observation each of the n random variable tuples (x_1, y_1), (x_2, y_2), …, (x_n, y_n). The variables x_i and y_i are not stochastically independent of each other, whereas the realizations of the pairs (x_i, y_i) are. All the random variables are continuous and measured on at least an ordinal scale. To derive the test statistic, the *absolute* differences $|d_i| = |x_i - y_i|$ of the values are determined first. If we assume for the moment that no zero differences $|d_i| = 0$ and no differences of equal size $|d_i| = |d_j|$ occur, we know that with, for example, $n = 3$ observations, we must obtain the ranks 1, 2, 3. If the actual difference between the sample values is positive (value of the first sample is greater than the value of the second sample), a plus sign is assigned to the rank and a minus sign in the reverse case. A sample size of $n = 3$ would result in $m = 2^n = 2^3 = 8$ possibilities of distributing the signs to the three ranks (see ◘ Table 4.7).

The rank sums of the positive ranks are denoted by W^+ and those of the negative ranks by W^-, where in all $m = 8$ cases

$$W^+ + W^- = \frac{n(n+1)}{2} = 6.$$

Since one rank sum can always be derived from the other, it does not matter which of the two values W^+ or W^- we use as the test statistic. In the following, we choose W^+ and all further steps are based on this choice. The null distribution for $n = 3$ is given in ◘ Table 4.8.

This shows, for example, that the probability of obtaining a value of W^+ less than or equal to 2 is $1/8 + 1/8 + 1/8 = 3/8 = 0.375$.

In PQRS, we can graph the sample distribution by selecting "Wilcoxon signed rank" under "Distribution" and entering the value 3 in the "*n*" field. After clicking on "Apply New Distribution", we obtain the image shown in ◘ Fig. 4.22.

It is easy to see in this figure that the smallest possible significance level at which we can test for $n = 3$ is $1/8 = 0.125 = 12.5\%$. If the sample size is increased, we gain more

◘ **Table 4.7** Eight different ways to assign two signs to three ranks

Rank	1.	2.	3.	4.	5.	6.	7.	8.
1	–	–	–	–	+	+	+	+
2	–	–	+	+	–	–	+	+
3	–	+	+	–	+	–	–	+
W^+	0	3	5	2	4	1	3	6
W^-	6	3	1	4	2	5	3	0

◘ **Table 4.8** Null distribution in the Wilcoxon signed-rank test

Poss. values W^+	0	1	2	3	4	5	6	
Absolute frequency	1	1	1	2	1	1	1	8
Probability	1/8	1/8	1/8	2/8	1/8	1/8	1/8	1

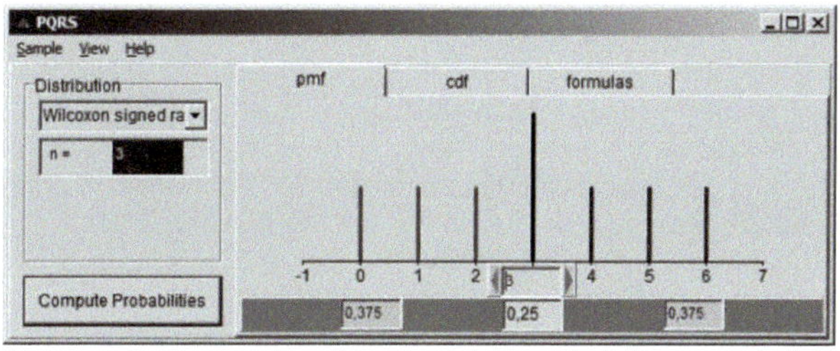

Fig. 4.22 Null distribution in the Wilcoxon signed-rank test

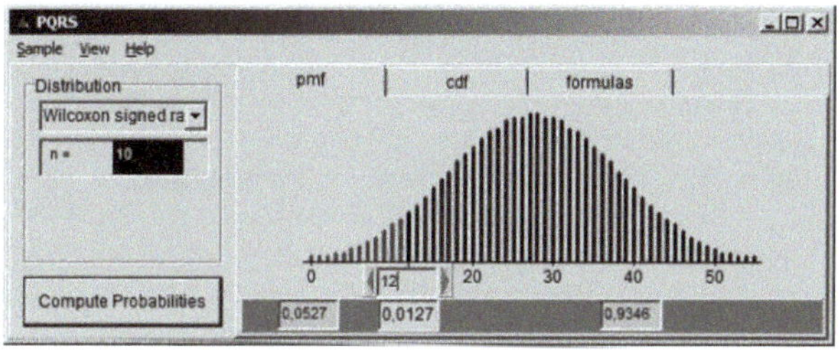

Fig. 4.23 Null distribution in the Wilcoxon signed-rank test with $n = 10$

possible values of W^+ and correspondingly finer increments in the null distribution. For n = 10, we obtain the distribution shown in Fig. 4.23.

Here, the probability of obtaining a value of W^+ of 12 or less is 0.0527 = 5.27%, which is almost the standard significance level of 5%. Furthermore, it is noticeable that the distribution is already very similar to a normal distribution. In fact, we could approximate the null distribution using a normal distribution with the parameters $\mu = n(n + 1)/4$ and $\sigma^2 = n(n + 1)(2n + 1)/24$. The normal distribution is, however, a continuous distribution, and the following continuity correction factor should therefore be made when standardizing the test statistic $(W^+ - \mu)/\sigma$: if $W^+ < \mu$, then add 0.5 to W^+ and if $W^+ > \mu$, then subtract 0.5 from W^+. In both cases, the distance from W^+ to μ, and thus the z-value of the samples, is reduced.

In closing, it should be mentioned that the more zero differences, $|d_i| = 0$, and ties, $|d_i| = |d_j|$, that occur, the more problematic the method described above becomes. Since the test assumes continuous variables, given sufficiently accurate measurements, neither of these variants can actually occur in theory. The continuity assumption therefore means that the probability of the occurrence of zero differences and ties should be zero, at least theoretically. In practice, however, the above ties occur more frequently because it is not always possible to measure the variables as precisely as desired. For example, amounts of money in euros are limited to two decimal places (the smallest unit is 1 cent or 1/100 euro) and it does not make much practical sense to refine this unit further. Several proposals in connection with this have been discussed in the statistics literature. In practice, though, it has generally become accepted to remove zero differences from the observations, resulting in a smaller sample size n^*, and to assign the common mid-rank to ties. At this juncture, it would be going too far to describe the precise demarcations of the various methods. Interested readers are referred to, for example, Conover (1973).

Example

The influence of group membership on the trustors' behavior in the trust game is to be investigated. In an experiment, $n = 12$ people (6 trustors and 6 trustees) are successively subjected to two different treatments. First, they play the trust game in pairs without information concerning the subject of study of their respective game partner ("without info"). In the game, the trustors decide whether they want to send 1, 2, ... or 10 euros to the trustee. The trustee then decides, using the strategy method, how much he wishes to send back to the trustor. The subjects do not receive any feedback about the behavior of their respective partner after completion of the first treatment. In the second treatment, the same subjects play the same trust game again, but they also receive information on the subject of study of their respective partner ("with info"). The behavior of the trustor in both treatments is reported in Table 4.9.

Since subject 4 shows no change between the two treatments, he is removed from the analysis, reducing the sample size to $n = 5$. (There are, however, also methods that take these zero differences into account; see, for instance, Marascuilo and McSweeney 1977).

Suppose it is assumed that the information about the partner's subject of study reduces the amount sent by the trustor. Then the hypotheses are:

- H_0: $E(d) = 0$. The expected difference in the amount sent by the trustor in the two treatments is zero, i.e. the information on the subject of study does not reduce the amount sent.
- H_1: $E(d) > 0$. The expected difference in the amount sent by the trustor in the two treatments is greater than zero, i.e. the information on the subject of study reduces the amount sent by the trustor.

The null distribution for $n = 5$ in PQRS is shown in Fig. 4.24.

Since we have a right-tailed test, we will consider the right end of the distribution. The probability of obtaining a test statistic greater than or equal to 14 is 0.0625 = 6.25%. Thus, our critical value is 14 and the corresponding significance level is 6.25%. In order to make a test decision, we only need the value of the test statistic from our data or the associated *p*-value.

First, we calculate the absolute values of the differences between the outcomes obtained "without info" and "with info". These are entered in column $|d|$. Ranks are then assigned to the $|d|$ values and entered in the "Rank |d|" column. Depending on whether the difference between the

Table 4.9 Data on trustors' behavior

	Without Info	With Info	$\|d\|$	Rank $\|d\|$	+ Rank	– Rank
1	4	6	2	2		−2
2	6	5	1	1	+1	
3	6	3	3	3	+3	
4	3	3	0	–	–	–
5	8	4	4	4	+4	
6	10	1	9	5	+5	
					$W^+ = 13$	$W^- = 2$

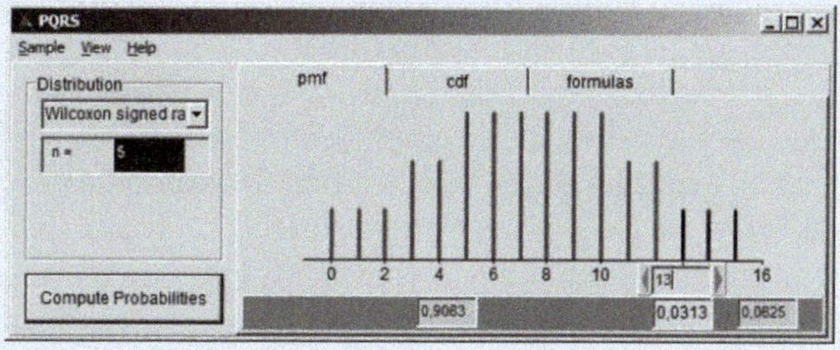

Fig. 4.24 Null distribution in the Wilcoxon signed-rank test with $n = 5$

"without info" and "with info" values were positive or negative, a plus (+) or minus (−) sign is placed before the values in the "+ Rank" or "− Rank" column. We call these numbers "signed ranks". From this, we calculate the absolute sum of the positive values W^+ and of the negative values W^-. We choose one of these two numbers as the test statistic W. We decide for $W = W^+$, although the same test result (with a slightly adapted test procedure) is also obtained with $W = W^-$. Our test statistic is then $W^+ = 1 + 3 + 4 + 6 = 13$. Since this value is to the left of our critical value of 14, we cannot reject the null hypothesis at a significance level of 6.25% by only a narrow margin.

The data therefore do not support the hypothesis that explicit information on the subject of study has a negative effect on the amount the trustor sends. This may come as a surprise, because a quick inspection of the data suggests that an effect is present. Four out of six subjects, after receiving the information, gave a lower amount. Subject 6 even showed the maximum possible change from 10 to 1. This shows the loss of information associated with the use of ordinal rank data. Subject 6 was assigned the highest possible rank of +5, but in relation to the cardinal change of 10–1 = 9 this appears "too small", whereas subject 5 was ranked +4 with a cardinal change of only 4. Under these conditions, and given the small sample size, a single negative rank of −2 is sufficient to make the change in the values insignificant.

4.6.10 The Binomial Test

Many variables in experiments have only two possible outcomes, such as "accept offer/reject offer" in the ultimatum game, "cooperate/defect" in the prisoner dilemma game, or "choose an even number/choose an odd number" in the matching pennies game. A coin toss with the results heads or tails can also be represented by such a dichotomous variable. We call the one-off performance of such an experiment a *Bernoulli trial* and the two results *success* and *failure*. The probability of one of the two results of a one-off Bernoulli trial is the probability of success or failure, which is 0.5 for flipping a fair coin, for example.

In a laboratory experiment involving decision-making, the probability of the subjects deciding on one or the other alternative action is generally not known in advance. Yet it is precisely this which is often of particular interest. If a theory specifies a particular value, the laboratory data and a suitable hypothesis test could be used to check whether the laboratory data statistically support the specific theoretical value or not. In the matching pennies game mentioned above, for example, game theory predicts an equilibrium in which both players play both alternatives with equal probability, i.e. with

p = P(choosing an even number) = 1 − p = P(choosing an odd number) = 0.5. If this game is played sufficiently frequently in the laboratory, a relative frequency for "even number" ("success") and "odd number" ("failure") is obtained by simply counting the respective realizations. This frequency is also referred to as the *empirical probability of success* $\hat{\pi}$. The binomial test examines whether the observed value of $\hat{\pi}$ is that which would be expected if it is assumed that in reality the probability of success takes on a specified value $p = p_0$, which in the case of the matching pennies game is $p = 0.5$. If the difference between π and p_0 is sufficiently large, then the null hypothesis is rejected, i.e. taking into account a given probability of error, the specified value p is not consistent with the observed sample. If, however, the null hypothesis cannot be rejected, the experimental data support the theoretical prediction.

4

The possible hypotheses in the binomial test are

	Two-tailed	Left-tailed	Right-tailed
H_0:	$p = p_0$	$p \geq p_0$	$p \leq p_0$
H_1:	$p \neq p_0$	$p < p_0$	$p > p_0$

The variable under consideration is either dichotomous, i.e. it can by definition only have two values, such as the result of a coin toss, or it is categorically scaled with 2 categories, e.g. the amounts given in the dictator game, which are "high" if they exceed a certain amount, and otherwise "low". Furthermore, the theory of the test requires that all n repetitions of the Bernoulli trial are stochastically independent of each other and the probability of success p remains constant over all trials. These two requirements may prove to be problematic in many experiments because they create a conflict. On the one hand, the runs should be stochastically independent of each other. This usually means that two consecutive repetitions of the experiment must not be performed by one and the same pair of subjects. On the other hand, the probabilities should remain constant over all repetitions. This, in turn, means that there should, first of all, be no feedback on the behavior of the subject's partner between the repetitions and, second, that the subject pairs should *not* be changed over several repetitions since each individual "brings along" his or her personal probability of occurrence from the outset. Yet even with within-subject designs, a sufficiently high number of repetitions may lead to learning or fatigue effects changing the probability of success. Without saying too much about the design of experiments at this point, a compromise could be to have the game played once by a large enough number of pairs and to ensure adequate randomization when recruiting them.

The test statistic B is the number of successes in a Bernoulli trial repeated n times. The null distribution of B, which is derived in the same way as other discrete density functions, is

$$\pi(x,n,p) = \binom{n}{x} p^x (1-p)^{n-x},$$

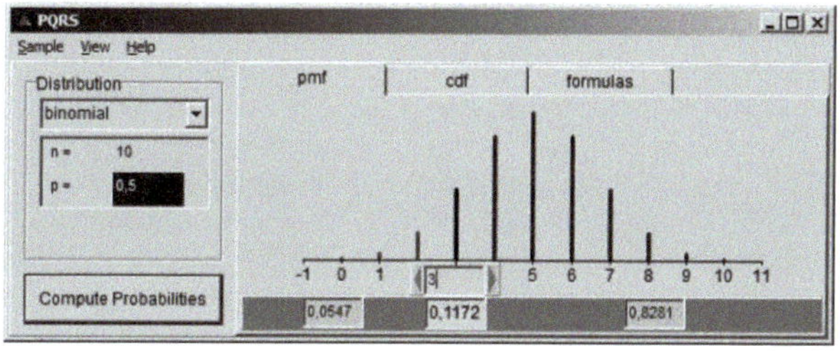

Fig. 4.25 Null distribution in the binomial test with $n = 10$ and $p = 0.5$

with this providing the probability of success x-times in n trials with a probability of success of p. For example, the density function value $\pi(3, 10, 0.5) = \binom{10}{3} 0.5^3 (1-0.5)^{10-3} = 0.1172 = 11.72\%$ indicates how likely it is that in 10 Bernoulli trials "success" is observed exactly three times, if both outcomes of the variables are equally probable.[10]

This distribution can be represented in PQRS as shown in Fig. 4.25.

The calculated probability of $0.1172 = 11.72\%$ can be read at position 3.

If the sample size n is "large enough", then the null distribution can be replaced by the normal distribution with $\mu = np_0$ and $\sigma^2 = np_0(1 - p_0)$. The standardized test statistic is then

$$z = \frac{x - np_0}{\sqrt{np_0(1-p_0)}} = \frac{p - p_0}{\sqrt{\frac{p_0(1-p_0)}{n}}}.$$

This test variant is also known as the *z-test for population frequency*. A rule of thumb for "large enough" is $np_0(1 - p_0) \geq 9$ (Bortz and Lienert 2008, p. 42). Since we are again mapping a discrete distribution with a continuous distribution, this means that particularly with a sample size of $15 < n < 60$, the resulting p-value tends to be too small due to an additional approximation error, resulting in the null hypothesis therefore being rejected too often. This error is corrected with the following continuity correction (Fleiss et al. 2003, p. 27):

$$z_{corr} = \frac{p - p_0 - \frac{1}{2n}}{\sqrt{\frac{p_0(1-p_0)}{n}}}.$$

10 An example is a coin toss, in which "heads", for instance, is defined as the case of success.

Example

A one-off first-price auction experiment is conducted with 10 subjects acting as bidders. Of the 10 bidders, $B = 7$ overbid the symmetrical Nash equilibrium prediction calculated under the assumption of risk neutrality, while 3 underbid this. The aim is to test whether overbidding and underbidding are equally likely, i.e. whether the probability of overbidding differs significantly from 50%. If we do not know in advance the direction of a deviation, we have to accept a higher Type II error or a lower power and test the two-tailed hypothesis H_0: $p = 0.5$ against H_1: $p \neq 0.5$. The null distribution in PQRS is shown in ◘ Fig. 4.26.

We can see from this that the probability of $B \geq 7$ or $B \leq 3$ is $(0.1172 + 0.0547) * 2 = 0.3438$.[11] This p-value is greater than the significance level of 5%, so we do not reject H_0: $p = 0.5$. The empirical probability of success of 0.7 is therefore not sufficiently different from the tested value of 0.5. In other words, if H_0 were true and the probability of success actually amounted to 0.5, we would observe a value $B = 7$ in 34.38% of all cases (with very many repetitions). If we had observed overbidding $B = 8$ times and had performed a right-tailed test, the p-value would have been reduced to only 0.0546 = 5.46%.

We have $10 * 0.5(1 - 0.5) = 2.5 < 9$ and therefore an approximation with the normal distribution is not suitable. Due to the small sample, however, the exact calculation of the p-value is also quite simple, making a simplifying approximate test unnecessary.

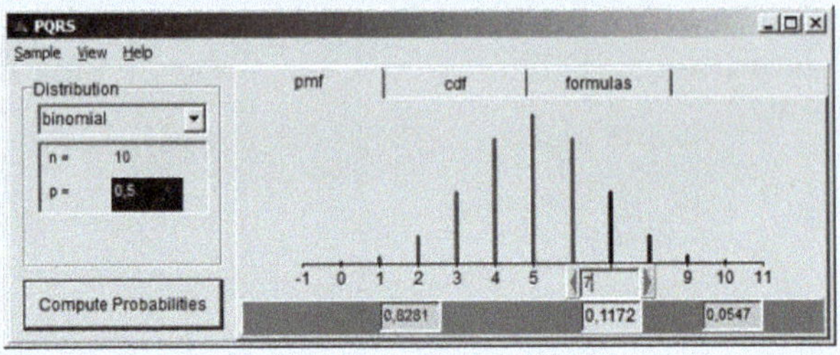

◘ **Fig. 4.26** Binomial test in the example

4.6.11 The Multinomial Test (1 × *k*)

The multinomial test is the generalization of the binomial test to categorical variables with $k > 2$ categories. For example, it might be desirable to classify amounts given in the dictator game not only in "high" and "low", but rather more refined in "high", "medium" and "low", which would correspond to a categorical variable with three categories. Otherwise, the test principle of the multinomial test is completely analogous to that of the binomial test. The test examines whether the empirical frequencies $\pi_1, \ldots, \pi_k$ of the k categories are those that would be expected on the premise that in reality the probabilities of success of the categories assume certain given values $p_1, \ldots, p_k$ (null hypothesis).

The test statistic in the multinomial test is the observed empirical frequency in all k classes. For example, suppose that the variable "hair color" has only $k = 3$ categories, "blonde", "black" and "other". From a given population (e.g. all Germans) we select $n = 20$ persons and assign them to the three categories according to their attribute "hair

11 It should be noted that the density function of the binomial distribution is only symmetrical when the probability of success or failure is 0.5. For all other values, the bars at the left and right end of the density function would have to be added up individually to obtain the p-value.

color". After counting out the classes, we get the absolute frequencies $x_1 = 5$, $x_2 = 8$ and $x_3 = 7$ and the empirical probabilities $\pi_1 = \frac{5}{20}$, $\pi_2 = \frac{8}{20}$ and $\pi_3 = \frac{7}{20}$.

If $\boldsymbol{x}$ is the vector of the absolute frequencies of a class and $\boldsymbol{p}$ is the vector of the true probability of being in a certain class, then the null distribution (multinomial distribution) is

$$\pi\left(n, \boldsymbol{x}, \boldsymbol{p}\right) = n! \prod_{i=1}^{k} \frac{\pi_i^{x_i}}{x_i!}.$$

For example, if in Germany exactly 30% of all people are blonde, 40% black-haired and 30% others, then out of a group of 20 randomly selected Germans, the probability of drawing exactly 5 blonde people, 8 black-haired people and 7 others is

$$\pi\left(20, \left(5;8;7\right), \left(0.3;0.4;0.3\right)\right) = 20! \frac{0.3^5}{5!} \cdot \frac{0.4^8}{8!} \cdot \frac{0.3^7}{7!} = 0.03475.$$

At this point it becomes clear that we are dealing with very large numbers (the number 20! corresponds to about 2432 quadrillion, i.e. a number with 19 digits). It becomes no less cumbersome if a p-value is calculated. First, all the k-dimensional vectors of the frequencies that might possibly be drawn would have to be determined. It can be shown that this number is

$$\binom{n+k-1}{k-1}.$$

In our case, the null distribution would already have 231 places for which we would have to calculate $\pi(n, \boldsymbol{x}, \boldsymbol{p})$. Then we would have to find all of the realizations that have a probability less than or equal to 0.03475 and add them up to obtain a single point on the distribution function, our p-value. Now imagine a survey with $n = 100$ people. It would then be necessary to calculate 5151 probabilities in which even the number 100! occurs. This alone would bring any normal pocket calculator to its knees. In other words, without a computer and the appropriate software, using the exact multinomial test does not make much sense.[12]

Fortunately, as is the case with the binomial test, there are also approximate alternatives for the multinomial test that provide good approximations if the sample size is "sufficiently large". If none of the expected frequencies is less than 5 in the multinomial test, the χ^2 goodness-of-fit test can also be used. Like the multinomial test, it examines whether or not an observed frequency distribution over k classes is consistent with a theoretical, expected distribution. The main difference to the multinomial test is that asymptotic properties of the differences of observed frequency and expected frequency are exploited for large samples. If e_i is the expected frequency of a class and b_i is the observed frequency, then it can be shown that for large samples the expression

$$z_i = \sqrt{\frac{\left(b_i - e_i\right)^2}{e_i}}$$

12 To perform a multinomial test in R, for example, we need the EMT (Exact Multinomial Test) package from Uwe Menzel, which is available on every CRAN server. The execution is then done using the command multinomial.test().

Fig. 4.27 Approximation of the multinomial test with the χ^2 distribution

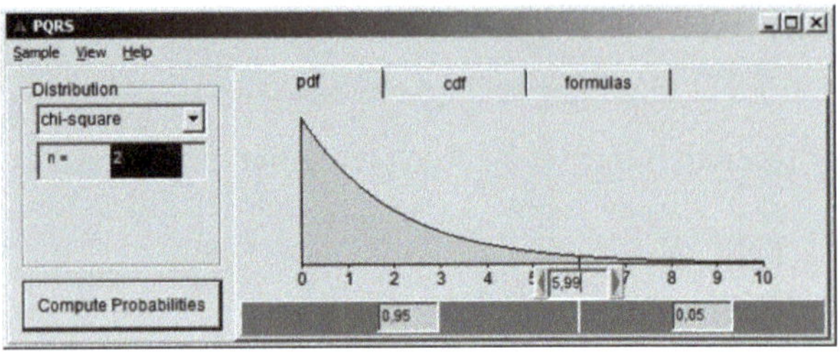

is approximately standard normally distributed. Therefore, for sufficiently large n, the test statistic

$$\chi^2 = \sum_{i=1}^{k} z_i^2 = \sum_{i=1}^{k} \frac{(\mathrm{b_i} - \mathrm{e_i})^2}{\mathrm{e_i}}$$

is χ^2 distributed with k degrees of freedom. Using this test statistic, the χ^2 goodness-of-fit test tests whether all the observed frequencies are those that would be expected or whether at least one observation deviates significantly from the respective expectation.[13] In contrast to the multinomial distribution, the χ^2 distribution is included in PQRS. With $v = k-1 = 2$ degrees of freedom, the null distribution in Fig. 4.27 results.

The following example not only fulfills the criterion $e_i > 5$ for all $i = 1 \ldots k$, but also has the special feature that the frequencies are uniformly distributed under the null hypothesis. The χ^2 test is extremely robust under this second condition, which means that it still produces useful results even if the first condition is not fulfilled or only just fulfilled (Zar 1999).

Example

Brosig-Koch, Helbach, Ockenfels and Weimann (2011) conducted a survey of a total of n = 144 subjects (students) before carrying out their actual experiment. Among other things, they asked how much money was available to the students each month. Classifying the answers in "poor", "standard" and "rich" resulted in the absolute frequencies $x_1 = 47$, $x_2 = 56$ and $x_3 = 41$ and the empirical probabilities $\pi_1 = 47/144$, $\pi_2 = 56/144$ and $\pi_3 = 41/144$. The aim was to test whether the available amount of money is uniformly distributed among the students, which corresponds to the null hypothesis $p_1 = p_2 = p_3 = 1/3$. If no computer were available, we would have to manually calculate the respective probability of occurrence for 10,585 possible empirical frequency distributions. In the statistics package R, the command multinomial.test() returns the output

Exact Multinomial Test, distance measure: p		
Events	pObs	p.value
10585	0.0018	0.3191

13 Because a single class ($k = 1$) does not provide any indication of a deviation between expectation and observation, the number of classes for the degrees of freedom is reduced by one.

"Events" is the number of null distribution values mentioned above, "pObs" is the probability of obtaining precisely the observed frequency distribution if the null hypothesis is true, and "p.value" is the sum of all the probabilities that are less than or equal to 0.0018 (p-value). We see that the null hypothesis $p_1 = p_2 = p_3 = 1/3$ cannot be rejected.

In order to carry out the χ^2 goodness-of-fit test, it is advisable to first compare the observed and expected frequencies as shown in ◘ Table 4.10.

These values are used to calculate the test statistic

$$\chi^2 = \sum_{i=1}^{k} \frac{(x_i - e_i)^2}{e_i} = \frac{1}{48} + \frac{64}{48} + \frac{49}{48} = 2.375.$$

Entering this value as a quantile in PQRS yields ◘ Fig. 4.28.

This shows that if the null hypothesis is correct, the probability of obtaining a more unusual sample than that observed is $p = 0.305 = 30.5\%$. This value is already very close to the exact p-value of the multinomial test ($p = 0.3191$). Thus, the null hypothesis is not rejected in this analysis either, because the p-value is greater than the probability of a Type I error. The given sample thus supports the hypothesis that all the classes are equally likely to occur.

◘ **Table 4.10** Data on the amount of money available

	"poor"	"standard"	"rich"
Observed frequencies x	47	56	41
Expected frequencies e	48	48	48

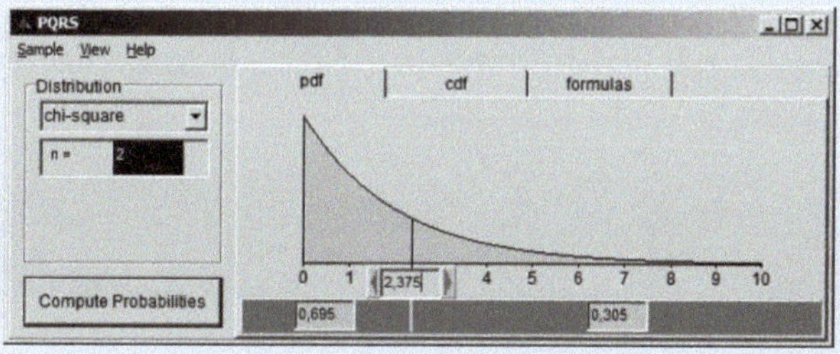

◘ **Fig. 4.28** Approximation of the multinomial test with the χ^2 distribution in the example

4.6.12 Fisher's Exact Test (2 × 2)

The multinomial test and its approximation, the χ^2 test in the $1 \times k$ variant, compared the frequencies of a single sample over k categories with the expected values of a reference distribution (e.g. uniform distribution over all k categories). If we now want to compare two independent, categorically scaled samples (or groups or treatments) with each other, then Fisher's exact test offers a good solution. As before, the observed frequencies are first calculated and summarized in a contingency table. The rows and columns of this table contain the respective values of the two categorical variables. The simplest case with only $k = 2$ categories of the variable measured results in ◘ Table 4.11 with the four values x_{ij}, where $i = 1 \ldots g$ represents the index of the group or sample and $j = 1 \ldots k$ the index of the categories.

Table 4.11 Four-field table for Fisher's exact test

		Measured categorical variable		
		category $j = 1$	category $j = 2$	
Group	control $i = 1$	x_{11}	x_{12}	n_1
	treatment $i = 2$	x_{21}	x_{22}	n_2
		N_1	N_2	N

4

Fisher's exact test now checks whether the frequencies x_{11} and x_{21} (or alternatively $x_{12} = n_1 - x_{11}$ and $x_{22} = n_2 - x_{21}$) are sufficiently different to indicate a significant difference between the groups. The null hypothesis assumes that the population frequencies $p_{1,j}$ and $p_{2,j}$ are equal or, alternatively, that the two samples originate from the same population. The research hypothesis can be formulated as left-tailed, right-tailed or two-tailed.

The data consist of two independent samples of sizes n_1 and n_2 relating to a nominally or ordinally scaled attribute with 2 categories. In order to obtain the sampling or null distribution, we first need a new random thought experiment. A population of size N with the two mutually exclusive attribute values $j = 1$ or $j = 2$ and the corresponding population frequencies N_1 and N_2, where $N = N_1 + N_2$, is given. Assuming we draw an element without replacement from this population exactly n times, what is the probability that we draw $j = 1$ exactly x times (and therefore $j = 2$ exactly $n - x$ times)? It can be shown that this probability is

$$P(N, N_1, n, x) = \frac{\begin{bmatrix} N_1 \\ x \end{bmatrix} \cdot \begin{bmatrix} N - N_1 \\ n - x \end{bmatrix}}{\begin{bmatrix} N \\ n \end{bmatrix}}.$$

This discrete density function is termed *hypergeometric.*

Example

To illustrate this, imagine that one year of a part-time continuing education course consists of $N = 50$ students, of whom $N_1 = 40$ have a high school diploma and $N - N_1 = 10$ have no high school diploma. We randomly select $n = 8$ students from this population and find that $x = 5$ of them have a high school diploma. The probability of this event is

$$P(50, 40, 8, 5) = \frac{\begin{bmatrix} 40 \\ 5 \end{bmatrix} \cdot \begin{bmatrix} 10 \\ 3 \end{bmatrix}}{\begin{bmatrix} 50 \\ 8 \end{bmatrix}} = 0.1471 = 14.71\%.$$

PQRS also contains the hypergeometric distribution. After entering the parameters $N = 50$, $N_1 = 40$ and $n = 8$ and clicking on "Apply New Distribution", we obtain Fig. 4.29.

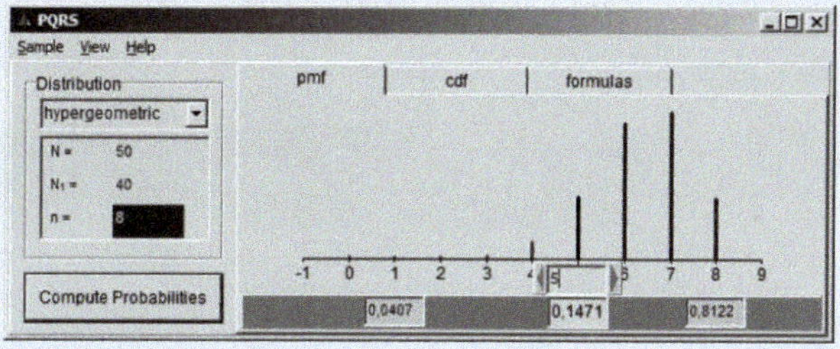

Fig. 4.29 Hypergeometric Distribution

Table 4.12 Four-field table for Fisher's exact test in the example

		Measured categorical variable		
		High school diploma	No high school diploma	
Sample	1	5	*3*	8
	2	*35*	*7*	*42*
		40	*10*	50

The value 0.1471 can be read directly below the quantile x = 5.

Now let us imagine that we also determine the attribute values "high school diploma" and "no high school diploma" among the remaining 42 students. Then it is clear that in this "complementary" sample, we will find 40–5 = 35 students with a high school diploma and 10–3 = 7 students without a high school diploma, with the probability of the occurrence of this "complementary" sample having been exactly the same as that of the first sample, since the population frequencies present mean one sample is always automatically a result of the realizations of the other sample. This can easily be checked using the summary in Table 4.12.

The marginal frequencies in bold and the underlined value 5 permit all the remaining numbers in the table to be determined unambiguously. So if the probability of obtaining the value 5 (top left) is 14.71%, then the probability of any combinations of all four values 5, 3, 35 and 7 must be the same (especially the probability for the realization of the entire table). In PQRS, for example, we can confirm that in a sample of size $n_2 = 42$ out of $N = 50$ students in which $N_1 = 40$ have a high school diploma, the probability of selecting $x_{11}=35$ students with a high school diploma is also 14.71% (cf. Fig. 4.30).

To calculate the probability 14.71%, the population frequency $N_1 = 40$ (or $N_2 = 10$) and $n_1 = 8$ (or $n_2 = 42$) must be known. The population frequencies are, of course, unknown in a typical comparison of two groups. For example, the population of the 50 students of one year could be extended to all the students in the country, from whom $n_1 = 8$ males and $n_2 = 12$ females are randomly selected. The values of "high school diploma" and "no high school diploma" are then determined in both groups and the (hypothetical) contingency Table 4.13 is be generated.

Although it is now possible to calculate the marginal frequencies, the population frequency of the "number of male students" or "students with a high school diploma" is still not known for the entire country. The trick in Fisher's exact test is to pretend that the combined sample of size N is a population from which samples of sizes n_1 or n_2 are taken. Under this assumption, it would be possible to use any value of the four-field table as the test statistic

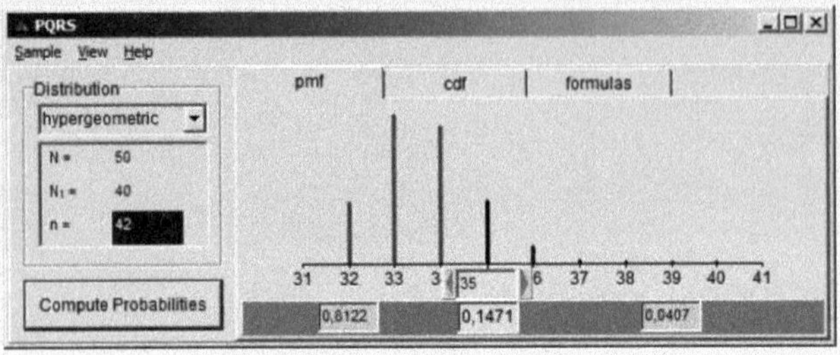

Fig. 4.30 Hypergeometric Distribution in the example

Table 4.13 Contingency table 1 for Fisher's exact test in the example

		Measured categorical variable		
		High school diploma	No high school diploma	
Gender	Male	$x_{11} = 2$	$x_{12} = 6$	$n_1 = 8$
	Female	$x_{21} = 9$	$x_{22} = 3$	$n_2 = 12$
		$N_1 = 11$	$N_2 = 9$	$N = 20$

Table 4.14 Contingency table 2 for Fisher's exact test in the example

		Measured categorical variable		
		High school diploma	No high school diploma	
Gender	Male	$x_{11} = 1$	$x_{12} = 7$	$n_1 = 8$
	Female	$x_{21} = 10$	$x_{22} = 2$	$n_2 = 12$
		$N_1 = 11$	$N_2 = 9$	$N = 20$

whose null distribution is the hypergeometric distribution. The probability of occurrence of the above contingency table (or a single value thereof) is, for example

$$P(20, 11, 8, 2) = \frac{\begin{bmatrix} 11 \\ 2 \end{bmatrix} \cdot \begin{bmatrix} 20-11 \\ 8-2 \end{bmatrix}}{\begin{bmatrix} 20 \\ 8 \end{bmatrix}} = 0.0367 = 3.67\%.$$

In order to calculate the *p*-value of a one-tailed null hypothesis, we now have to determine the probabilities for "more extreme" values of the contingency table that favor the alternative hypothesis. To do this, we make the values x_{11} and x_{21} even more dissimilar by reducing the smaller value by one and increasing the larger value by one. This results in the contingency table in Table 4.14

Table 4.15 Contingency table 3 for Fisher's exact test in the example

		Measured categorical variable		
		High school diploma	No high school diploma	
Gender	Male	$x_{11} = 0$	$x_{12} = 8$	$n_1 = 8$
	Female	$x_{21} = 11$	$x_{22} = 1$	$n_2 = 12$
		$N_1 = 11$	$N_2 = 9$	$N = 20$

Note that with the same marginal frequencies, the values in the second column "no high school diploma" also become more dissimilar. The probability for this more extreme contingency table is

$$P(20,11,8,1)=\frac{\begin{bmatrix}11\\1\end{bmatrix}\cdot\begin{bmatrix}20-11\\8-1\end{bmatrix}}{\begin{bmatrix}20\\8\end{bmatrix}}=0.0031.$$

The "most extreme" of all the contingency tables in the sense of the alternative hypothesis would ultimately be Table 4.15.

The probability of occurrence is

$$P(20,11,8,0)=\frac{\begin{bmatrix}11\\0\end{bmatrix}\cdot\begin{bmatrix}20-11\\8-0\end{bmatrix}}{\begin{bmatrix}20\\8\end{bmatrix}}=0.0001.$$

The probability that the observed or a more extreme distribution of frequencies will occur is therefore the sum $p = 0.0367 + 0.0031 + 0.0001 = 0.0399$. At a significance level of 5%, the null hypothesis, "Gender has no influence on school education", would thus be rejected in favor of the alternative hypothesis, "Women achieve a higher level of school education than men".

4.6.13 χ^2 Test ($2 \times k$)

Fisher's exact test, discussed in the last section, quickly becomes impractical when the number of classes of the categorical variable or the number of observations increases. The χ^2 test offers a simplifying approximation for these cases and is also one of the tests that allow a statistical conclusion based on the sample (observed) frequencies to be drawn concerning the population (expected) frequencies. To do this, the χ^2 test compares the actually realized values in a contingency table with those that could be expected if the null hypothesis is true. If all the differences are sufficiently large, this does not support the null hypothesis. The $2 \times k$ variant of this test involves the comparison of 2 independent samples in relation to an attribute with k categories. The null hypothesis is then, "Both samples come from the same population", or, "The population frequencies of the two samples do not differ". In a control/treatment comparison, it is also possible say, "The treatment has no influence on the population frequencies of the attribute's classes".

The data consist of two independent samples involving a nominally or ordinally scaled attribute with k mutually exclusive categories. The samples are of size $n_1$and n_2. The test statistic only approximately, i.e. for sufficiently large samples, follows a χ^2 distribution. A common rule of thumb is therefore: at least 80% of all the expected frequencies in the contingency table must be greater than 5 and the remaining 20% greater than 1 (Bortz and Lienert 2008).

In order to obtain the test statistic in the χ^2 test $(2 \times k)$, first the *expected frequencies* e_{ij} of the contingency table are calculated using "column total multiplied by row total divided by N".

For the purposes of illustration, we again look at the 2×2 contingency table in ◘ Table 4.11. If we had a population of size N, in which N_1 subjects have a value of 1 of the attribute and N_2 subjects have a value of 2 of the attribute and n_1 subjects belong to group 1 and n_2 subjects belong to group 2, then the probability that a randomly selected subject comes from group 1 *and* has an attribute value of 1 would be

$$\frac{n_1}{N} \cdot \frac{N_1}{N}.$$

If we now randomly selected a subject (with replacement) N times, we would expect to select

$$e_{11} = \frac{n_1}{N} \cdot \frac{N_1}{N} \cdot N = \frac{n_1 N_1}{N}$$

subjects who belong to group 1 and have an attribute value of 1. In general, we can write the expected number of subjects as

$$e_{ij} = \frac{n_j N_i}{N}.$$

These expected frequencies are now compared with the actually observed frequencies. The greater the differences between "control" and "treatment" with respect to the attribute, the greater the difference between the expected values and the observed values in a cell.[14] The latter differences therefore form the basis for a test statistic that can be used to decide whether there is a significant difference between the "control" and the "treatment" with regard to the variable being assessed.

The normalized, squared difference between x_{ij} and e_{ij}, for the observed values x_{ij} and the expected values e_{ij} for all i and j, contribute to χ^2. The sum of all these differences is the χ^2 test statistic

14 The example in this section will illustrate this using a numerical example.

$$\chi^2 = \sum_{i=1}^{2}\sum_{j=1}^{k}\frac{\left(x_{ij} - e_{ij}\right)^2}{e_{ij}},$$

which, at fixed marginal frequencies, approximates a χ^2 distribution with

$$(k-1)(2-1) = k-1$$

degrees of freedom. In a 2 × 2 table, for example, only one of the four possible values (or summands in the test statistic) can vary freely; the remaining three can always be calculated from the given marginal frequencies. Therefore, in this case, the degree of freedom is (2−1)(2−1) = 1.

Example

The aim is to test whether alcohol changes responsiveness. For this purpose, the attribute reaction time is measured in two groups of subjects (no alcohol/alcohol, size per sample: $n_1 = n_2 = 100$) using an ordinal scale (fast/slow). The observations are presented in ◘ Table 4.16.

We see that 90% of the participants in the group without alcohol show a fast reaction, in contrast to only 20% of the participants in the group with alcohol. In view of this large difference, we could now already surmise that administering alcohol has an influence on responsiveness. It does not matter whether the (relative) frequencies of the fast or the slow participants are compared. Similarly, an influence of alcohol on responsiveness could be concluded if we observe that only 10% of the participants in the group without alcohol have a slow reaction, but 80% in the group with alcohol. In both cases, a significant difference between the frequencies might be inferred if the frequencies in one of the columns are sufficiently different. If we wanted to make the two frequencies in one column more similar, the other column would also adjust in the same way given the same group sizes. For the purpose of clarification, let us imagine that instead of measuring the alcohol content in blood, the intelligence quotient of the participants in both groups is measured on an ordinal scale (high/low). We might surmise that this attribute has no influence on the reaction speed. The results in ◘ Table 4.17, for example, would therefore be conceivable.

The frequencies in the left column have decreased by 29 or increased by 38. With the fixed group size of 100 each, the frequency increases by 38 at the top right and decreases by 29 at the bottom right. In this case, the frequencies within a column are very close to each other and it is not expected that the frequencies differ significantly between the groups, i.e. IQ has no significant influence on responsiveness.

◘ **Table 4.16** Contingency table 1 for the χ^2 test in the example

		Reaction time		
		Fast	Slow	
Alcohol	No	90	10	100
	Yes	20	80	100
		110	90	200

Table 4.17 Contingency table 2 for the χ^2 test in the example

		Reaction time		
		Fast	Slow	
IQ	Low	52	48	100
	High	49	51	100
		101	99	200

Table 4.18 Contingency table 3 with expected frequencies in the example

		Reaction time		
	Exp. frequency	Fast	Slow	
Alcohol	No	55 (22.27)	45 (27.22)	100
	Yes	55 (22.27)	45 (27.22)	100
		110	90	200
	Exp. frequency	Fast	Slow	
IQ	Low	50.5 (0.045)	49.5 (0.045)	100
	High	50.5 (0.045)	49.5 (0.045)	100
		101	99	200

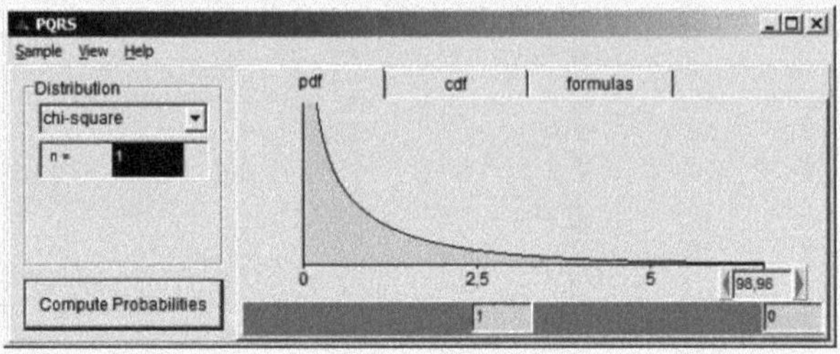

Fig. 4.31 Null distribution in the χ^2 test

To perform the χ^2 frequency test, we first calculate the expected frequencies and the resulting contributions to χ^2, as shown in Table 4.18 (values in parentheses).

The test statistic is the sum of all the contributions to χ^2. In the first case, $\chi^2 = 22.27 + 22.27 + 27.22 + 27.22 = 98.98$ and, in the second case, $\chi^2 = 0.045 + 0.045 + 0.045 + 0.045 = 0.18$. Figure 4.31 shows the null distribution for the first case with $k - 1 = 1$ degree of freedom as it is presented in PQRS.

This shows a p-value close to zero, which indicates a significant influence of alcohol on responsiveness (rejection of the null hypothesis). In the second case, the null distribution is as shown in Fig. 4.32.

Here, a p-value of 67.14% shows no significant influence of IQ on responsiveness (non-rejection of the null hypothesis).

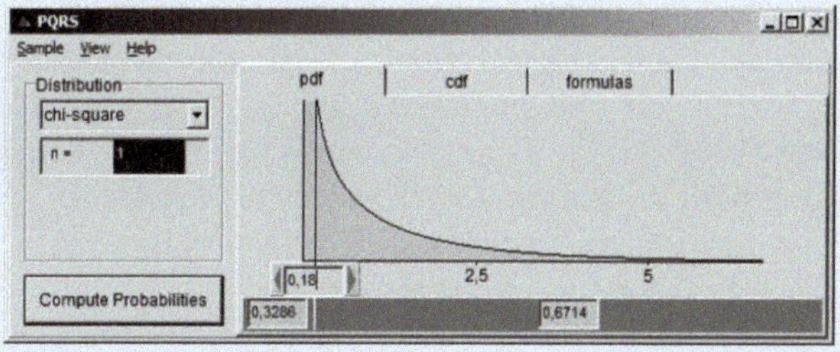

Fig. 4.32 Null distribution in the χ^2 test

4.6.14 McNemar Test

The McNemar test is suitable for comparing two related or dependent samples with respect to a dichotomous attribute. Typically, the experimental design is a within-subject design in which the same group of subjects undergoes two different experimental treatments. Since each individual subject then makes two successive decisions, these two decisions will depend on each other.

The basis for the McNemar test is again a 2 × 2 contingency table, which we present in slightly modified form (Table 4.19).

Frequency a indicates how many subjects have the attribute value $j = 1$ both before and after the treatment, i.e. no change. Likewise, the subjects in cell d show no change, but this time with the attribute value $j = 2$. The changes from $j = 1$ to $j = 2$ are displayed in cell b and the changes from $j = 2$ to $j = 1$ are displayed in cell c. The basic test principle is the same as for the 2 × 2 χ^2 test. The observed frequencies are compared with the expected frequencies and if the differences are sufficiently large, the null hypothesis, "There was no change in the attribute between the two samples", is rejected. The main difference to the comparison of two independent samples is the way in which the expected frequencies are calculated.

The data consist of two dependent samples of an attribute measured on a nominal or ordinal scale with 2 mutually exclusive categories. The test statistic follows a χ^2 distribution when the samples are sufficiently large. The expected frequencies at the bottom left and top right of the contingency table must be equal and both greater than 5 (Bortz and Lienert 2008). If this condition is not met, a binomial test is used.

If the null hypothesis is valid, the frequencies before and after the treatment are not expected to change. The marginal frequency $a + b$ represents the number of

Table 4.19 Contingency table 1 in the McNemar test

		Treatment		
		attribute $j = 1$	attribute $j = 2$	
Control	attribute $j = 1$	a	b	$a + b$
	$j = 2$	c	d	$c + d$
		$a + c$	$b + d$	N

subjects with $j = 1$ *before* the treatment (of whom b changed to $j = 2$ after the treatment). The marginal frequency $a + c$ represents the total number of subjects who showed $j = 1$ *after* the treatment (of whom c changed to $j = 1$ after the treatment). Therefore, if the null hypothesis applies, it is expected that $a + b = a + c$ or $b = c = (b + c)/2$. The expression $b = c$ means that if there were changes, they were equal in both directions. The expected frequencies $e = (b + c)/2$ are now compared with the observed frequencies, as in the χ^2 test, by adding up the normalized, squared differences. This results in the test statistic

$$\chi^2 = \frac{(b-e)^2}{e} + \frac{(c-e)^2}{e} = \frac{(b-c)^2}{b+c},$$

which is χ^2 distributed with one degree of freedom. Since the χ^2 distribution is again only a continuous approximation for what is in fact a discrete null distribution, this approximation can be improved with the following continuity correction:

$$\chi^2 = \frac{(|b-c|-1)^2}{b+c}.$$

If the expected frequencies are less than 5, an exact binomial test can also be used. If the null hypothesis is valid, the probability of a change from $j = 1$ to $j = 2$ is the same as a change from $j = 2$ to $j = 1$, i.e. 0.5. The smaller of both frequencies, $x = \min(b, c)$, is used as the test statistic. This is binomially distributed (as is the other "frequency of change") with the null distribution $B(b + c; 0.5)$.

Example

We would like to re-examine the impact of information about the subject of study on the amounts sent by the trustor in the trust game, but this time in a variant of the game in which the trustor can only choose between a "high" or a "low" amount to give to the trustee. In addition, this time $n = 104$ individuals (52 trustors and 52 trustees) are successively subjected to two different experimental treatments. First, the subjects play the trust game in pairs without knowing the subject of study of their game partner ("without info", control). In the second treatment, the same subjects play the trust game again, but they also receive information about the subject of study of their game partner ("with info", treatment). The data lead to ◘ Table 4.20.

Assuming that there is some theoretical justification for information about the subject of study only having a negative influence, if any, on the amount sent, then we can formulate the one-tailed research hypothesis: "Among all the changes, the probability of changing from "high" to "low" (cell *b*) is greater than the probability of changing from "low" to "high" (cell *c*)".

First using the exact binomial test, we obtain the test statistic $c = 6$ and the parameters of the null distribution are $n = 15 + 6 = 21$ and $p = 0.5$. After entering these parameters in PQRS, we obtain the density function in ◘ Fig. 4.33.

This shows the *p*-value of the one-tailed hypothesis $p = 0.0133 + 0.0259 = 0.0392 = 3.92\%$ $< 5\%$. The influence of information on the subject of study thus has a significant negative effect on the amount sent. In a two-tailed hypothesis, the *p*-value would double to $p = 0.0784$, which suggests that the effect is not significant.

Table 4.20 Contingency table 2 in the McNemar test

		With info		
		High	Low	
Without info	High	21	15	36
	Low	6	10	16
		27	25	52

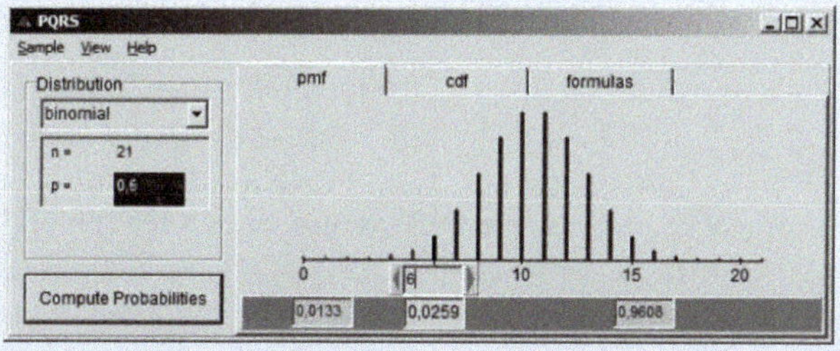

Fig. 4.33 Binomial distribution in the McNemar test

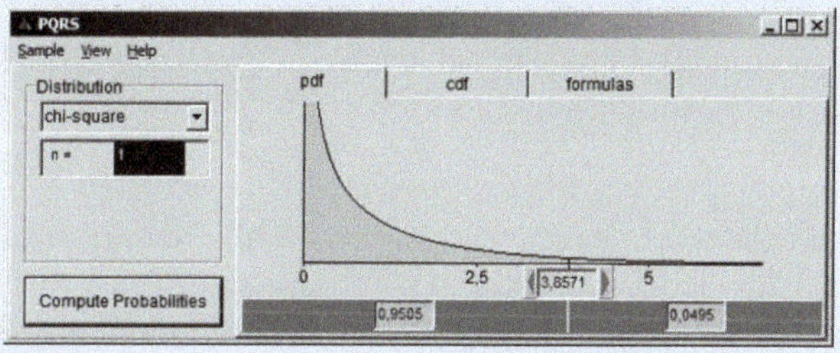

Fig. 4.34 Chi-square distribution in the McNemar test

We will now perform the χ^2 approximation for a two-tailed research hypothesis to demonstrate the influence of the continuity correction and for the sake of completeness. The test statistic without continuity correction is

$$\chi^2 = \frac{(b-c)^2}{b+c} = \frac{(15-6)^2}{21} = 3.8571.$$

The null distribution is a χ^2 distribution with one degree of freedom (see Fig. 4.34).

The p-value is thus $p = 0.0495 = 4.95\% < 5\%$ for a *two-tailed* hypothesis. The exact two-tailed p-value without correction was $p = 0.0784 = 7.84\% > 5\%$, which shows that a χ^2 approximation without continuity correction can certainly lead to wrong decisions.

With continuity correction, we obtain the test statistic

$$\chi^2 = \frac{(|b-c|-1)^2}{b+c} = \frac{(|15-6|-1)^2}{21} = 3.0476,$$

which in PQRS leads to a more accurate p-value of $p = 0.0809 = 8.09\% > 5\%$ and to the "correct" test decision.

4.7 Statistical Models

4.7.1 The Fundamentals

Testing the statistical significance of a treatment effect in the form of a hypothesis is not the only reason behind many experimental studies. Collecting data on variables experimentally in order to estimate relationships between the variables is another. For this second purpose, a *statistical model* is developed in order to explain the data obtained as well as possible. This model can be used to answer further questions, such as:

- How can the attributes of a variable be explained using other variables and how well can this be done?
- What value would a variable presumably have if it were influenced by the attribute of another variable that was not elicited in the experiment?

The starting point for a statistical model is the desire to model the changes in an experimentally *observed* variable y or to at least *explain* these changes with the help of a model. The variable y is therefore also called the variable *to be explained* or the *endogenous variable*. The information we use to explain the endogenous variable originates from one or more *explanatory* variables (*exogenous variables*). The basic assumption of each statistical model is that there is a true relationship between the two variables, but this is unknown. In particularly simple cases, it may be appropriate to assume a true, *linear* relationship between the endogenous variable y and *exactly one* exogenous variable x. This would then have the form

$$y = a + bx,$$

where the constant parameters a and b of this line are unknown.

If this model were left as it is, it would certainly not model the data actually observed in the best possible way. For the sake of clarification, suppose we have a treatment variable x with the values $x = (1, 2, 3, 4)$ for 4 subjects in the experiment. This could be, for example, a given initial endowment or a given cost in an economics game. Depending on these values for x, the continuous variable y is measured in the experiment with the values $y = (7.6, 12.2, 11.1, 14.8)$. Drawing the data points (x_1, y_1), (x_2, y_2), (x_3, y_3), (x_4, y_4) in an x-y graph results in the points shown in Fig. 4.35.

On the basis of this graph, we can see that our previous model does not do justice to these observations for two reasons. First, the observations are not a continuous function, but only 4 discrete points, each consisting of x- and y-coordinates. For this reason, we discretize the variables x and y of the model using an observation index $i = 1, 2, 3, 4$ and obtain the new model

$$y_i = a + bx_i.$$

Each individual observation y_i is thus explained deterministically by a linear transformation of a single observation x_i. Therefore, the right side of this equation is also called the (deterministic) *linear predictor*.

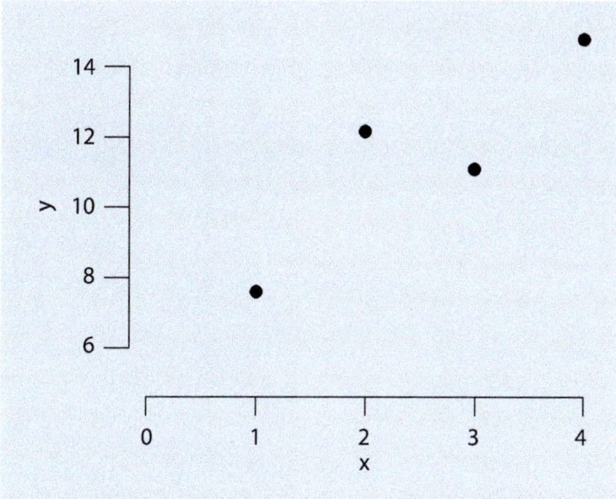

■ **Fig. 4.35** Representation of the four hypothetical observations of the variables *x* and *y* in the *x*-*y* graph

Second, it can be seen that the linear predictor for the constant parameters *a* and *b* cannot fully map or explain *all* the observations. In other words, there is no single straight line on which all four observations lie. No matter how we try to draw a straight line in the graph, some observation points are always above or below this straight line. This deviation of an actual observation y_i from its linear predictor $a + bx_i$ (true straight line) is corrected by a random error term, or random disturbance, u_i, with the result that the following applies

$$u_i = y_i - (a + bx_i)$$

or

$$y_i = a + bx_i + u_i.$$

As we can see in ■ Fig. 4.36, this disturbance is sometimes positive and sometimes negative, and sometimes strong and sometimes weak, since an observation at different distances is sometimes above and sometimes below the true dashed line. In our stochastic model, we therefore assume that the subjects are subject to a random effect with regard to the variable y_i, which sometimes causes them to deviate upwards and sometimes downwards from the true, deterministic value $a + bx_i$.

In terms of an experiment, this random deviation can be explained in two ways. If we carry out the experiment with the same four subjects several times in succession, each person is subject to a new random influence each time. For example, variable *y* could measure the amount given in the ultimatum game. Then it would be expected that on a day when student 1 happens to be in a particularly good mood (he has slept well, the weather is fine, he has received a good exam result, etc.), the amount given, y_1, also happens to be particularly high. Although the variable "mood" remains unobserved in the experiment and therefore cannot be controlled, it too has an influence on

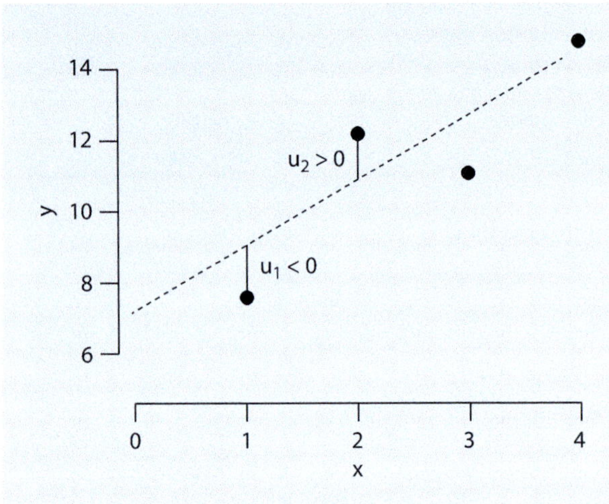

Fig. 4.36 Random disturbances in the four observations

the endogenous variable y.[15] In the same way, of course, a particularly low value can be explained if it is assumed that the students happen to be in a bad mood some other day.

Alternatively, we can invite, or "draw" from a total population, groups of students to participate in the experiment several times in succession, with four *new* students *each time*. In this case, we are not modeling a random effect within-subject over time, but rather between-subject across different subjects. This is because every individual subject also possesses individual character traits that we are generally not able to observe and control, as is the case with "mood".

According to our stochastic model, the random variable y_i is thus subject to two additive effects. First, the fixed, non-random effect b, which represents the slope of the "true" relationship. The larger this value is, the greater the average increase in the y_i values with an increase in x_i.[16] A horizontal line would therefore represent a fixed effect of zero, where the variable y_i remains unaffected by a change in x_i. Second, the variation of y_i is affected by a random effect that does not result from the variation of the explanatory variable. At first, we will make it very easy for ourselves and simply summarize all the conceivable random influences on y_i in this effect. In this case, we call the model an *econometric model* (cf. von Auer 2016).[17]

In the simplest case, the random effect is formally represented by a normally distributed random variable u_i with an expected value $\mu = 0$ and a constant variance σ^2, or more compactly

$$u_i \sim N\left(0,\ \sigma^2\right).$$

15 Unobservable variables are sometimes called *latent* variables.

16 Of course, the intercept *a* also has an influence on the value of *y*. However, this influence is the same for all *x* values and only determines the overall level of the relationship. Therefore, *a* is not considered an effect.

17 Later we will refine this overall random influence and explicitly model random effects.

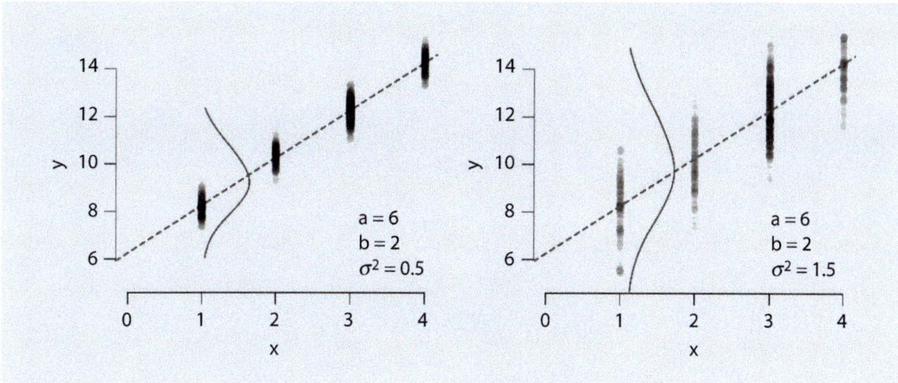

Fig. 4.37 Weak versus strong random effect in a simulated experiment with 200 repetitions

The expected value of zero means that, with very many repetitions of the experiment, the positive and negative values of the random disturbances balance out exactly on average. It follows that the expected value of the variable y_i is equal to to the true value of the linear predictor,

$$E(y_i) = a + bx_i.$$

The variance σ^2 represents the strength of the random effect. The greater the dispersion of the random disturbance, the stronger the impact of the random effect on the endogenous variable's possible magnitude. In Fig. 4.37, the two hundred-fold repetition of our hypothetical experiment with the controlled values $x = (1, 2, 3, 4)$ and the fixed effect $b = 2$ with the intercept $a = 6$ has been simulated. On the left of Fig. 4.37, a comparatively weak random effect of $\sigma^2 = 0.5$ has been used as a basis, with the result that there is little fluctuation in the observations y_i around the true value. The right-hand side of Fig. 4.37 shows the stronger random effect of $\sigma^2 = 1.5$, which causes a correspondingly larger spread. In both cases, it is easy to see the normal distribution of the disturbance variable, with the highest probability density or density of points located around the true y value. Upwards and downwards, however, the density of the points becomes less and less. In the left picture, the density function is therefore narrow with a high peak, whereas in the right picture it is wide with a flatter peak.

The fixed effect b, the random effect σ^2 and the true intercept a of the relationship are always *unknown* in practice. The aim of a *regression* is to obtain the best possible *estimators*, $\hat{b}$, $\hat{\sigma}^2$ and $\hat{a}$, for these unknown parameters.[18] In the statistical sense, "best possible" means that (i) the mean values of the estimates should correspond to the true values of b, σ^2 and a (*unbiased estimation*) over many repetitions and that (ii) of all the unbiased estimators, the ones with the lowest deviation or highest estimation accuracy are selected (*efficiency*). In the course of this section, we will explore different models, each of which requires different estimators. An unbiased and efficient estimator in the

18 In the following, we will always indicate estimated values with a "hat".

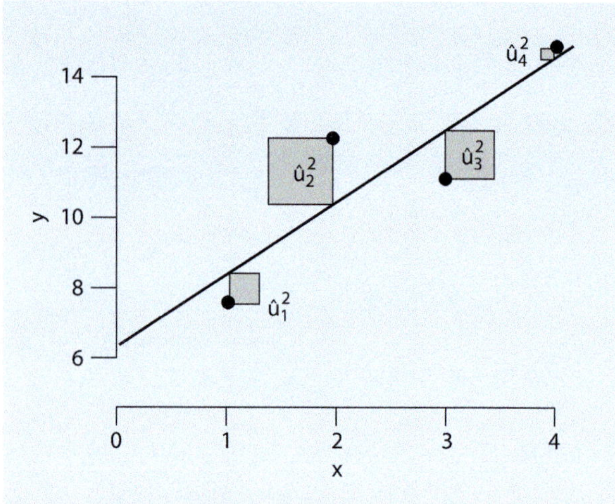

Fig. 4.38 Residual squares in four observations

4

case of our simple linear model is the *least squares estimator*. It determines the values of a and b that lead to a line that has the minimum possible distance to all the data points.

Suppose the observations of our dependent variable are 7.5, 12, 11 and 14.5. Then this estimator supplies the values $\hat{a} = 6.25$ and $\hat{b} = 2$. The resulting regression line, $\hat{y} = \hat{a} + \hat{b}x = 6.25 + 2x$, is shown in Fig. 4.38. The distance of an observation y_i from the estimated value $\hat{y}_i$ on the regression line is called the *residual* $\hat{u}_t$. It represents the part of the observation y_t that cannot be explained by our estimated model.[19] In the diagram, the squared residuals are represented by gray squares. The least squares estimators $\hat{a}$ and $\hat{b}$ minimize the sum of these squares. Thus, there is no other line leading to a smaller sum of all the gray areas than $\hat{y} = 6.25 + 2x$. In our example, this total area is 5.25. In order to estimate the random effect σ^2, we calculate the *average* residual square. However, this residual sum of squares is not divided by the number of observations T, but by the number of observations that provide information content regarding deviation, i.e. $T - 2$.[20] We obtain $\hat{\sigma}^2 = 5.25/(4-2) = 2.625$ as an estimate for the random effect.

Once a model has been estimated, making a *prediction* is very simple. To do this, we insert a new x-value into the estimated equation and obtain a predicted y-value without random influence. Of course, it is important not to extrapolate too much and use x-values that are very far away from the available x-values. The further away the x-values are from the data points used to estimate the model, the less reliable the predictive power of the estimated model.

---

19 Therefore we often refer to the "unexplained residual".

20 Two observations are always "useless" in this sense, since a regression line with only two observations will always go exactly through these two observed points, with the result that the residuals assume the value zero and, consequently, do not provide any information content with regard to a deviation.

Box 4.4 Correlation Versus Causation

The strength of association between two variables can be considered from a quantitative and qualitative point of view. In qualitative terms, the strongest relationship is a causal one. This means that the value of one variable causes a change in the value of another variable. For example, the force transmitted from one foot to a football is one of the reasons why the football flies so far.

A correlation between variables is a qualitatively weaker form of relationship and exists when it can merely be observed that the increase of one variable is accompanied by an increase or decrease of the other variable. Even a perfect correlation does not necessarily mean that the variables are also causally related. For example, it may be observed that the amount of hair men have on their head and their respective income are inverse to each other, i.e. the less hair men have, the higher their income. If there were a causal relationship here, all men would probably shave off their hair in the hope of becoming richer. The actual causal relationship can be established easily if a third variable, age, is included. The older a man is, the more professional experience he has and, therefore, the higher the average income he earns. At the same time, it is in the nature of things that hair loss in men is also age-related. Age therefore has a causal effect on both the amount of hair and the average income of working men. Causation always means a correlation, but not every correlation means causation. To put it another way, if two variables are not correlated, there cannot be a causal relationship either. However, even if no causal relationship exists, there may well be a correlation.

A simple statistical model such as the one described above merely measures the strength of a relationship and therefore provides purely quantitative information on the relationship between the variables. Whether one variable is responsible for the change in the other variable is not the subject of the statistical model (at least not in the form shown here). The relationship quantified by a statistical model is only ever causal to the extent that the experimental design, which was carried out in advance, has allowed it. The three factors of control, repetition and randomization discussed earlier are decisive for the causality in an experiment. Experiments in which randomization is not possible are called quasi-experiments. It is much more difficult to derive causal relationships in such experiments, but there are special statistical models and estimation methods that facilitate the determination of causalities (regression discontinuity designs). These include, in particular, instrumental variables estimation and the differences-in-differences (DiD) method.

4.7.2 Using Statistical Models

In the simplest case, statistical models consist of only a single equation. This equation explains the variation of an observed dependent variable y (left-hand side of the equation) using a functional term f with K explanatory variables x and a random influence u_i (right-hand side of the equation). It is generally written

$$y = f\left(x_1, x_2, \ldots, x_K, u_i\right).$$

Such a model is always subject to certain assumptions. The more assumptions the model requires and the stronger they are, the simpler it becomes, and vice versa. The simple linear model already mentioned in ▶ Sect. 4.7.1 is very easy to understand and to handle, but it requires very restrictive assumptions that often cannot be fulfilled in experimental practice. The most important of these are:

1. There are no relevant exogenous variables missing in the econometric model and the exogenous variables used are not irrelevant.
2. The true relationship between the exogenous variable and the endogenous variable is linear.
3. The intercept and slope parameters are constant for all the observations, i.e. they have no index t or i.
4. The disturbance is normally distributed with $u_i \sim N(0, \sigma^2)$ for all the observations i and the disturbances of all the observations i are statistically independent of each other.
5. The values of the independent variable x are statistically independent of the disturbance variable u.

Violating some of these assumptions can have consequences of varying severity. In some cases, the least squares estimation only needs to be adjusted slightly, while in others completely new models with special estimation methods have to be developed. For example, if it can be shown that a *univariate* model with only one explanatory variable x is underspecified, further variables must be added. These models with more than one variable are called *multivariate* or *multiple regression models*. In this case, the least squares estimate can continue to be used under otherwise identical conditions. If f is linear in the parameters, then the model is called a *linear* model, if not, it is called a *non-linear* model. Non-linearities in the exogenous variables can usually be linearized by means of transformation, but non-linearities in the parameters cannot. *Dummy variable regression models* and *structural break models* are suitable for modeling coefficients that change abruptly over two or more sections of the observations (e.g. two different time periods). If they vary randomly across individuals, *multi-level models* (also random coefficient models) can be used.

Another central feature used to classify regression models is the nature of an explanatory variable. In particular, if the disturbance variable is no longer normally distributed with an expected value of zero and constant variance, this will have a direct effect on the distribution of y. Is y a continuous, discrete or categorical variable? Is its variance constant? Does its expected value correspond to the linear predictor? A large class of non-normally distributed endogenous variables can be modeled with the *generalized linear model*. Often an upper or lower bound for the endogenous variable, which by definition cannot exceed the values of this variable, is also introduced. This is referred to as *censored* or *truncated endogenous variables*, each of which also requires its own estimation method.

An explanatory variable, x, always has a strong ability to predict well explanatory content if it correlate strongly with the variable to be explained, y. However, if a variable x is now simultaneously correlated with the disturbance variable u, it then immediately follows that there is a correlation between the variable to be explained, y, and the disturbance variable, u. This correlation is problematic because it leads to biased (and inconsistent) estimators. A way out of this so-called endogeneity problem is to use what is known as *instrumental variables* estimation. The conventional least squares estimation cannot be used here either, but a two-stage variant (two-stage least squares) is commonly employed.

In addition, the structure of the data also plays a role in the correct choice of a model: were several individuals measured at a single point in time (*cross-sectional data*), was one individual measured at several points in time (*time series data*) or were several individuals measured at several points in time (*panel data*)?

As we can see, there are several possible deviations from the standard "simple linear model" case. Discussing all these cases in detail and presenting both the consequences of a violation and possible remedial measures would go beyond the scope of the book. Therefore, we refer readers to a number of good textbooks (von Auer 2016; Griffith et al. 1993; Kennedy 2008; Gujarati and Porter 2008).

Instead, in this section we will concentrate on four special cases that are particularly common in connection with behavioral data collected experimentally:

1. The dependent variable is not continuous, but has only a countable number of values;
2. The dependent variable is not normally distributed;
3. The observations of the dependent variable are not statistically independent of each other;
4. The dependent variable is truncated from above or below.

In the following, we will show how each of these four cases can be modeled with a suitable methodology and how a typical computer output of the respective estimate looks. To this end, we use simple hypothetical examples. To better illustrate the differences between these models and the classic linear model, we will first present the linear model in a notation compatible with the other models and then discuss the extended models.

4.7.3 The Linear Model (LM)

The linear model relates a continuous endogenous variable y_i to a linear predictor η consisting of K continuous and/or categorical, exogenous variables x_{1i} … x_{Ki}. All the random influences on y_i are modeled by the random variable u_t. We thus have

$$y_i = \eta + u_i$$

$$u_i \sim N\left(0, \sigma^2\right)$$

with

$$\eta = b_0 x_{0i} + b_1 x_{1i} + \ldots + b_K x_{Ki}.$$

The value of the artificial "variable" x_{0i} is always 1, so b_0 is the intercept.

If we were to repeatedly draw a random sample with fixed values of the exogenous variables and thus constantly obtain new values of the random variables u_i and y_i, then with a great many repetitions we would on average reach a point on the "true" line or level. In the following, this expected value of the endogenous variable, which is part of the expression x_i, is termed μ_i. In Fig. 4.36, for example, $\mu_1 = 8$ is at position $x = 1$ and $\mu_2 = 10$ at position $x = 2$. Since the disturbance variable has an expected value of zero, the following generally holds

$$E(y_i) = \mu_i = \eta$$

4

and therefore

$$y_i \sim N(\eta, \sigma^2).$$

This means that the expected value of the endogenous variable (or the "true" y_i without random influence) is predicted by the linear predictor or a linear combination of the given exogenous variables. In the example in Fig. 4.38 in Sect. 4.7.1, this linear combination is $\eta = 6 + 2x$ ("true" straight line). We see that a constant change in the value an exogenous variable in this type of model also leads to a constant change in the expected value of the endogenous variable. In other words, the linear model is called "linear" because it is linear in the $N = K + 1$ parameters $b_0 \ldots b_K$. The exogenous variables may well be transformed in a non-linear way, but it is still a linear model. For example, the model

$$y_i = b_0 + b_1 \ln(x_{1i}) + b_2 \sqrt{x_{2i}} + u_i$$

is a linear model, whereas

$$y_i = b_0 + x_{1i}^{b_1} + e^{b_2} x_{2i} + u_i$$

is not.

The linear model from Sect. 4.7.1 with only one independent variable was $y_i = b_0 + b_1 x_i + u_i$. With the specified data, the model can be estimated using a computer program, with a typical computer output possibly as shown in Table 4.21.

This table is divided into two parts, the regression table and the other key indicators listed below it.

In our case, the regression table consists of four columns. The first column contains the estimated values for the parameters: the y-axis intercept of the regression line $\hat{b}_0$, denoted by the line name (intercept), and $\hat{b}_1$, the slope of the regression line, denoted by the line name x, which is name of the associated exogenous variable. The column std. err is the estimated dispersion of both estimators, i.e. an estimated measure of their accuracy when attempting to hit the true value of b_0 or b_1. In concrete terms, one could imagine a precise and an imprecise estimator as a professional biathlete and an ordinary person trying to hit the black region of a target with a rifle. If each shoots 20 times under the same conditions, the dispersion of the biathlete's 20 bullet holes will be significantly smaller than the normal shooter's. Similarly, for a precise estimator with a small

Table 4.21 Typical computer output of a linear regression

	coef	std.err	t.value	p.value
(Intercept) X	6.2500	1.9843	3.1497	0.0877
	2.0000	0.7246	2.7603	0.1100
Number of observations: 4				
Number of coefficients: 2				
Degrees of freedom: 2				
R-squ.: 0.7921				
Adj. R-squ.: 0.6881				
Sum of squ. resid.: 5.25				
Sig.-squ. (est.): 2.625				
F-Test (F-value): 7.619				
F-Test (p-value): 0.11				

dispersion, the average estimate is closer to the true value than for an imprecise estimator with a large dispersion.

The sample t-value of a t-test with the null hypothesis "true parameter = q" can be determined from the estimated value and the estimated dispersion. The standardization formula for this is

$$t = \frac{\text{coef} - q}{\text{std. err}}.$$

What is most frequently checked is whether a parameter is significantly different from zero. In this case, $q = 0$ and the values of the t-value column are obtained by dividing the values of the first column by those of the second column. For the null hypothesis $b_0 = 0$, for example, the corresponding t-value is calculated from (2–0)/0.72 = 2.76. If we now compare the sample t-values with the critical t-values, which would have to be determined separately depending on the estimated significance level, we could make a test decision. But it is faster to look at the corresponding p-values of the last column p.value. If these are lower than the estimated significance level (e.g. 5%), the null hypothesis "true parameter = 0" is rejected and the parameter is statistically significant different from zero.[21]

21 The wording "different from zero" is often omitted and then we only say, a parameter is "statistically significant".

The parameters below the regression table are partly self-explanatory. The first three values are the number of observations, the number of estimated model parameters and the difference between the two, also called the "degrees of freedom" of the model. The latter is necessary, for example, to determine the critical value in a hypothesis test for this model.

The fourth number, R-squ., is known as the *coefficient of determination*. It measures the proportion of the total variance in the observations y_i that can be explained by the estimated model. If this proportion is very low, the model cannot explain very much. This is especially the case when the dispersion of the observations around the regression line and thus the sum of the residual squares, or the "unexplained" dispersion, is very high. The determined regression line then has little value insofar as a completely different straight line would have been equally realistic. A high coefficient of determination, on the other hand, suggests a low dispersion of the disturbance, meaning that a representative line has probably been determined and this should not change too much if the experiment is repeated. A perfectly linear relationship, therefore, has a coefficient of determination of 1 or 100%, because the exact position of the observations is fully explained. The square root of the coefficient of determination is called the correlation coefficient and measures how linear the relationship is. A value of -1 stands for a perfectly linear relationship with negative slope, while a value of $+1$ for a perfectly linear relationship with positive slope.[22] If there is no correlation at all between the endogenous and the exogenous variables, both the correlation coefficient and the coefficient of determination are zero. In this case, the least squares estimator also returns the value of zero for the slope. Increasing the explanatory variable by one unit would therefore have no influence on the variable to be explained and the suitability of this variable in the model should be called into question.

The coefficient of determination is only conditionally suitable for the purpose of specifying the model since its value can never decrease if another variable is included in the model. Thus a coefficient of determination close to 100% can easily be generated artificially by simply adding enough variables to the model – whether they make sense or not. The corrected coefficient of determination takes this fact into account by including the number of variables negatively in its calculation. An additional variable can therefore lead to a lower corrected coefficient of determination. A disadvantage of this measure is that it can no longer be interpreted as a proportion, since it may also take negative values. The corrected coefficient of determination therefore has very little meaning in itself. And even for the purpose of comparing alternative models, it is generally not the corrected coefficient of determination but other parameters, such as the Akaike information criterion, that are normally used.

The two values that next appear in the computer output are the sum of squared residuals (Sum of squ. resid.) and the estimated random effect, or disturbance variance estimator, (Sig.-squ.(est.)).

Finally, the last two values are the test statistic and the p-value for a simultaneous significance test of all the slope parameters in the model. Since only a single slope parameter is included in our example model, the *F*-test and the *t*-test in the regression table are identical. Both provide the same *p*-value and the *F*-statistic corresponds to the squared *t*-statistic.

---

22 Of course, it is not possible to draw any conclusions about the sign from the coefficient of determination, since both signs would be possible when calculating the square root.

4.7.4 Models for Discrete and/or Non-Normally Distributed Dependent Variables

The linear model discussed in the previous section is one of the most frequently used statistical models. This model is based on a true linear relationship that provides the expected value of the endogenous variable for a given value of the exogenous variable, i.e.

$$E(y_i) = \mu_i = b_0 x_{0i} + b_1 x_{1i} = \eta.$$

This relationship $\mu_i = \eta$ could also be written in a somewhat extended form as

$$g(\mu_i) = \eta,$$

where $g(z) = z$ is the transformation function "without any effect", namely the line of equality. Of course, such a transformation function only makes sense if it does not correspond to the line of equality – otherwise it could simply be omitted. In the following, we will proceed from precisely such cases. The main feature of function g is that it must be invertible and differentiable. Apart from that, it can in principle take any linear or non-linear form.

The basic idea of the *generalized linear model* (GLM) developed by Nelder and Wedderburn (1972) is to transform non-normally distributed dependent variables in such a way that they can always be explained with the help of the linear predictor. Using g, a link between a non-normally distributed and/or discontinuous dependent variable and a linear combination of independent variables can be established. For this reason, function g is also called the *link function*. For example, the scope of application of the GLM is extended to the modeling of *categorical* quantities that have a countable number of values (e.g. acceptance/rejection in the ultimatum game, three types of behavior "self-interested, positive reciprocal, negative reciprocal") or frequencies (how often was a public good created in the public-good game?). The "costs" of these generalizations are essentially reflected in a more complicated estimate of the model. Instead of a simple least squares estimation, an iterative, weighted variant is used that leads to maximum likelihood estimators (McCullagh and Nelder 1989). Since this method is included in most statistical programs, we will save ourselves the technical details and concentrate on the functionality and application of possible models.

To give a concrete example to illustrate the flexibility of GLM, we assume that our variable to be explained can only take two values. These values follow a Bernoulli distribution, whereby the probability of success p is unknown and is to be estimated. For example, in the ultimatum game, we want to explain the acceptance or rejection behavior y_i of the second mover on the basis of the amount x_i proposed by the first mover. It should be immediately clear that a linear model of the type $y_i = b_0 + b_0 x_i + u_i$ would make little sense, since the right-hand side of the equation would also represent values between "yes" and "no". As a result, we would get statements like "with $x_i = 30$ %, the second mover would respond with a 73% acceptance". There is, however, no 73 percent acceptance – either the proposal is rejected or not. Instead, it makes more sense here to model the *probability* of an acceptance, which means that the above statement reads "with $x_i = 30$ %, the second mover would accept with a probability of 73%". Instead of a binary variable to be explained, the left-hand side of our model equation contains a probability with a value range between zero and one. This value range must now be

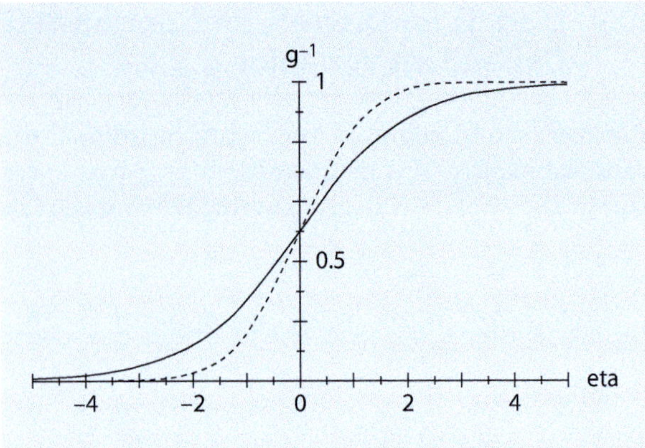

Fig. 4.39 Two possible sigmoid functions (S-curves) for modeling probabilities: the expit function (solid line) and the distribution function of the standard normal distribution (dashed line)

defined with the aid of a suitable link function. The inverse function that follows from $g(\mu_i) = \eta$ is

$$g^{-1}(\eta) = \mu_i$$

and from μ_i, the mean y-value, we know that it should be between zero and one. This means we are looking for a function that maps the real numbers to the interval (0, 1), that is, $g^{-1}: R \rightarrow (0, 1)$. Two very popular candidates are

$$g_1^{-1}(\eta) = \frac{e^{\eta}}{e^{\eta}+1} = expit(\eta) = p$$

and the distribution function of the standard normal distribution $g_2^{-1}(\eta) = p$, which, however, is much more difficult to use than an explicit term (which is why we do not use it here). Both functions are shown in Fig. 4.39. Taking the inverse function of both functions again, we get

$$g_1(p) = \ln\left(\frac{p}{1-p}\right) = logit(p) = \eta$$

$$g_2(p) = probit(p) = \eta.$$

Both equations establish a relationship between the (transformed) probability p and the deterministic linear predictor $\eta = b_0 + x_{1i}$. The first equation is the model for *logistic regression* (also *logit regression*) and the second is the model for *probit regression*. Both hardly differ in their nature and also in practice deciding for one model over the other does not play a major role.

In the logit variant, our model is

$$ln\left(\frac{p}{1-p}\right) = b_0 + b_1 x_{1i}.$$

Solving the equation for p would result in the (non-linear) regression equation. The parameters b_0 and b_1 are estimated in all GLMs using the maximum likelihood method and its variations.

Example

We have the data set shown in Table 4.22.

The maximum likelihood estimate of our model leads to the estimates $\hat{b}_0 = -5.45$ and $\hat{b}_1 = 23.04$. The estimated linear relationship between the logits and the linear predictor is shown in Fig. 4.40 (left).

Table 4.22 Hypothetical data set on giving behavior in the ultimatum game with 20 observations, with "offer" denoting the proportion of one's own endowment to be shared and "acc" representing the acceptance (value 1) or rejection (value 0)

offer	*acc*	*offer*	*acc*	*offer*	*acc*	*offer*	*acc*	*offer*	*acc*
0.11	0	0.32	1	0.21	0	0.44	1	0.32	1
0.2	1	0.29	1	0.24	0	0.15	0	0.43	1
0.16	1	0.33	1	0.47	1	0.18	0	0.36	1
0.05	0	0.42	1	0.21	0	0.27	0	0.35	1

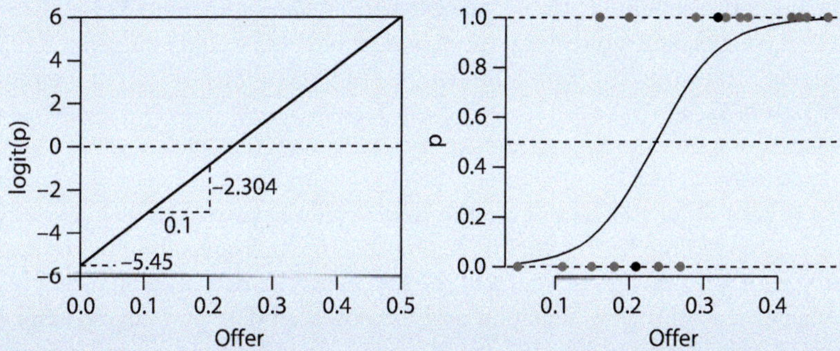

Fig. 4.40 Two representations of the estimated equation of a logistic regression. In the right illustration, the 20 observations are also shown as points. The darker a point is, the more observations that are superimposed

From a purely formal point of view, we can say that an increase of 0.1 in the share of levies would increase the logits by $\hat{b}_1/10 = 2.304$. However, this finding can hardly be interpreted in any meaningful way. Therefore, logits are usually transformed back into the original unit "odds" $(p/1-p)$[23] via the exponential function. Then the estimated

23 Odds represent the relationship between the probabilities of two opposing events. For example, p = 0.2 is the probability that a horse will win a horse race and 1−p = 0.8 is the probability that it will not win. Then the odds are 0.2/0.8 = 1/4 which are shown as odds 4:1 in continental European horse races. In case of a win you would get 4 Euro for every Euro you bet.

Table 4.23 Typical computer output of a logistic regression

	Estimate Std.	Error z value		Pr(>|z|)
(Intercept)	−5.4522	2.4868	−2.1925	0.02834*
Offer	23.0400	9.9453	2.3167	0.02052*
Null deviance: 26.9205 on 19 degrees of freedom				
Residual deviance: 14.2276 on 18 degrees of freedom				
AIC: 18.2276				

equation could be solved for p and expressed in a non-linear but more interpretable form (see Fig. 4.40, right). It then becomes clear that the marginal effect of the offer variable on the probability (p) is not constant. Furthermore, the predicted probability of an acceptance of a given amount can be read off. For example, it is almost certain that the responder will accept if the offer is 40% of the initial endowment of the first mover. If the same estimation were performed with the *probit* link function, the z-values of the standard normal distribution would be obtained on the ordinate in Fig. 4.40 (left) instead of the logits, and a direct interpretation would also be difficult. After a retransformation to p, however, the estimation curve would be almost exactly the same as in the right part of Fig. 4.40.

A typical computer output of a logistic regression with the data from our example is shown in Table 4.23.

What is new about this output are the values null deviance, residual deviance and AIC. Broadly speaking, the deviance in generalized linear models – similar to the R^2 in linear models – is a measure of how well the model fits the data. The smaller this number, the better the estimation equation and the observed data fit. Null deviance is the index for what is known as the null model, which does not contain any independent variables and only the y-axis intercept is estimated. Residual deviance, on the other hand, refers to our actually estimated model, i.e. with just one independent variable. The difference between the two values shows how much the inclusion of the "offer" variable in the model improves the fit to the data. In our case, we get a value of $26.92 - 14.23 = 12.69$. Whether this number is significantly different from zero can be determined with a χ^2 test. The degrees of freedom of the null distribution of the χ^2 test correspond exactly to the difference of the degrees of freedom of both models, i.e. in our case $19 - 18 = 1$, resulting in the null distribution shown in Fig. 4.41, from which we can read a p-value of 0.0004. This means that our model makes a statistically significant contribution to explaining the data compared to the null model.

The value AIC stands for Akaike Information Criterion and represents the counterpart to the corrected coefficient of determination in linear models. This indicator can be used to (re-)specify generalized linear models. If two models are identical, except for a single variable that is present in one model and not in the other, the model with the smaller AIC value is preferable.

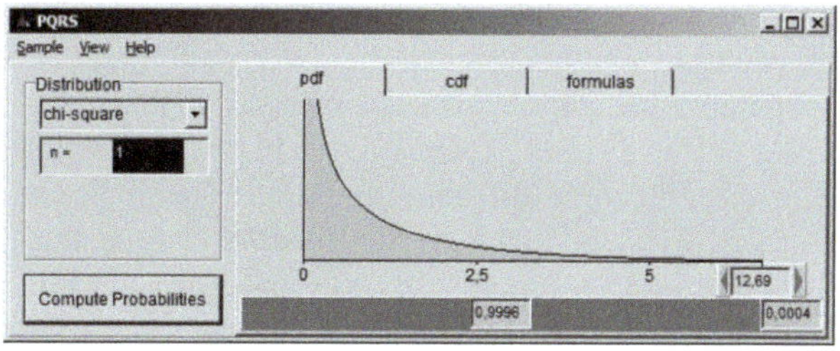

Fig. 4.41 Null distribution in the test to explain the logistic regression

Another type of dependent variable that is quite common in experimental economics is the count variable. For example, we may want to model how often subjects free ride in a repeated public-good game and what factors influence this number. Possible values of this variable to be explained would then be 0, 1, 2, 3,... and it would follow a Poisson distribution. This type of distribution can also be modeled very easily with a GLM. The only significant change required to estimate a Poisson model is to transform the link function into the natural logarithm.

It is of course not possible to present the full scope of GLMs in this section. The interested reader is referred to the original book by McCullagh and Nelder (1989). The main advantages of a GLM are:

- Many common types of variables (continuous, normal. non-normal, discrete and categorical) can be modeled.
- A non-normally distributed dependent variable does not need to be transformed into a normally distributed variable to be modeled.
- It has a high degree of flexibility in modeling due to the almost free choice of the link function.
- It is based on a maximum likelihood estimate and therefore has statistically desirable properties of the estimator.
- It is a standard feature in statistics programs, it is easy to carry out the estimation.

Important

If the variable to be explained is not continuous and/or not normally distributed, the generalized linear model (GLM) can be used. It covers a wide range of models and is included as a standard feature in all major statistics programs.

We have thus solved the first two of the four problems we mentioned earlier. However, a GLM cannot solve the third problem, that of dependent observations in multiple measurements, because it assumes that the observations are statistically independent of each other. This is the problem we will now turn our attention to.

4.7.5 Models for Statistically Dependent Observations

Statistically dependent observations are very common in experimental economics. Let us imagine that the entirety of all the observed decisions of a single subject represents a cluster or a class of decisions. In general, two types of dependence can then occur. The first is dependence *between* classes (between-class dependence, or inter-subject

4

dependence). This means that a player's decision is influenced by another player's past decisions. In a public good game, for example, this can occur quite quickly. Repeated cooperative behavior by one one member of the group can lead to another member of this group also playing cooperatively. In this case, this dependence possibly represents a concrete object of investigation and would have to be explicitly modeled. The models and methods discussed in this section, however, assume that there is no dependence between the subjects. This requirement can easily be met by a suitable experimental design. Hoffmann et al. (2015), for example, use the round-robin arrangement discussed earlier (▶ Sect. 2.8.1). Each player makes a decision several times in succession, but plays against new anonymous players in each round, meaning that the decisions between the classes are not expected to be dependent. When each player is informed about this design, the games of a session represent a sequence of independent one-shot games.

The second type of dependence occurs between the observations *within* a class (within-class dependence, or intra-subject dependence). By definition, this type of dependence cannot be avoided in repeated games. Once a subject plays a game a second time, whether against a new anonymous opponent or not, the second observation inevitably depends on the first, since it is one and the same person with the same characteristics and preferences who makes both decisions. The main problem of dependent observations within a class is that a player's subsequent decisions have less "exclusive" or "delimiting" information than the first. The more intra-class decisions are treated as statistically independent (although they are not), the greater the extent to which information that is not available is mistakenly taken into account and the more standard errors in these observations are systematically underestimated. Biased estimators of standard errors lead to biased estimators of test statistics and thus to false inferences. From a statistical point of view, therefore, within-class dependences represent a serious problem that must be solved.

The simplest approach to avoiding within-class dependence is to aggregate the data within a class by simply using the group averages as data. This inevitably leads to a loss of information about the dynamics of behavior, although it is generally precisely this information that is of particular interest when subjects are measured repeatedly. Alternatively – instead of aggregating the data directly – it is possible to correct only the standard errors according to their correlation within the class (usually upwards) (e.g. Hilbe 2009). Many software packages offer a "cluster" option for this purpose in their regression commands, with several variants with different levels of correction available:

1. *None*: No clustering, all observations are treated as independent.
2. *Individual*: A person is treated as a cluster of all x decisions made.
3. *Session*: A session is treated as a cluster of y people making x decisions.

Such adapted standard errors are called Huber-White, sandwich or empirical standard errors. When the standard errors are clustered, the regression coefficients remain unchanged and it is only the inferences that change if the correction is large enough.

In a sense, both of these approaches only represent "repair jobs". Using statistical models that explicitly incorporate the dependence of observations into the model from the outset is more elegant and consistent. The two best-known approaches are *multilevel models* (MLM) and *generalized estimating equations* (GEE). Both approaches will now be presented briefly.

Multilevel models are generally suitable for modeling grouped data. These models often have different names depending on the grouping and the science in which they are used. Sociologists, for example, often examine hierarchically arranged groups. A classic subject of research here is the connection between socioeconomic status and the school performance of pupils. A special school class then represents the lowest hierarchical level, while the next higher level is school, region, state, country, etc. Multilevel models that take this hierarchical data structure into account are called *hierarchical linear models* (HLM).

Hierarchical data structures are rare in experimental economics. It is much more common to observe changes in individual behavior over time in different treatments (*longitudinal data*). In this case, each individual subject represents a group of observations over time. Multilevel models are also suitable for this, although they will then bear different names, such as multilevel model for change (MMC) (Singer and Willett 2003). Such a model can be applied to various forms of longitudinal data. Time can be measured in any unit and the time interval between the measurements can either be fixed (each person is measured at equal time intervals) or different (each person has a different schedule). Even the number of measuring points need not be the same for all individuals.

In the simplest case, a multilevel model is used to model two levels, each representing different views of the data:

1. Level 1 (within-person, within-individual or within-subject): The relationship between time and the dependent variable within each individual person is considered. Typical questions at level 1 are: Are we observing a person with an increasing correlation between the person's responses over time? Are there people for whom this correlation is decreasing? Is the relationship between the person's responses linear? The subject of a Level 1 analysis is always the individual trajectory of a subject. A typical example is a person's contribution behavior in a public good game over a certain number of repetitions of the game.
2. Level 2 (between-person, between-individual or between-subject): In this view, the relationship between time and the dependent variable *between* the individuals is of interest. For example, a typical level 2 question is: Why does one trajectory start at a low level and another at a high level? Can we explain observed differences between the trajectories' intercepts and/or slopes using a different variable? In the public good game, for example, are women's (declining) contributions higher than men's declining contributions? Does the MPCR influence the intercept or slope of the trajectories?

The concrete way a multilevel model works will now be illustrated by means of a greatly simplified example.

Example

We have a data set of a hypothetical public good game as shown in ◘ Table 4.24. Two groups of 4 people each (2 men and 2 women) were given the opportunity to contribute to a public good over 5 rounds. The contributions were elicited every 2 minutes from the beginning of the game on. The first group's game had an MPCR of 0.3 (coded "low"), while the second group's game had an MPCR of 0.4 (coded "high"). Men were coded with the number 0 and women with the number 1.

4

Table 4.24 Hypothetical data set for a public good game

			Time of Measurement (minutes)					
subj	group	mpcr	2	4	6	8	10	gender
1	1	Low	52	48	35	36	25	1
2	1	Low	44	47	30	22	11	0
3	1	Low	35	22	15	14	8	0
4	1	Low	49	51	44	32	21	1
5	2	High	65	55	53	56	34	0
6	2	High	70	70	73	65	55	1
7	2	High	75	70	73	68	53	1
8	2	High	60	58	54	49	30	0

Before we set up a specific model, we need to choose the unit of measure of the temporal predictor in such a way that parameters estimated later can be interpreted in a meaningful way. This applies in particular to the intercept since this always represents the corresponding estimated value of the dependent variable if all the independent variables are zero. For example, obtaining an intercept of 80 in the current "minute" unit would mean that the players would contribute 80 on average at minute 0. However, no contributions were made at minute 0, since the first contributions were not elicited until minute 2. So we simply shift the time by two units by subtracting the two from all the values of the variable Minute. This variable is called Time in the following. It is also measured at two-minute intervals, but its zero point is at the time of the first contribution.[24]

The left part of Fig. 4.42 shows the course of the dependent variable over time, henceforth called the *trajectory*, for all 8 subjects. There is a downward trend in the contributions across all the individuals. Furthermore, it can be seen that the contributions between the individual subjects vary greatly, as the trajectories are at very different levels. Within an individual, however, the observations spread more or less equally, since the observations along a trajectory show approximately the same upward and downward deviations.

The first "naive", or extreme approach, to a statistical model is a simple linear model with pooled data. It has the form

$$contrib_t = a + b \cdot time_t + u_t$$
$$u_i \sim N\left(0, \sigma^2\right).$$

24 An even clearer example would be a regression of "wage level" on "age", with the first observations expected from an age of 16 years at the earliest. Without zero centering, a positive intercept of 300 euros, for example, would mean that newborns would receive an average wage of 300 euros.

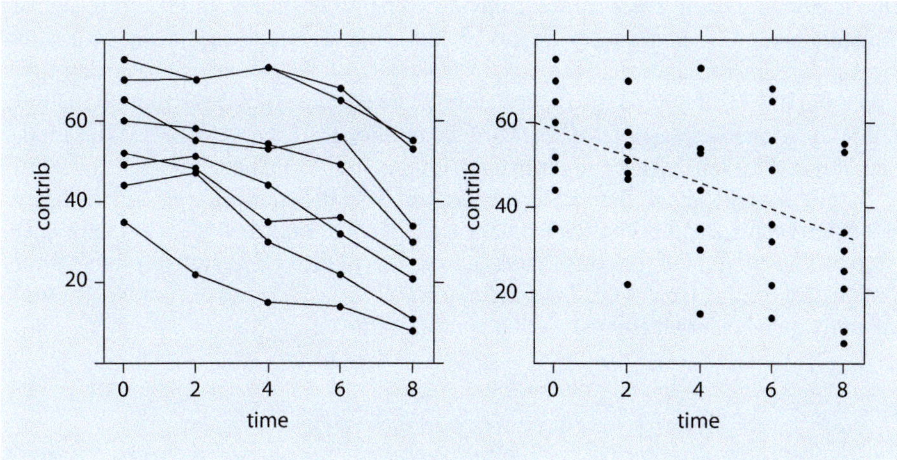

Fig. 4.42 Data of the example displayed graphically. The left figure shows the contributions of all 8 subjects over time (trajectories). In the right figure, the data was pooled and a naive least squares estimation was performed (dashed line)

Table 4.25 Naive least squares estimation with pooled grouped data

	Coef	std.err	t.value	p.value
(Intercept)	58.3000	4.6516	12.5333	<0.0001
Time	−3.1562	0.9495	−3.3241	0.002
Number of observations: 40				
Number of coefficients 2				
Degrees of freedom: 38				
R-squ.: 0.2253				
Sum of squ. resid.: 10962.9625				
Sig.-squ. (est.): 288.499				

For each of the values $time_1 = 0$, $time_2 = 2$, … $time_5 = 8$, there are now 8 *contrib* values of the 8 subjects. Since the variables no longer have an index *i*, the fact that the 8 observations at a certain point in time originate from different individuals is ignored. Implicitly, we assume that we are not dealing with individuals with different characteristics, but completely uniform individuals who start at the same level and make contributions at the same rate over time. Differences in the level of contributions are regarded as completely random. From a purely technical point of view, such a least squares estimation can be performed effortlessly and would lead to the results in Table 4.25.

The regression line for this model is the dashed line in the graph on the right-hand side of Fig. 4.42. Although this model estimates the average downward trend fairly reliably, the estimation of the intercept, in particular, is subject to a large standard error.

This is not surprising since all the individual characteristics of the subjects are not explicitly taken into account by the model, but rather are implicitly incorporated into the random disturbance u_t. The large dispersion of the observations around the regression line is, of course, also reflected in a low coefficient of determination of only 22.5% and a very large estimated spread of the random disturbance of 288.5. This model also does not explain *why* the levels of contributions vary so much.

In our example, the individual downward trends of the contributions and the aggregated downward trend of the pooled data are fairly well matched. If we are only interested in a mean trend of the contribution amounts, a pooled OLS estimate often delivers quite useful results. However, a brief excursion will make it clear that completely ignoring individual heterogeneity may well be problematic.

Box 4.5 Simpson's Paradox: It Depends on How You Look at It

Let us assume we want to estimate the relationship between the blood alcohol content (in per mille) and driving ability. Driving ability is measured in the distance in meters covered without collision in a vehicle in an obstacle course. We invite three participants, Anne, Bob and Cindy, who each complete the course with 5 different blood alcohol levels. The data set pooled across all three individuals can be seen in the left part of ◘ Fig. 4.43. If we carried out a linear regression, this would result in a rising regression line, which suggests that roadworthiness increases with increased alcohol consumption.

This false conclusion results from the failure to take account of the heterogeneity between individuals. In fact, all three subjects have a negative correlation between alcohol and driving ability (see the right-hand side of ◘ Fig. 4.43). It is only by chance that the subjects have particular individual characteristics that cause slight differences in the negative slopes and greater differences in the intercepts. For example, Cindy might have a particularly efficient liver, which ensures that she is still relatively roadworthy even with a relatively high level of alcohol consumption. Therefore, her regression line is at a particularly high level in comparison to the others. What is also not observable but nevertheless relevant for the individual intercepts is what the subjects ate before the experiment and how much. Anne might have had an empty stomach during the test, for example, whereas Cindy may have eaten a tuna pizza with double cheese beforehand. It is therefore possible that strong heterogeneity between the groups may lead to aggregated and individual views of the data resulting in contradictory statements. This phenomenon goes back to Simpson (1951) and is therefore called Simpson's Paradox.

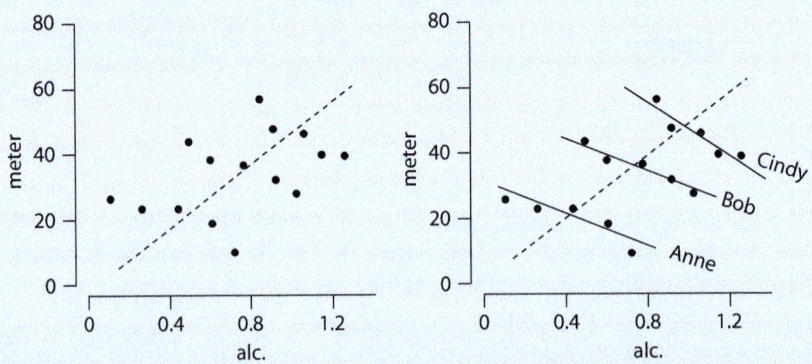

◘ **Fig. 4.43** Illustration of Simpson's paradox. Pooled data are on the left and individual regressions on the right

We have seen that the (naive) least squares method with pooled data ignores the differences between individuals and thus the additional information that can be used for a regression, while Simpson's paradox shows that this can possibly lead to wrong conclusions. The other extreme of modeling is to fully model the differences between individuals. In this case, a separate linear model would be estimated for each individual, i.e.

$$contrib_{it} = a_i + b_i \cdot time_{it} + u_{it} \quad \text{(Model FE)}$$

$$u_{it} \sim N\left(0, \sigma_u^2\right).$$

There now exists a variable $time_{it}$ for each player i at any time t. The model is based on the assumption that the true relationship between time and the amount contributed is linear for each player $i = 1 \ldots 8$. The parameters a_i and b_i are the intercept and slope of individual i, respectively. They include all the subjects' time-constant, individual characteristics that are not explicitly controlled for using a corresponding variable. Examples of unobservable characteristics are intelligence or political sentiments. Age (in years) or gender can also be regarded as constant over time, but can in principle be observed and should also be included in the model if they have an informative value. u_{it} is a random error that can happen to each player to varying degrees in each period, with a positive deviation from the true value being possible in one period and a negative one in another.

It is possible to estimate this type of fixed-effects model using *dummy variables*. These binary variables can be used to represent the heterogeneity between individuals or groups, both in terms of intercept and slope. The basic procedure is to define one group as the "base group" and estimate the differences of all other groups from this base group using a least squares estimate. The difference between two groups in intercept or slope represents the estimated parameter of the corresponding dummy variable. Among economists, this model is also known as the least squares dummy variable (LSDV). Since we only need this model as an intermediate step, we will not go into the details of an estimate here. This model belongs to the within-individual level, Level 1, mentioned earlier.

As shown in ◘ Table 4.26, each individual now has his own regression line with its own estimated intercept $\hat{a}$ and its own estimated slope $\hat{b}$. The respective regression lines of all 8 individuals can be seen in the left part of ◘ Fig. 4.44.

Since the total dispersion of the observations is now distributed over 8 individual regressions, each regression in itself provides a considerably higher amount of information for a single individual, with the coefficients of determination lying between 60% and 90%. The estimated standard deviations of the estimators, however, have not really become smaller. This is because the same total number of observations must be used when estimating a significantly larger number of parameters. For 8 individuals there are already $8 \cdot 2 = 16$ parameters. This reduces the degrees of freedom and thus the accuracy of the estimators in the form of higher estimated standard deviations. Another problem of this purely Level 1 model is that we only gain information about specific individuals. We cannot draw any conclusions about any relationship between the contribution made and time at the population level. What's more, we cannot explain the differences between

Table 4.26 Least square estimates for grouped data for each group individually. The estimated standard deviations of the estimators are shown in brackets

	Individual							
	Contrib							
	(1)	(2)	(3)	(4)	(5)	(6)	(7)	(8)
Time	−3.300***	−4.550**	−3.100**	−3.750**	−3.050*	−1.750	−2.300*	−3.450**
	(0.542)	(0.810)	(0.594)	(0.810)	(1.100)	(0.810)	(0.872)	(0.934)
Constant	52.400***	49.000***	31.200***	54.400***	64.800***	73.600***	77.000***	64.000***
	(2.653)	(3.967)	(2.912)	(3.967)	(5.387)	(3.967)	(4.271)	(4.576)
Observations	5	5	5	5	5	5	5	5
R2	0.925	0.913	0.901	0.877	0.719	0.609	0.699	0.820
Std. Error (df = 3)	3.425	5.122	3.759	5.122	6.955	5.122	5.514	5.908

Note: *p<0.1; **p<0.05; ***p<0.01

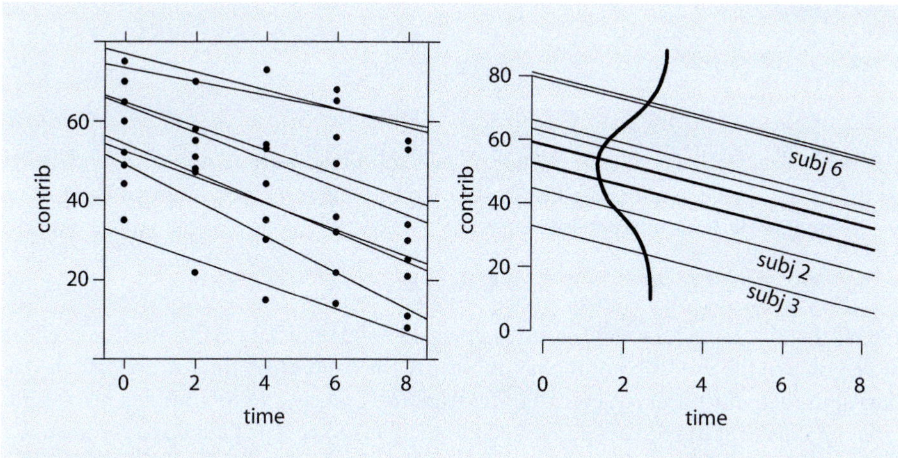

Fig. 4.44 Estimation results of the Level 1 model (left) and modeling the contribution amount using a random effect (right)

the individuals. For example, we can see that individual 5 contributes at a considerably higher level than individual 3, but we do not know how this difference comes about.

A multilevel model is a compromise between the two extreme variants of the modeling mentioned above. It allows the basic logic of individual regressions to be applied without actually carrying out these regressions (Bliese and Ployhart 2002). On the one hand, it is based on the pooling solution, since even in a multilevel model an attempt is made to keep the degrees of freedom as large as possible and to estimate as few parameters as possible. On the other hand, it does not ignore the heterogeneity between the subjects, but considers them as an explicit part of the model. It can explain not only the observations *within* an individual at level 1, but also the differences between the trajectories and the intercept and/or slope of the individuals at level 2. These differences can be traced back to a change in the level 1 parameters a_i and b_i between the individuals. The trick is to explain the differences between individuals with respect to these parameters using a separate model with random disturbance. Even if such a level 2 model is limited to linear approaches, the design of the entire model is very flexible. We will now go through all its possible variations step by step.

To begin with, we assume that a, representing all the contributions of the single individuals, is randomly spread around a constant expected value γ_a. At level 2, a new individual-specific random disturbance $\epsilon_{ai} \sim N(0, \sigma^2_{\epsilon a})$ must therefore be introduced to explain the individual slopes. The model equation describing the different intercepts is $a_i = \gamma_a + \epsilon_{ai}$. We further assume slope b is not subject to any influence and that it therefore assumes a constant value $b_i = \gamma_b$ for all individuals. The remaining part v_{it} of the total spread varies within an individual over time and is therefore the level 1 random disturbance. Overall, we then obtain the following multilevel model:

Level 1 (within-subject):

$$contrib_{it} = a_i + b_i \cdot time_{it} + v_{it}$$

(Model A)

Level 2 (between-subject):

$$a_i = \gamma_a + \epsilon_{ai}$$

$$b_i = \gamma_b.$$

This multilevel form can also be brought into a combined form (composite model). To do this, the level 2 equations are simply inserted into the level 1 equation yielding

Composite (within- and between-subject):

$$contrib_{it} = \gamma_a + \gamma_b \cdot time_{it} + \epsilon_{ai} + v_{it}.$$

We see that the composite view of the multilevel model differs from the pure fixed-effect model only in the modeling of the random part. The fixed-effect model models a deviation of the actually observed contribution from the expected value completely by using a random disturbance u_{it} that individual i was subject to at time t alone. In contrast, the multilevel model (model A) divides this deviation into a "between" component and a "within" component. The component ϵ_{ai} is responsible for random differences between the individuals with respect to the intercepts, but is the same for a given individual in all periods. Since this effect on the intercept is random, it is called the (individual-specific) *random effect*. The component v_{it} is the remaining part of the total dispersion that causes random deviations over time within a given individual i.

To illustrate this relationship, we look at the regression results of model A in ◘ Table 4.27 and ◘ Fig. 4.45.

In the column Random effects of ◘ Table 4.27, we can see that the random effect of subject 3 is exactly −26.36. In ◘ Fig. 4.45, this value corresponds to the vertical distance

◘ **Table 4.27** Estimation results of model A

Random effect parameters:						Random effects:	
Groups	Name		Variance	Std.Dev.		(Intercept)	
subj	(Intercept)		283.115	16.8260		1	-6.35
Residual			27.735	5.2664		2	-14.59
						3	-26.36
Number of obs: 40, groups: subj, 8						4	-6.15
						5	6.79
Fixed effect (parameters):						6	20.52
	Est.	SE	df	t	Pr(>|t|)	7	21.70
(Intercept)	58.30	6.12	7.55	9.52	<0.0001***	8	4.44
time	-3.16	0.29	31.00	-10.72	<0.0001***		

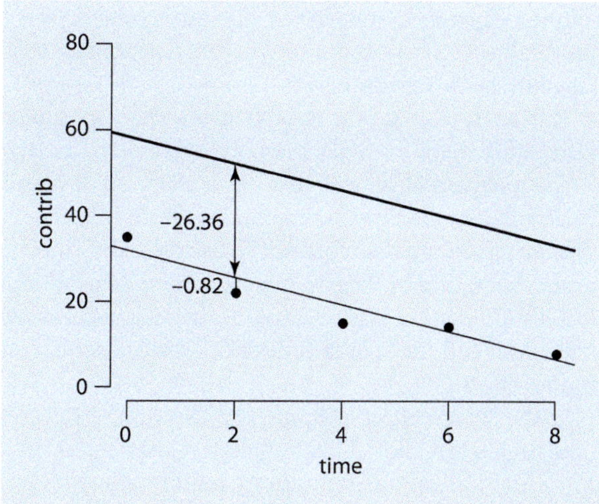

Fig. 4.45 Splitting the total deviation of subject 3 at time 2 into a between-group part and a within-group part

between the average population regression line of the model (bold) and the individual straight line of the subject $i = 3$ (thin). This difference is due to a randomly occurring, time-constant characteristic of this subject. As a result, he randomly contributed 26.36 less ($\epsilon_{a3} = -26.36$) over all periods than the mean of all the subjects. This part of the total deviation of the actual observation from the mean regression line corresponds to the random between-group deviation that each individual is subject to independent of time. On the other hand, the deviation within this individual $i = 3$ at time $t = 2$ was $v_{32} = -0.82$, resulting in a total deviation of $u_{32} = -26.36 - 0.82 = -27.18$ at time $t = 2$.

Random effects can in principle arise from variables whose observed values represent only a small part of a larger population (e.g. Bolker et al. 2009). For example, a time-independent random disturbance might happen to be a subject's particularly good mood, which happens to ensure that this individual's level of contributions is higher than usual over all rounds. However, in order to be considered a variable that generates a random effect, it would certainly have to be able to assume more than the two values "good" and "bad", according to the above condition, since its influence would otherwise be modeled as a fixed effect. Which variable is actually used is ultimately of no importance for the model. A random effect can generally incorporate the influences of all those variables that have a random and time-constant impact on the individual parameters a_i or b_i, and that are at the same time either not completely observable or not a direct object of investigation.

Having chosen to model the different contribution levels using a random effect would mean that repeating the experiment with new subjects would randomly rearrange the respective trajectories in a vertical direction. To illustrate this, we look at the right part of Fig. 4.44, where we see the entire regression result presented graphically. Using the bold population regression line as a starting point, each subject's individual regression line deviates up or down by the respective value of the random effect. In this case, subject 3 would have started the day particularly badly and subject 6 would have started particularly well, thus achieving low or high contribution levels, respectively, over all the rounds. Carrying out the experiment again would be equivalent to a new random occurrence of deviations for each subject. It is then quite possible that in this round subject 6's contributions will be particularly low while subject 3's contributions at a medium level over time.

This analysis makes it clear that the specific value of a random effect is not very meaningful. If, for example, subject 6 makes particularly high contributions over all the rounds today because he is in a particularly good mood, then this does not help us in explaining the observed contributions, since this effect was positive purely by chance. Random effects are therefore not characterized by their specific values at a certain point in time, but rather by the parameters of their distribution. The expected value, for example, tells us whether positive and negative deviations currently balance each other out on average, and the standard deviation provides information about the strength of the random effect.

In the "Fixed effect (parameters)" section of ◘ Table 4.27, we find the data on the population regression line. It has a y-axis intercept of $\hat{\gamma}_a = 58.3$ and a slope of $\hat{\gamma}_b = -3.16$. Thus, this variant of a multilevel model provides the same regression line as the naive least squares estimation with pooled data. So what is the difference between the two models? To understand this, we recall how the respective models model a random disturbance. The naive least squares estimate with pooled data assumed that at a time t there was a random error u_t. All the factors influencing a contribution amount that we could not control for were incorporated into this random disturbance – including the various personal characteristics of the subjects. What our grouped data structure provides, however, is in fact information about the *values* of the different personal characteristics of the subjects. For example, a subject $i = 1$ happens to contribute at a relatively high level over time, while another subject $i = 2$ happens to make a relatively low level of contributions. This information can be explicitly incorporated into the modeling of the random disturbance. In concrete terms, this is achieved by dividing the disturbance variable into two components, as mentioned earlier. Model A splits the total random influence u_t, which does not differentiate between individuals, into u_{it}, which applies specifically to individual i at time t (within-group) and an independent part ϵ_i, which impacts on individual i at all times (between-group). Since both components add up to u_t again, there is no quantitative difference between the two models in the estimation of the intercept and slope parameters at the population level.

The difference between the two models is more a matter of a *qualitative* nature. The naive least squares estimation is "naive" because it is based on the independence of the observations, although they are not independent. In other words, the naive least squares estimation assumes that the subsequent observations of an individual provide as much information for an estimate as the first observation. In fact, the amount of new information provided by observations within an individual becomes less and less over time. This error is inevitably associated with a systematic underestimation of the standard errors of both the estimators intercept and slope. The value of 4.65 for the intercept provided by the naive model must therefore be regarded as too small. Model A avoids this error by means of a more differentiated modeling of the disturbance variable and therefore shows a higher, unbiased value of 6.12. Specifically, model A introduces the time-independent disturbance ϵ_{ai}, whose impact on the contributions is the same in all periods. Thus the disturbances of one individual are "linked" over time, so that the disturbance of one period can no longer be independent of the disturbance of another.

However, precisely this *autocorrelation* is the property we need for modeling with grouped data and which a naive least squares estimation does not provide.

In this context, the question that arises is how to judge whether the use of a differentiated disturbance variable instead of a normal disturbance variable is justified. To this end, we look again at the random effects of the eight subjects in ◘ Table 4.27. The random vertical deviation of the trajectory, for example, of subject 1 from the regression line is $\epsilon_1 = -6.35$ units, that of subject 2 $\epsilon_2 = -14.59$ units and that of subject 3 $\epsilon_3 = -26.36$ units. The greater these deviations are on average across all subjects, the greater the variance between the groups and the greater the strength of the random effect. In our example, we obtain a value of $\sigma_{\epsilon a}^2 = 283.12$. The remaining share of the total variance, which cannot be explained by a random dispersion of individual characteristics between the subjects, is $\sigma_v^2 = 27.74$. Thus 283.12/(283.12 + 27.74) = 91.08% of the total dispersion can be explained by a dispersion between the groups. This number is called the intraclass correlation coefficient (ICC). It provides information about how strongly the observations within a group or an individual are correlated, and thus also about how large the error would be if a non-differentiated disturbance variable were used. In our case, the correlation of observations within the groups is extremely high. This can be seen, for example, in the observations of individual 3 in ◘ Fig. 4.44 (left). The first observation is clearly below the regression line, as are the second, third, fourth and fifth. For another individual, on the other hand, all five observations are clearly above the regression line. In order for the ICC to become smaller, the observations within an individual would have to scatter more and "cross" the regression line from time to time. In the (theoretical) extreme case of an ICC = 0%, there would be no difference between an ordinary least squares estimation and model A.

To conclude the interpretation of model A, we will look at the estimated standard deviation of the slope and compare it with that of the pooling model. It is striking that the value of 0.29 is significantly lower than the value of 0.95 in the naive model. The reason is that model A is based, first of all, on different individuals and, second, on the same "true" slopes of all the individuals. If the experiment were to be repeated frequently, new estimated slopes would always result, which would, however, be of equal magnitude between the subjects. Under these conditions, the true mean slope can be estimated more precisely than if the slope between the individuals after each repetition were also different. In another model, we will examine to what extent this assumption is justified. But if we look at the individual trajectories of the subjects, it is already possible to say that it is at least not completely unrealistic.

We will now assume that the levels of the contributions no longer vary completely randomly, but can at least partly be explained by another variable. For example, if we assume that men and women generally contribute different amounts, then one explanatory variable for the difference in levels in the individual trajectories could be sex or the factor variable gender. This results in a complete model, which is again composed of two different levels:

Level 1 (within-subject):

$$contrib_{it} = a_i + b_i time_{it} + v_{it}$$

$$(\text{Model B})$$

Level 2 (between-subject):

$$a_i = \gamma_a + \delta_a gender_i + \epsilon_{ai}$$

$$b_i = \gamma_b.$$

γ_a is the true mean male contribution (i.e. *gender* = 0) in the first round of the game (i.e. *time* = 0). The parameter δ_a specifies the *change* in this value when *gender* changes from 0 to 1. Thus, $\gamma_a + \delta_a \cdot 1$ is the true mean female contribution in the first round of the game. The actual amount a_i contributed by a player i in the first round (be it that of a man or woman) randomly differs from the true value by $\epsilon_{ai} \sim N\left(0, \sigma_\epsilon^2\right)$. We thus obtain the combined form

$$contrib_{it} = \gamma_a + \gamma_b time_{it} + \delta_a gender_i + \epsilon_{ai} + v_{it}.$$

The individual contribution level a_i is now not only explained using the random effect ϵ_{ai}, but also using the *fixed effect* δ_a. Fixed effects are characterized by the fact that their values have been observed and are of direct interest to the experimenter. In our case, we have a particular assumption about the effect of gender on contribution levels and would like to quantify this effect. In addition to other collected variables with only few characteristics, such as "gender", "origin" or "foreign languages spoken", treatment variables, such as the MPCR specified by the experimenter, are classic examples of variables with a fixed effect. A linear model in which level 1 parameters are modeled using both a random effect and a fixed effect is called the linear mixed (effect) model (LMM).

▫ Figure 4.46 illustrates how this model works. The dashed lines show the regression line at the population level (thick) and the random effect deviations (thin) of the women, while those of the men are solid. We see that the women's estimation line is above that of men. This suggests that women on contribute more than men. If we were to once again repeat the experiment with new subjects, the trajectories would not change completely randomly in a vertical direction, in contrast to a pure random effects model at

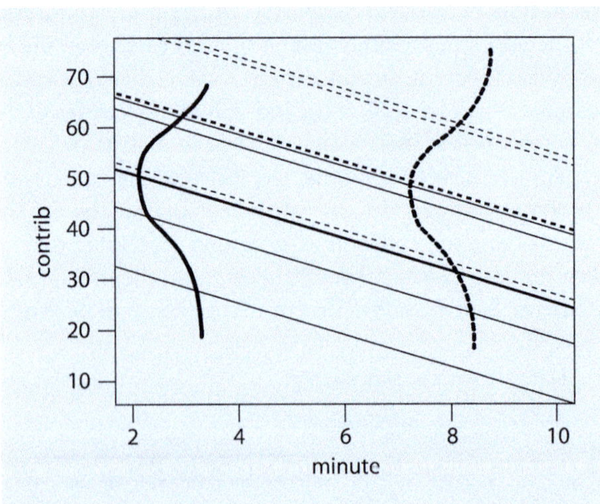

▫ **Fig. 4.46** Modeling the differences in level of the contributions using the fixed effect gender and a random effect (model B). The trajectories of the women are shown dashed and the trajectories of the men are solid

level 2. Instead, the fixed effect ensures that the trajectories of women in repeated experiments lie in an area above the trajectories of men. The still existing random effect does, however, cause the vertical arrangement of the trajectories *within* the group of men (lower area) and the group of women (upper area) to be random, as before.

More detailed information on both effects using the estimation of model B is provided in ◘ Table 4.28. It is hardly surprising that the estimate for the slope is the same as in model A, because we are still only modeling the variations in the intercepts not the slope. The intercept of 58.30 is modeled in a more differentiated way. The intercept for men is $\hat{\gamma}_a = 50.73$ and the estimated fixed effect of the variable gender when changing from male (gender = 0) to female (gender = 1) is $\hat{\delta}_a = 15.15$. Thus the estimated intercept for women is 50.72 + 15.15 = 65.87.

It is striking that the effect of gender is not significant, although the difference in the mean contributions between men and women is over 15 percentage points! Here again, the difference between economic and statistical significance can be clearly seen. After all, such a large effect is economically significant in the sense of "relevant". Situations such as this always tend to cause interpretation and specification errors if there is too much focus on the *p*-value and nobody questions what this number means or how it came about. ◘ Figure 4.46 clearly shows that it is not only the fixed effect of 15 percentage points that is very large (vertical distance between the two bold lines), but also the dispersion within the two groups "men" and "women". Large random effects within the groups mean that their values can easily overlap. Then some dashed straight lines of the women (upper area) fall into the lower area of the men (solid straight lines) and vice versa. However, if the two areas between men and women cannot be distinguished clearly enough because they overlap too much, it is difficult for a significance test to identify an actual existing

◘ **Table 4.28** Estimation results of model B

Random effect parameters:						Random effects:	
Groups	Name	Variance	Std.Dev.			(Intercept)	
subj	(Intercept)	254.718	15.9599			1	−13.75
Residual		27.735	5.2664			2	−7.14
Number of obs: 40, groups: subj, 8						3	−18.89
Fixed effect (parameters):						4	−13.55
	Est.	SE	df	t	Pr(>|t|)	5	14.19
(Intercept)	50.73	8.15	6.26	6.22	0.0007***	6	13.07
time	−3.16	0.29	31.00	−10.72	<0.0001***	7	14.24
gender1	15.15	11.41	6.00	1.33	0.2324	8	11.84

effect between these groups. In this case, the power of a test that we discussed earlier is small and its alternative term "sensitivity" takes on a clear meaning.

If we now equated economic significance with statistical significance, we could be tempted to remove the variable gender from the model. However, this would be a clear specification error because, as we will see in a moment, this variable is both economically and statistically significant. We already know that the wide dispersion within the two groups ensures a low power and thus a large Type II error. But why is the dispersion so large? Well, it takes in differences between individuals that are not explicitly specified by their own variable, i.e. for which there is no control. Thus, if the dispersion within the groups is unexpectedly large relative to the effect between the groups, this indicates that the model is under-specified. We therefore must not take variables or information out of the model, but precisely the opposite. To show this, we will now explain the differences in the intercept using another fixed effect variable.

The most obvious variable to be included in the model is mpcr. After all, it is not "only" a covariable of the subjects, such as gender, but a treatment variable explicitly determined by the experimenter. We are therefore particularly interested in the effect of this variable. One research question could be: "Can a difference in contribution levels (partially) be explained by a difference in MPCRs? We add the variable mpcr at level 2 and obtain the model:

Level 1 (within-subject):

$$contrib_{it} = a_i + b_i time_{it} + u_{it}$$

$$(\text{Model C})$$

Level 2 (between-subject):

$$a_i = \gamma_a + \delta_a gender_i + \tau_a mpcr_i + \epsilon_{ai}$$

$$b_i = \gamma_b.$$

The main technical difference between model C and model B is that another parameter τ_a is estimated at level 2. The results are shown in ◘ Table 4.29.

The effect of including mpcr in the model is very obvious: changing it from "high" to "low" is accompanied by an average decrease of 27.25% points in the amounts contributed (for both men and women). The variance of the random effect falls from 254.72 to 9.74. The resulting higher power can be clearly seen in ◘ Fig. 4.47 (left). All 4 groups (men with low/high MPCR and women with low/high MPCR) are clearly distinguishable. No straight line of one group projects into the area of another group. Thus, even a hypothesis test has no problems showing an actual effect. Both the effect of gender and mpcr is statistically significant with p-values substantially less than 5%.

Important

***p*-values are not suitable for specifying models. "Non-significance" does not necessarily mean "irrelevance".**

Table 4.29 Estimation results of model C

Random effect parameters:						Random effects:	
Groups	Name		Variance	Std.Dev.		(Intercept)	
subj	(Intercept)		9.746	3.1219		1	−0.27
Residual			27.735	5.2664		2	4.03
Number of obs: 40, groups: subj, 8						3	−3.62
Fixed effect (parameters):						4	−0.14
	Est.	SE	df	t	Pr(>|t|)	5	0.56
(Intercept)	64.35	2.67	7.64	24.11	<0.0001***	6	−0.18
time	−3.16	0.29	31.00	−10.72	<0.0001***	7	0.59
gender1	15.15	2.77	5.00	5.48	0.0028**	8	−0.97
mpcrlow	−27.25	2.77	5.00	−9.85	0.0002***		

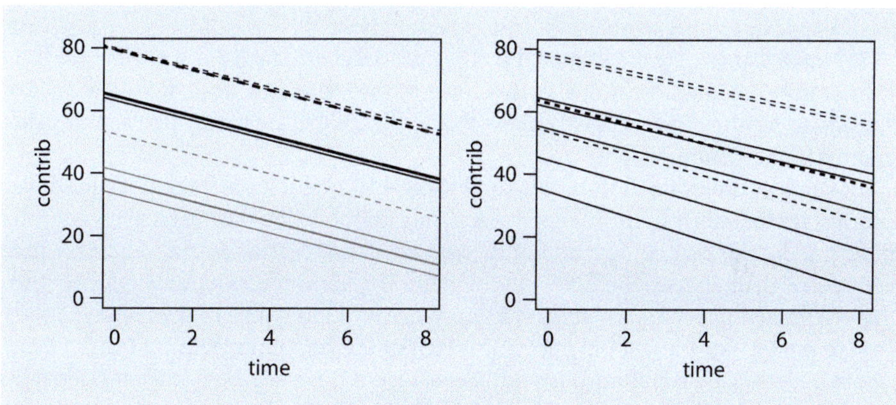

Fig. 4.47 Graphical representation of model C (left) and model D (right). In model C, the treatment "MPCR low" is gray and the treatment "MPCR high" is black

Last but not least, we might wonder whether the variations in the slopes also need to be explained. To this end, we will introduce another random effect, using model B as a starting point, to explain the slope at level 2. We thus have model D:

Level 1 (within-subject):

$$contrib_{it} = a_i + b_i time_{it} + u_{it} \qquad \text{(Model D)}$$

Level 2 (between-subject):

$$a_i = \gamma_a + \delta_a gender_i + \epsilon_{ai}$$
$$b_i = \gamma_b + \epsilon_{bi}.$$

γ_b is the true mean rate at which the amounts an individual contributes change over time and $\epsilon_{bi} \sim N(0, \sigma_{\epsilon b}^2)$ is the time-independent, random deviation from this value. We obtain the following in the combined form

$$contrib_{it} = \gamma_a + \delta_a gender_i + \gamma_b time_{it} + \epsilon_{ai} + \epsilon_{bi} time_{it} + u_{it}.$$

The first line shown is the deterministic part of the model and the second line is the random part. A new component is noticeable in the random part: $\epsilon_{bi} time_{it}$ is a random variable that cannot vary completely freely, but whose value depends on the period we are in. Such a structure of values can only be generated by a random variable whose distribution has a non-constant variance. Thus, compared to model B, model D is not only able to model autocorrelation, but also *heteroskedasticity* – another property that is not available in a conventional least squares estimation.

Figure 4.47 (right) shows model D. The only difference to Fig. 4.44 or model B is that not only the intercepts but also the slopes of the individual straight lines may vary at random. Although we explicitly allow this in the model, we hardly perceive any differences in the slopes.

Looking at the estimation results in Table 4.30, we first notice that we have two columns for the values of the random effects, one for modeling the intercept and one for modeling the slope. A comparison of the two columns reveals that the values are perfectly correlated (also see the Corr column in the "Random effects parameters" section). This phenomenon occurs when there is too little variation in one of the parameters we want to model using a random effect. This lack of information means that both effects cannot be estimated simultaneously and the statistics program then generates perfectly correlated values of the random effects. In fact, model D is "overparameterized"; with the introduction of the random effect for the slope, we are trying to find an explanation for an effect that probably does not exist. Our data set suggests a constant rate of contributions across all individuals. In this case, a pure random intercept model, such as model C, would be preferable.

For the sake of completeness, the variation of the slope can also be modeled using a mixed effect. However, since the same problems would then arise as in model C, we will not pursue this idea further at this point.

Table 4.30 Estimation results of model D

Random effect parameters:						Random effects:		
Groups	Name	Variance	Std.Dev.	Corr			(Intercept)	time
subj	(Intercept)	195.02	13.97			1	−8.91	−0.33
	time	0.25	0.50	1.000		2	−9.49	−0.36
Residual		25.07	5.01			3	−19.32	−0.72
Number of obs: 40, groups: subj, 8						4	−8.84	−0.33
Fixed effect (parameters):						5	9.29	0.35
	Est.	SE	df	t	Pr(>|t|)	6	14.61	0.55
(Intercept)	54.28	7.04	6.52	7.71	0.0002***	7	15.50	0.58
time	−3.16	0.34	10.65	−9.38	<0.0001***	8	7.17	0.27
gender1	8.05	9.63	6.54	0.84	0.43			

Box 4.6 Interpreting the Coefficients: Marginal Versus Conditional

There are two different ways of interpreting estimated parameters in multilevel models. In order to better differentiate these from each other, we first look at the simple linear model $y_i = a + bx_i + u_i$ or $y_i = \eta + u_i$ with $u_i \sim N(0, \sigma^2)$. Since the disturbance variable has an expected value of zero, the expected value of the observations y_i corresponds to the linear predictor, i.e. $E(y_i) = \mu_i = \eta = a + bx_i$. An estimated parameter of the model always applies to the estimation of the "true values" a and b in this mean value. For example, a coefficient $\hat{b} = 3$ means that the expected value increases by three units if x (ceteris paribus) is increased by one unit. Thus the estimated value $\hat{b} = 3$ represents the slope of the regression line $E(y_i) = \hat{a} + \hat{b}x_i$. The concept of the *marginal effect* was developed from this interpretation of a slope since the slope of a function at one point, especially in the case of non-linear functions, is specifically the ratio of a marginal change in the y-value to a marginal change in the x-value. If a model has such an interpretation, we often use the term *marginal model*. Another characteristic of the simple linear model is that the marginal effect provides a prediction at the population level and not for a specific individual or group. In this case, this is referred to as a *population-average model*.

Since the multilevel model is more nuanced than the simple linear model, the interpretation of the coefficients must also be more nuanced. For example, we consider the simplest variant $y_{it} = \gamma_a + \gamma_b x_{it} + \epsilon_{ai} + v_{it}$, in which the intercept a is modeled using a random effect ϵ_{ai} (model A). The expected value is then $E(y_{it}) = \gamma_a + \gamma_b x_{it} + \epsilon_{ai}$. We see that the expected value does not correspond to the linear predictor, as it does in the simple linear model. Rather, the actual value depends on the value of the random effect ϵ_{ai}. In contrast to a population-average model, the slope can therefore only be interpreted conditionally for a given individual with a given random effect ϵ_{ai}. Graphically, this means interpreting the straight line of a particular individual. The population-averaged view is obtained by once again forming the expected value via the random effect. Then the expected value $E(E(y_{it})) = \gamma_a + \gamma_b x_{it}$ corresponds to the linear predictor with reference to the population regression line. Multilevel models therefore include both a conditional, or subject-specific, and a marginal, or population-averaged, interpretation. Some statisticians see this as an advantage over pure population-average models (e.g. Lee and Nelder 2004).

Population-averaged models are often estimated using the *generalized estimating equations* (GEE) method (Liang and Zeger 1986). The idea behind this is to extend the estimation of generalized linear models (GLMs) so that it can also be applied to longitudinal data and the associated within-subject dependence of the observations. For this purpose, an a priori correlation structure of the observations between two points in time is defined, with this describing the dependency of the observations over time as well as possible. Assuming we consider three points in time t over which an individual was measured, and ρ denotes the correlation measure between the observations of two points in time, then different correlation structures can be depicted as a matrix:

1. *Independent:* The observations of any two points in time have no correlation.

$$\begin{pmatrix} 1 & 0 & 0 \\ 0 & 1 & 0 \\ 0 & 0 & 1 \end{pmatrix}$$

2. *Exchangeable:* The observations of any two points in time have the same correlation.

$$\begin{pmatrix} 1 & \rho & \rho \\ \rho & 1 & \rho \\ \rho & \rho & 1 \end{pmatrix}$$

3. *m-dependent:* The observations with an interval of one period have the same correlation ρ_1, with an interval of two periods they have the same correlation ρ_2, and with an interval of m periods they have the same correlation ρ_m. In this case: 1-dependent:

$$\begin{pmatrix} 1 & \rho_1 & 0 \\ \rho_1 & 1 & \rho_1 \\ 0 & \rho_1 & 1 \end{pmatrix}$$

4. *Autoregressive:* The dependencies between two observations decrease exponentially with each period that lies between them.

$$\begin{pmatrix} 1 & \rho^1 & \rho^2 \\ \rho^1 & 1 & \rho^1 \\ \rho^2 & \rho^1 & 1 \end{pmatrix}$$

5. *Unstructured:* Each pair of observations has its own correlation.

$$\begin{pmatrix} 1 & \rho_1 & \rho_2 \\ \rho_1 & 1 & \rho_3 \\ \rho_2 & \rho_3 & 1 \end{pmatrix}$$

It is remarkable that the selection of the correlation structure has no influence on the consistency of the estimators, but only on their efficiency. With a sufficiently large number of observations, therefore, the decision for a particular correlation structure plays a subordinate role. Nevertheless, GEEs with different correlation structures are estimated in practice and a sensitivity analysis is carried out with the individual estimates. It should also be mentioned that the estimators in a GEE are determined using the quasi-likelihood method and a GEE does not always have an inner solution.

If we look at the case of a linear model, we notice that the estimators and their interpretation of an LMM and a GEE do not differ. For example, if we estimate the generalized linear model with the linear predictor $\mu_i = a + b_1 time_t + b_2 gender_t + b_3 mpcr_t$ and the link function $g(z) = z$ in the GEE (with the "exchangeable" correlation structure), we get the results shown in ◘ Table 4.31. The coefficients correspond to those of the multilevel model C we have already discussed.

4

Table 4.31 Estimation results of the generalized estimating equations method

Mean Model:

Mean Link: identity

Variance to Mean Relation: gaussian

Coefficients:

	estimate	san.se	wald	p
(Intercept)	64.35	1.94	1096.75	<0.0001
time	-3.16	0.28	124.35	<0.0001
gender1	15.15	2.19	48.03	<0.0001
mpcrlow	-27.25	2.19	155.38	<0.0001

Box 4.7 The "Best of Both Worlds": Longitudinal Data with a Discrete Variable to be Explained

Let us summarize briefly. Generalized linear models can model data with non-normally distributed discrete endogenous variables. Linear multilevel models and (linear) generalized estimating equations, on the other hand, are able to adequately model dependent observations within a group. The question now arises as to what happens if both special cases occur simultaneously. How, for example, would the dichotomous acceptance behavior of the responder have to be modeled in the ultimatum game if the same person has to decide on acceptance and rejection of the offer over several rounds? This step is particularly simple for generalized estimating equations since they are already based on the generalized linear model, which in turn is designed from the outset to be able to explain non-normally distributed endogenous variables. Instead of selecting the identity function as a link function, we could simply use a parameter (e.g. link = "logit") to instruct a GEE estimation command to use the logit function as a link function. In this way, we can quickly and easily perform a logistic regression with dependent observations.

One disadvantage of this approach is that with a GEE we always "only" get a marginal, population-averaged model. If we want an analog *conditional* model, there are two possibilities. One possibility is to extend the generalized linear model again – in this case by adding individual-specific effects. Thus a GLM (generalized linear model) becomes a GLMM (generalized linear MIXED model). The second possibility is to modify the multilevel model so that it can also be applied to variables that are not normally distributed. This is mainly done by loosening the assumption of normally distributed disturbance variables and introducing a link between the expected value and the linear predictor. In this case, we generalize the LMM (linear mixed model) to the GLMM (generalized linear MIXED model). If OR denotes the odds ratio $p/(1-p)$, then in the simplest case we obtain the *conditional* model with only one random effect ϵ_{ai} for the intercept a

$$OR = \exp\left(a + bx_{it} + \epsilon_{ai}\right).$$

In this model, the odds ratios depend on what the individual-specific value of the random effect ϵ_{ai} is. In terms of the expected value, the "mean" OR in the population-averaged interpretation of the conditional model is then

$$E\left(OR\right) = \exp\left(a + bx_{it}\right) \cdot E(\exp\left(\epsilon_{ai}\right))$$
$$= \exp(\eta_{it}) \cdot E\left(\exp\left(\epsilon_{ai}\right)\right).$$

In the GEE estimate, however, the corresponding *marginal* model is

$$OR = \exp(\eta_{it}).$$

We can see from this that the mean estimate of the odds in GLMM and GEE differ by the factor $E(\exp(\epsilon_{ai}))$. As we already verified using an example, both methods lead to equivalent estimates of the parameters if the link function is a linear function lin(·). In this case, the mean OR in the conditional model $E(OR) = E(\text{lin}^{-1}(a + bx_{it} + \epsilon_{ai})) = \text{lin}^{-1}(a + bx_{it}) + dE(\epsilon_{ai}) = \text{lin}^{-1}(\eta_{it})$, which would also correspond to the OR in the marginal model. However, estimating ORs or probabilities with a linear model is rather unusual because this can lead to difficulties in interpretation such as probabilities greater than 1 or less than 0.

Discussing GLMMs in detail at this point would again go beyond the scope of the book. Interested readers are once more referred to the relevant literature. For example, Rabe-Hesketh and Skrondal (2010) provide a fine and compact review article. The same authors have also written a textbook, which is also a good source for more in-depth information on GLMMs (Rabe-Hesketh and Skrondal 2004).

4.7.6 Models with Limited Dependent Variables

Limited variables have a natural or defined upper or lower limit (or both). For example, the number of days an employee works has the lower limit zero and the upper limit 365. Such variables are particularly common in economic contexts since many economic variables can only be interpreted for a limited interval (demand, production quantity, capacities, etc.). A basic distinction is made between *censored* and *truncated* data. ▫ Table 4.32 summarizes the main differences.

▫ **Table 4.32** Comparison of censored and truncated data

	Censored	Truncated
Boundaries	Permitted and observed	Not permitted (even if observed)
No. of observations	n	$<n$
Information loss	Moderate	Greater
Mean	Tends to boundary	Doesn't tend to boundary

Example
A lecturer would like to know whether there is a connection between the lateness y_t of his 50 students in minutes and the distance of their commute to college x_t in meters. The actual data of all 50 students are shown in ▫ Fig. 4.48.

The only problem is that the lecturer himself is regularly a good 5 minutes late for his own lecture. Therefore, of the students already present he does not know whether, and if so, by how much, they were late. He can only record the times of arrival of the students from the time of

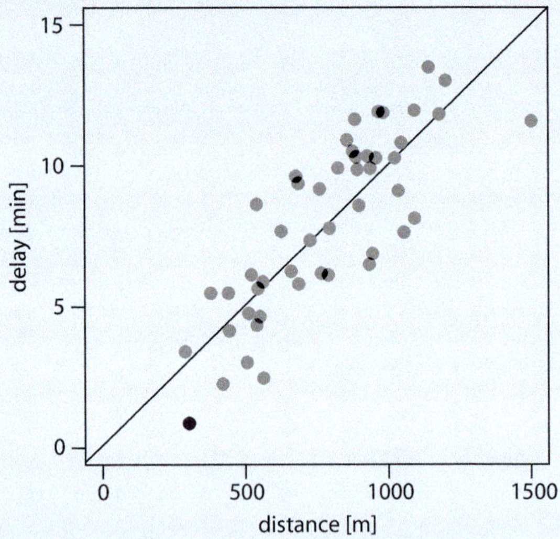

Fig. 4.48 Non-bounded data of the example with least squares regression line

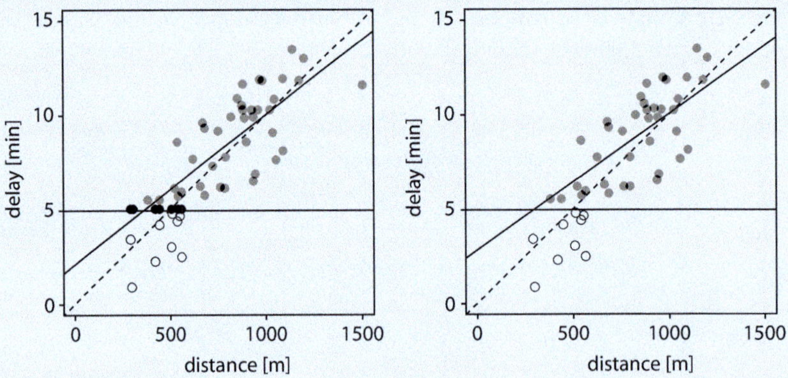

Fig. 4.49 Censored (left) versus truncated data (right). In the case of censored data, all the observations below the horizontal line are raised from their original position (grey circle) to the level $y_i^{\min} = 5$. Information about the explanatory variable (commute to college) is retained, whereas information about the variable to be explained is lost. In the case of truncated data, all the observations below the horizontal line are removed from the data set, i.e. information on both variables is lost. The dashed line shows the regression line that would have resulted from an unbounded endogenous variable

his own arrival. So how does he deal with the students already present? Now, he can at least use the little information he has on these students. He knows, for instance, that none of the students present is more than 5 minutes late. The lecturer therefore makes no distinction between students who are completely punctual ($y_i = 0$) and those who are late by 5 minutes ($y_i = 5$) or less. All these students are assigned the lower lateness value $y_i^{min} = 5$. This process corresponds to a censorship of the "outlying" data. Sometimes this is referred to as bottom-coding (or top-coding for an upper limit). Graphically, this means a vertical shift of all unobserved data points with a value of $y_i < 5$ up to the level of $y_i^{min} = 5$ (see ▫ Fig. 4.49, left). All 50 students are still included in the data set, but the exact y_i value of some of them is not known – only that it is a maximum of 5 minutes.

Another possibility would be to remove the students already present from the data set. In this case, the data set would be "cut off" and a *truncated data set* would result (see ▫ Fig. 4.49, right).

▫ Figure 4.49 clearly shows that an ordinary least squares estimator would be biased for both censored and truncated data. The intercept would be systematically overestimated and the slope parameter systematically underestimated. If the lecturer in the above example were to have the exact times of the students already present, it would therefore be best to use them. Whenever the endogenous variables have been collected for all the observations and these values are basically realistic, there is of course no point in censoring or truncating them afterwards. In the example, the instructor could simply avoid the problem of limited variables by arriving punctually himself.

In other cases, however, circumventing this problem is not so easy. Imagine, for example, ticket sales for the final of the soccer World Cup.

The stadium has a "natural" upper capacity limit y^{max}; if the actual demand for tickets is higher than this limit, then for this observation it is only known that the demand is higher y^{max}, but not exactly how much higher. In this case, the observation would have to be either censored or truncated to y^{max} or removed. The demand for tickets is a *latent* variable in this case. It corresponds to the number of tickets sold up to the capacity limit, but is no longer observable above this limit.

Another example is the limitation imposed by the state on the speedometers in all US vehicles of the 1980s to 85 mph. No matter how fast the vehicle could actually go, while driving at the actual maximum speed it was impossible to observe a speed greater than 85 mph. The actual speed in this case is the latent variable that is not fully observable. If the data were to be censored, this would mean that the study would be carried out with all vehicles, but a large number of observations (especially of sports cars and motorcycles) would accumulate at the 85 mph limit and by definition cannot exceed it. Truncating the data, on the other hand, means that the test is only carried out using the vehicles that actually have a maximum speed of less than 85 mph.

It is important to note that special regression methods do not have to be applied if a lower and/or upper limit theoretically exists, but this is either never or hardly ever reached. For example, if the worst possible score in a math test is 0 and the best 100, and no student reached these extremes at the same time, the estimators are not biased. The more data points lie at the limits (in relation to the other data points), the more serious the adverse consequences are if the problem is ignored.

Models that can handle limited data are generally called limited dependent variable models. The best-known model suitable for censored data is the *Tobit model*. It requires non-negative observations with one or two recognized data limitations and, like the logit and probit models, can be formulated as a latent variable model. In our lecturer example, the dependent variable y is assigned the value of the partially observed *latent variable* y^* (actual lateness), if this could be observed, i.e. if $y^* > 5$, then

$y = y^*$ (fully observable). In contrast, if $y^* \leq 5$ then $y = y^{min} = 5$ holds, which represents a partial observation, since we still have the information about the commute of all the students. In short, this means

$$y = \max\left(y^{min}, y^*\right).$$

In the case of the latent variable y^*, we assume the linear relationship

4

$$y^* = a + bx_i + u_i.$$

The disturbance variable u_i is bound to the same conditions as in the simple linear model. However, it can be shown that the Tobit estimate is more sensitive to violations of these assumptions than the least squares estimate in the linear model (especially a violation of homoskedasticity). Checking the disturbance assumptions is therefore even more important in the case of Tobit models than it already is in the case of the normal linear model.

Table 4.33 shows three columns with one regression each. The first column shows an ordinary least squares estimate with complete information, i.e. we can fully observe all the data points (lecturer arrives on time). This estimate is the reference. The second column shows the least squares estimate for the censored data set. Even if the slope differs by "only" 0.002 in absolute terms, this difference has a strong effect on the relationship due to the measurement of the commute to college in meters (and the associated extreme scaling). This can be seen by looking at the difference between the estimated intercepts. The biased value is almost twenty times greater than the unbiased value. The Tobit model is shown in the third column of Table 4.33. It estimates the slope param-

Table 4.33 Comparison of least squares estimates and a Tobit regression

	OLS		Tobit
	(1)	(2)	(3)
x	0.010***	0.008***	0.010***
	(0.001)	(0.001)	(0.001)
logSigma			0.542***
			(0.114)
Constant	0.110	2.060***	0.539
	(0.775)	(0.667)	(0.873)
Observations	50	50	50
Residual Std. Error (df = 48)	1.783	1.534	

Note: *p<0.1; **p<0.05; ***p<0.01

eter in almost the same way as the reference model, resulting in a significantly smaller deviation in the intercept.

The estimated parameters of a Tobit estimate always refer to the marginal effect of x on the uncensored *latent* variable y^* and not on y. In the lecturer example, we could say that the estimated additional lateness is 0.01 minutes per additional meter of commute to college.

4.8 Statistics Software

There is too much data in practically all laboratory experiments to be evaluated by hand within a reasonable time. In addition, every type of experimental software used already supplies data in an electronic form. For example, the experimental software *zTree* is able to store data in xls or csv format. Therefore, the first question we have to ask ourselves in the context of statistical evaluation is which software package is suitable for our purposes.

Traditionally, students do their first data analyses with Microsoft Excel. The big advantage of this software is that the user becomes familiar with it in a relatively short time and can quickly obtain his own initial results. Excel is particularly suitable for the initial preparation or graphical presentation of data. However, if the data sets are very large or special procedures such as nonparametric tests are required, the limits of this program are soon reached. Experimenters who already know that they will be working in an experimental science for a longer period of time are therefore better advised to use a more flexible, more powerful and more efficiently programmed software solution. Common examples are SPSS (▶ ibm.com), SAS (▶ www.sas.com), S+ (▶ spotfire.tibco.com), GAUSS (▶ www.aptech.com) or MATLAB (▶ www.mathworks.com). STATA (▶ www.stata.com) is very popular among economists. The statistical scope of these software packages far exceeds that of Excel. However, most of them have their own programming language, of which it is necessary to learn at least the basics and this can initially deter many people. In addition, the purchase costs and license fees are sometimes considerable. There are free alternatives to all the commercial products. Free variants of Excel are Calc (▶ www.openoffice.org) or SOFA (▶ www.sofastatistics.com), SPSS can largely be replaced by PSPP (▶ www.gnu.org/software/pspp), MATLAB by Octave (▶ www.gnu.org/software/octave) and S+ by R (▶ www.r-project.org). An overview of non-commercial statistical software can be found at statpages.org/javasta2.html.

If only specific tools for specific analyses are needed, and not complete packages, it is also possible to make use of powerful and free programs. PQRS (members.home.nl/sytse.knypstra/PQRS) provides the density and distribution functions for all conceivable probability distributions. All the parameters can be freely varied so that critical values and p-values of hypothesis tests in particular can be determined quickly and easily. The tool is available in three variants. Version 2 can be run immediately on any Windows computer without installing third-party software. Newer versions (currently version 2.7) are executable Python programs, so this programming language must be installed to run PQRS. In addition, an app that runs smoothly on all mobile Android devices is available.

To perform a power analysis of statistical hypothesis tests, the free program G*Power (▶ www.gpower.hhu.de) is also available. This is also extremely helpful for fast and straightforward analyses.

References

Bliese, P., & Ployhart, E. E. (2002). Growth modeling using random coefficient models: Model building, testing, and illustrations. *Organizational Research Methods, 5*(4), 362–387.

Bolker, B. M., Brooks, M. E., Clark, C. J., Geange, S. W., Poulsen, J. R., Stevens, M. H. H., & White, J. S. S. (2009). Generalized linear mixed models: A practical guide for ecology and evolution. *Trends in Ecology & Evolution, 24*(3), 127–135.

Bortz, J., & Lienert, G. A. (2008). *Kurzgefasste Statistik für die klinische Forschung: Leitfaden für die verteilungsfreie Analyse kleiner Stichproben. 3. Auflage*. Heidelberg: Springer.

4

Box, G. E. P., Hunter, J. S., & Hunter, W. G. (2005). *Statistics for experimenters: design, innovation, and discovery* (2nd ed.). New Jersey: John Wiley & Sons.

Brosig-Koch, J., Helbach, C., Ockenfels, A., & Weimann, J. (2011). Still different after all these years: Solidarity behavior in East and West Germany. *Journal of Public Economics, 95*(11–12), 1373–1376.

Cohen, J. (1988). *Statistical power analysis for the behavioral sciences* (2nd ed.). Hillsdale: Lawrence Erlbaum Associates.

Conover, W. J. (1972). A Kolmogorov goodness-of-fit test for discontinuous distributions. *Journal of the American Statistical Association, 67*(339), 591–596.

Conover, W. J. (1973). On methods of handling ties in the Wilcoxon Signed-Rank test. *Journal of the American Statistical Association, 68*(344), 985–988.

Conover, W. J. (1999). *Practical nonparametric statistics* (3rd ed.). New York: John Wiley & Sons.

Cox, J. C., Friedman, D., & Gjerstad, S. (2007). A tractable model of reciprocity and fairness. *Games and Economic Behavior, 59*, 17–45.

Cumming, G. (2013). The new statistics: why and how. *Psychological Science, 25*(1), 7–29.

Davis, D. D., & Holt, C. A. (1993). *Experimental economics*. Princeton: Princeton University Press.

Dufwenberg, M., & Kirchsteiger, G. (2004). A theory of sequential reciprocity. *Games and Economic Behavior, 47*, 268–298.

Ellis, P.D. (2010). The essential guide to effect sizes. Cambridge et al.: Cambridge University Press.

Falk, A., & Fischbacher, U. (2006). A theory of reciprocity. *Games and Economic Behavior, 54*(2), 293–315.

Fleiss, J. L., Levin, B., & Paik, M. C. (2003). *Statistical methods for rates and proportions* (3rd ed.). New York: John Wiley & Sons.

Griffiths, W. E., Hill, R. C., & Judge, G. G. (1993). *Learning and practicing econometrics*. New Jersey: John Wiley & Sons.

Gujarati, D., & Porter, D. (2008). *Basic econometrics* (5th ed.). Boston: McGraw-Hill.

Hilbe, J. M. (2009). *Logistic regression models*. London: Chapman & Hall/CRC.

Hoenig, J. M., & Heisey, D. M. (2001). The abuse of power: The pervasive fallacy of power calculations for data analysis. *The American Statistician, 55*(1), 19–24.

Hoffmann, S., Mihm, B., & Weimann, J. (2015). To commit or not to commit? An experimental investigation of pre-commitments in bargaining situations with asymmetric information. *Journal of Public Economics, 121*, 95–105.

Kanji, G. K. (2006). *100 statistical tests* (3rd ed.). London: Sage Publications Ltd.

Kennedy, P. (2008). *A guide to econometrics* (6th ed.). Malden: Wiley-Blackwell.

Lee, Y., & Nelder, J. A. (2004). Conditional and marginal models: another view. *Statistical Science, 19*(2), 219–228.

Lenth, R. (2000). Two sample-size practices that I Don't Recommend. Comments from panel discussion at the 2000 Joint Statistical Meetings in Indianapolis, http://www.stat.uiowa.edu/~rlenth/Power/2badHabits.pdf.

Lenth, R. (2001). Some practical guidelines for effective sample size determination. *The American Statistician, 55*(3), 187–193.

Lenth, R. (2007). Post Hoc power: Tables and commentary. Technical Report No. 378, University of Iowa, Department of Statistics and Actuarial Science, http://www.stat.uiowa.edu/files/stat/techrep/tr378.pdf.

Leonhart, R. (2008). *Psychologische Methodenlehre/Statistik*. München: UTB.

Levine, D. K. (1998). Modeling altruism and spitefulness in experiments. *Review of Economic Dynamics, 1*(3), 593–622.

Liang, K.-Y., & Zeger, S. L. (1986). Longitudinal data analysis using generalized linear models. *Biometrika, 78*, 13–22.
Marascuilo, L. A., & McSweeney, M. (1977). *Nonparametric and distribution-free methods for the social sciences*. Monterey: Brooks/Cole.
Massey, F. J. (1951). The Kolmogorov-Smirnov test for goodness of fit. *Journal of the American Statistical Association, 46*, 68–78.
McCullagh, P., & Nelder, J. A. (1989). *Generalized linear models* (2nd ed.). Boca Raton: Chapman and Hall.
Morris, M. (2010). *Design of experiments – an introduction based on linear models*. London: Chapman and Hall.
Murphy, K. R., Myors, B., & Wolach, A. (2014). *Statistical power analysis: A simple and general model for traditional and modern hypothesis tests* (4th ed.). New York: Routledge.
Nelder, J. A., & Wedderburn, R. W. M. (1972). Generalized linear models. *Journal of the Royal Statistical Society, Series A (General), 135*(3), 370–384.
Nowak, M. A., & Siegmund, K. (2005). Evolution of indirect reciprocity. *Nature, 437*, 1291–1298.
Rabe-Hesketh, S., & Skrondal, A. (2010). Generalized linear mixed models. In P. Peterson, E. Baker, & B. McGaw (Eds.), *International encyclopedia of education* (pp. 171–177). Amsterdam: Elsevier.
Rabe-Hesketh, S., & Skrondal, A. (2004). *Generalized latent variable modelling: Multilevel, longitudinal, and structural equation models*. Boca Raton: Chapman & Hall/CRC.
Rabin, M. (1993). Incorporating fairness into game theory and economics. *American Economic Review, 83*, 1281–1302.
Sheskin, D. J. (2000). *Parametric and nonparametric statistical procedures*. Boca Raton: CRC Press.
Siegel, S., & Castellan, N. J. (1988). *Nonparametric statistics for the behavioral sciences* (2nd ed.). New York: McGraw-Hill.
Simpson, E. H. (1951). The interpretation of interaction in contingency tables. *Journal of the Royal Statistical Society. Series B (Methodological), 13*(2), 238–241.
Singer, J. D., & Willett, J. B. (2003). *Applied longitudinal data analysis – modeling change and event occurrence*. Oxford: Oxford University Press.
von Auer, L. (2016). *Ökonometrie: Eine Einführung. 7. Auflage*. Berlin, Heidelberg: Gabler Verlag.
Wu, C. F. J., & Hamada, M. S. (2009). *Experiments: Planning, analysis, and optimization* (2nd ed.). New York: John Wiley & Sons.
Zar, J. H. (1999). *Biostatistical analysis* (4th ed.). Upper Saddle River: Prentice Hall.

Supplementary Information

J. Weimann, J. Brosig-Koch, *Methods in Experimental Economics*, Springer Texts in Business and Economics, https://doi.org/10.1007/978-3-319-93363-4

Appendix

A.1 A Basic Game-Theoretical Concepts

A.1.1 Game Theory

Game theory is concerned with the analysis of strategic interactions. Such an interaction exists when actors mutually influence the utility they can gain in a decision-making situation through the decisions they make. We will consider a strategic interaction between two players A and B as an example. Both have a predefined number of possible actions that they can take in a particular situation. Both players want to maximize their own utility by choosing the best action they can take, but the question of what action is best for player A depends on what action player B takes, and the best action for player B depends on what A does. This interdependence is the strategic interaction. For example, A and B could be companies operating on the same market. The question of which price A should choose to maximize its profit depends on which price B chooses. Similarly, the choice of the maximum profit price of B depends on which price A sets.

Game theory analyzes such strategic interactions from the perspective of a normative theory, i.e. it does not ask how real persons will probably behave in a strategic interaction, but rather assumes that the players behave strictly rationally and (expected) utility-maximizing, and examines which equilibria (see ► Sect. A.1.3) arise under this premise. The behavioral assumptions that game theory makes are very extensive. Not only does it require players to always choose the action that gives them the highest payoff (measured as utility), it also assumes that players behave strategically. This means that they are fully aware of the strategic interaction and fully rationally consider the strictly rational considerations of the other players. For this to happen, it is necessary to assume that the rationality of the players is common knowledge. This means that all the players know that all the players behave rationally and strategically, and everyone knows that this is known to everyone, and everyone knows that everyone knows that this is known to everyone, and so on.

The formal analysis of strategic interactions is made possible by representing them as a "game" (see ► Sect. A.1.2). The description of this game formalizes the interaction such that it is then open to analytical treatment. Identifying an equilibrium in a game described in this way provides a prediction of how individual players will act if they comply with the behavioral assumptions of the theory.

Within experimental economic research, it is mainly strategic interactions of relatively simple structure (this is important to ensure that the subjects' behavior is not distorted by fundamental problems of understanding) that are considered, with the result that the theory is particularly easy to test. Theoretical research, on the other hand, also considers very complex games. Complexity arises, for example, by allowing mixed strategies of the players. This means that the players do not decide which action they choose, but only the probability of choosing the various possible actions. Games can be static (all the players decide simultaneously) or dynamic (players decide sequentially). If at least one player is not exactly informed about another player's preferences regarding the possible outcome of the game, this can also lead to strategic situations becoming complex. These types of games are called games with incomplete information. Under certain conditions such games can be transformed into games with imperfect information, i.e. games in which not all moves of the players can be observed. The advantage of the transformation is that for games with imperfect information, there are solutions that allow behavior to be predicted.

A.1.2 The Description of a Game

In some respects, the description of a **game** is similar to the descriptions found in board games. First of all, it is specified who is a player, i.e. who can make decisions during the game (only these are players). In principle, the number of players is not restricted, apart from the fact that there must be at least two, otherwise no strategic interaction can arise. For the sake of simplicity, we will first introduce some notation.

We number the players $i = 1, 2, \dots, n$. For each of the players what is termed the strategy space S_i is defined. This is the total set of possible strategies s_i available to player i, i.e. from which he can choose. The strategy space therefore defines the options for action available to the player. These are fixed; the player cannot change them and create new strategies. A strategy profile $s = (s_1, s_2, \dots, s_n)$ contains a combination of n strategies of n players. $s_{-i} = (s_1, \dots, s_{i-1}, s_{i+1}, \dots, s_n)$ denotes a strategy profile that contains the strategies of the n-1 other players from player i's point of view. $S = S_1 \times S_2 \times \dots \times S_{n-1} \times S_n$ denotes all the possible strategy combinations that are possible in the game. The payoff function is defined on this set of possible game outcomes, because it assigns a real number to each element of u_i: $S \rightarrow$ R, $i \in \{1, \dots, n\}$. This function is simply a valuation of each individual game outcome by each player. The **payoff function** is therefore nothing more than a mapping of the preferences that the players have concerning the possible outcomes of the game.

The description of a game includes not only information about the players, their strategy spaces and payoff functions, but also information about which player can make a decision when and what information he has at the time of that decision. In static games, all the players decide simultaneously, i.e. they do not know the strategies chosen by the other players at the time of the decision. In a dynamic game, the players decide one after the other (similar to chess). With perfect information, players can observe the moves made by the players who moved before them. But this observability can be limited, i.e. it is possible that certain moves cannot be observed. These are games with imperfect information.

Static games are specified using their so-called normal form. This includes the strategy spaces and payoff functions of all the players. With two players and a discrete number of strategies, this can take the form of a matrix. The following Table A1

Table A1 Normal form of a 2×2 game

Player B ↓	Player A→	a_1	a_2
b_1		π_{B11}, π_{A11}	π_{B12}, π_{A12}
b_2		π_{B21}, π_{A21}	π_{B22}, π_{A22}

shows the normal form of a 2×2 game, i.e. for two players (A, B) who each have two strategies available: (a_1, a_2) for player A and (b_1, b_2) for player B.

The values in the cells of the table indicate the payoffs to the two players for the corresponding strategy combination. For example, π_{B11} is the payoff to player B for the strategy combination (a_1, b_1).

In a dynamic game, the information is given in the so-called extensive form. This is a game, or decision, tree whose nodes indicate the points at which a player can make a decision and whose branches indicate the actions that are available at a particular node. At the terminal nodes are the payoffs that are achieved if the player reaches that terminal node at the end of the game. Figure A1 shows a game tree for a dynamic game between two players who can choose between two actions at each node:

A **strategy** in a dynamic game tells the player which action to choose at which node. All the possible nodes are included, thus making a strategy a comprehensive plan that tells the player what to do no matter where in the game he is. Since player 2 in the example has two nodes at which he has to decide between two actions, a total of $2^2 = 4$ strategies are available to him.

A.1.3 The Nash Equilibrium

How do we move from the pure description of a game to a solution or to an analysis of the strategic interaction? The basic idea is to look for situations – or rather strategy combinations – that represent an equilibrium in the sense that when players are in this equilibrium, they no longer have any reason to change their behavior. The most important equilibrium concept used in this context is that of the Nash equilibrium, which dates back to John Nash (1950). The original article in which the equilibrium concept is presented and the Nash theorem is proven comprises only one page, but it revolutionized economic theory and earned John Nash the Nobel Prize.

Before explaining the Nash equilibrium, it is a good idea to explain what a "**best response**" is. In a two-player game, this is a strategy that maximizes one player's payoff for a given strategy of the other player. If there are more than two players, the definition is analogous, except that the strategies of all other players are

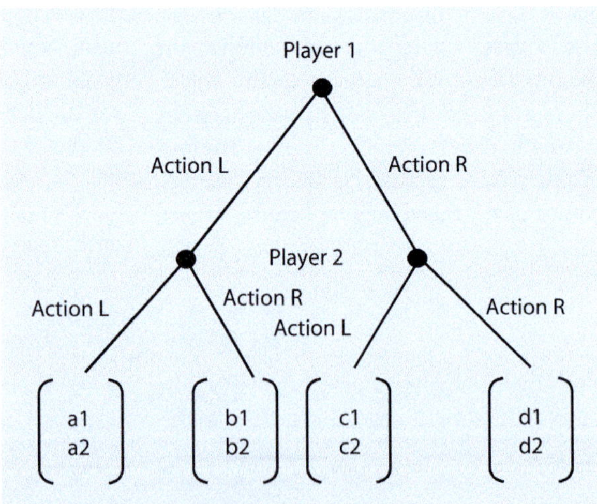

Fig. A1 Extensive form of a 2×2 game

taken as given. A Nash equilibrium is nothing more than a combination of strategies, where each strategy in the combination is also the best response to the other strategies in the combination. A Nash equilibrium therefore consists of the best mutual responses of all the players involved. The great significance of this concept of equilibrium can be explained by the fact that John Nash demonstrated that every game with a finite number of players and a finite number of strategies per player has at least one such equilibrium. This makes it clear that a majority of the strategic interactions also have a solution in the form of a Nash equilibrium. This is an extremely advantageous situation for theorists. When they embark on the analysis of such a game, they can be sure that there is a solution – all they have to do is find it. However, the Nash equilibrium does not always yield plausible predictions, which is why further solution concepts have been developed over time. We will look at one of them in the next section.

Nash's proof does not say that there is exactly one equilibrium for every game, but that there is at least one equilibrium. This means that there can be several. If this is the case, then the question arises, which of the equilibria will be achieved in the end or what does game theory predict? This problem of choosing the right equilibrium long occupied game theorists without a general solution being found. Frequently it is not possible to highlight specific equilibria as "special", but rather one has to be content with the fact that there simply are several equilibria to be found. Nash's proof also requires that mixed strategies be allowed, i.e. probability distributions across the player's strategy space. A good example of this is the penalty shoot-out in soccer. No shooter chooses a pure strategy (e.g. "always shoot to the right"). The shooters mix their strategies, sometimes choosing the left corner, sometimes the right.

A.1.4 The Extensive Form and Subgame Perfection

As in static games, Nash equilibria also exist in dynamic games. The problem is that there are too many of them. It is possible that a strategy combination forms a Nash equilibrium that is "on the way" so to speak, i.e. on a path through the game tree, and it is possible that this Nash equilibrium allows that one of the players does not play his best response. Such an equilibrium is hardly plausible.

In order to eliminate the implausible Nash equilibria, we must first introduce the term subgame. In a game tree, a subgame consists of a node and all its successor nodes. ▪ Figure A2 shows a simplified extensive form,[1] which has a total of four subgames (without the entire game).

Now that we know what a subgame is, we can introduce the "subgame perfect equilibrium", which goes back to Reinhard Selten (1965) and which also earned its discoverer the Nobel Prize. A subgame perfect equilibrium is one in which a Nash equilibrium exists in all the subgames and in the whole game as well. As a result, strategies that do not yield the best responses are no longer candidates for a subgame perfect equilibrium and the problem described above has been solved.

To determine subgame perfect equilibria, we apply the principle of backward induction. To do this, we first identify the Nash equilibria for all the subgames at the end of the game tree. We then replace each of these subgames with the payoffs that

1 The payoffs and the names of the players at the nodes are omitted for the sake of simplicity.

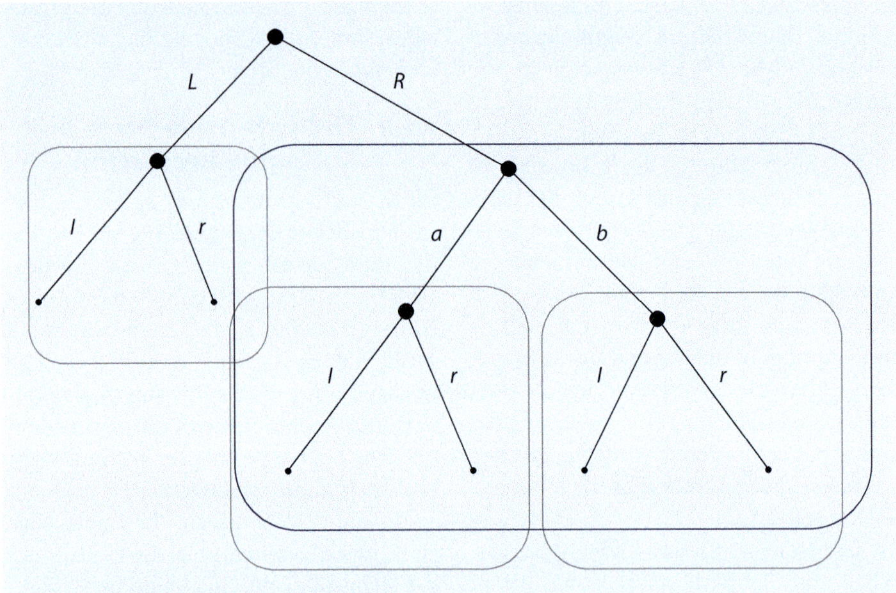

Fig. A2 Subgames in an extensive form

result in the equilibrium of this subgame (if there are several equilibria, one of them has to be selected). Then we repeat both steps for the game reduced in this way, continuing this procedure until we have determined all the moves in the entire game and the subgame perfect Nash equilibrium has been found. The procedure must be repeated for each of the Nash equilibria identified in its course.[2]

Our remarks so far have focused on games with complete information. There is no reason to hide the fact that there are also specific solution concepts for games with incomplete information, known as Bayesian equilibria. For more information on this subject, the reader is referred to relevant textbooks on game theory. The solution concepts outlined above should be sufficient to understand the main strategic incentives in the experimental studies described here.

2 Brosig-Koch et al. (2015) investigate the extent to which this ability is pronounced in different age groups. They observe that people find it difficult to apply the principle of backward induction. However, these difficulties decrease with age. Teams are also better able to induce backwards than individual decision-makers (Brosig-Koch et al. 2014).

A.2 Important Experiments

In this appendix, we will introduce important types of experiments in experimental economic research. The presentation is confined to the essentials and is intended to provide an insight to readers who are not familiar with the particular setups. We will describe the basic design of each experiment without going into the many variants that exist for these. In addition, we will briefly explain the economic background of the experiments and list what we consider to be the most important findings.

A.2.1 The Prisoner's Dilemma Game Experiment and the Public Good Experiment

The prisoner's dilemma is a game with which a fundamental and, in economics, very important conflict between individual and collective rationality can be modeled. We could also say that the prisoner's dilemma represents a kind of "cooperation paradox", which can be summarized as follows. The very fact that the players pursue the goal of maximizing their own payoffs completely rationally (i.e. without making a mistake) puts them in a situation where there is an alternative in which all the players could achieve a higher payoff if they had refrained from rationally pursuing their own interests.

The prisoner's dilemma game shown in ◘ Table A2 is a two-person game in which both players have two strategies and choose between them simultaneously. These are denoted D (for defection) and C (for cooperation).

Both players have a **dominant strategy**. Regardless of whether the other player chooses C or D, it is always best for the player to choose D. Thus, the Nash equilibrium of this game is (D_A, D_B) and both players gain a payoff of 3. The cooperative solution (C_a, C_B) would give them a payoff that is twice as high, but that is not accessible to rational and strictly selfish players under the rules of the game because C is never the best response. Cooperation could benefit both players, but this cooperation is not rational from the point of view of the individual payoff-maximizing player.

Larger groups with n players can also find themselves in such a cooperation paradox. The public good game is used in experimental economic research to model the phenomenon of public goods. The standard procedure is to apply the voluntary contribution mechanism (VCM), which dates back to Isaac & Walker (1988). In the VCM game, each player has an initial endowment of z_i. Two investment vehicles are available in which any portion of z_i can be invested. The first is a private investment vehicle, which provides a payoff p per unit invested to a player who invested in it, and the second a public project, in which each player receives a payoff of a/n multiplied by the sum of the contributions invested. Parameter a describes the marginal productivity of the public good. Denoting b_i the investment of player i in the public project, we then obtain the following the payoff function for this game:

$$\pi_i = (z_i - b_i)\,p + \frac{a}{n}\sum_{i=1}^{n} b_i$$

with $a > p > \frac{a}{n}$.

Given these parameters, it is a dominant strategy for rational and selfish players in the simultaneous public good game not to make any investment in the public

◘ **Table A2** Normal form of a prisoner's dilemma; payoffs: (A, B)

Player A ↓	Player B→	C_B	D_B
C_A		6, 6	1, 7
D_A		7, 1	3, 3

good. If the players choose their dominant strategy, they will receive a payoff of pz_i. If all the players were to invest all their endowment in the public project, the payoff to each player would be $az_i > pz_i$. This gives the same result as in the two-person prisoner's dilemma. Strictly rational pursuit of self-interest leads to a payoff that is smaller than that which can be achieved if all players refrain from pursuing their self-interest.

The public project in the game meets the conditions that characterize a public good: there is no exclusion of consumption and there is no rivalry in consumption. Public goods play an extremely important role in modern societies. Climate protection, national defense or the provision of environmental goods are prominent examples of public goods. Analogous to the prisoner's dilemma, a public good game is referred to as a situation in which the players find themselves in a social dilemma.

The experiments on the public good game are usually repeated, often over 10 rounds. The aim is to investigate whether the subjects are able to overcome the dilemma and achieve an efficient solution by acting cooperatively. In the unmistakable subgame perfect equilibrium of this repeated game, strict free-riding should be observed in each round. There are many experiments whose reproducible results can be outlined as follows. The cooperation rate in the first few rounds is about 40% to 50% of the efficient level, but then drops to about 10% in the course of the experiment.[3] Although this disproves the prediction of strict free-riding in these experiments, the observed differences between the efficient and the realized overall payoffs are nevertheless very high, although the subjects (often economics students) are well aware of the dilemma and know that they could substantially increase their overall payoff by cooperating. The provision of public goods is thus not an easy problem to solve, even under ideal laboratory conditions.

A.2.2 The Ultimatum Game Experiment

The ultimatum game models a negotiation situation between two people in a very simple way. It involves dividing a predetermined amount of money of value x. The two players have different roles. The so-called proposer makes the first move. He can make an offer to the second player – the responder – by offering him a share ax ($0 \leq a \leq 1$) and keeping $x(1\text{-}a)$ for himself. During the second stage of the game, the responder must decide whether or not to accept the proposed split. If he accepts, the money is divided accordingly and the game is over. If he rejects the offer, the game also ends and both players receive a payoff of zero. The proposer's offer is therefore an ultimatum that cannot be further negotiated (hence the name of the game).

The subgame perfect equilibrium of this game is determined by backward induction. At the final stage, the responder will accept any offer that makes him better off than a rejection of the offer. Since a rejection will result in a payoff of zero, he will accept any offer with a payoff larger than zero. The proposer anticipates this. His best response to the responder's strategy is to make him an offer that is only just better than rejection. He will therefore offer the responder the smallest possible share of x and the responder will accept this offer because it is his best response.

The equilibrium of the ultimatum game results in an extremely unequal split.

3 For an early survey see Ledyard (1995), for a more recent selective survey see Chaudhuri (2011).

While the proposer gets almost everything, the responder gets almost nothing. The ultimatum game was first played in experiments by Güth et al. (1982). The aim was to check whether the game-theoretical prediction was correct, although it would impose a very unequal allocation. It was found that the responders did not agree with the equilibrium payoffs and therefore often rejected low offers, although this worsened their position compared to an acceptance. Proposers anticipated this behavior and usually offered significantly higher shares than the equilibrium foresees. Often a 50-50 split was offered and this was always accepted. If the proposer demanded a larger share for himself, such as 80:20 or 70:30, he had no choice but to face rejection by the responder. Since then, these results have been confirmed time and again in a large number of experiments on the ultimatum game. The ultimatum game is rightly considered one of the best-studied games in experimental economic research.[4]

A.2.3 The Dictator Game Experiment

The dictator game is similar to the ultimatum game, but it does not allow the responder to reject the proposer's offer. The responder is thus reduced to the role of the so-called receiver, or recipient. This makes the proposer the "dictator", who alone can determine how the amount x is divided between the two players. Strictly speaking, the dictator game is not a game in the game-theoretical sense, as there is no strategic interaction with the receiver. It is the absence of strategic interaction, however, that makes the "game" so interesting, because we can assume that the dictator's decision is not influenced by expectations regarding the receiver's behavior. This means, however, that his decision only expresses his preference for possible payoff allocations. Therefore, the dictator game experiment can be used to gain information about this kind of preferences.

It is only natural to compare the results of the ultimatum game experiment with those of dictator game experiments (see e.g. Forsythe et al. 1994). This shows that, on average, dictators' allocations are significantly lower than those of proposers in ultimatum game experiments. This suggests that some of the offers observed in the ultimatum game experiment are driven by the expectation that excessively low offers could be rejected. Nevertheless, relatively high allocations to the receiver are also evident in dictator game experiments. However, it has also been observed that this giving behavior is highly sensitive to individual elements of the design. We report on these effects in more detail in ▶ Chap. 2 of the book.

A.2.4 The Trust Game Experiment and The Gift-Exchange Experiment

The trust game experiment (or sometimes also called the investment game) was introduced into the literature by Berg et al. (1995). It is a sequential two-person game in which both subjects first receive an initial endowment A. The first player (the trustor) has the option of giving any share $0 \leq \alpha \leq 1$ to the second player (the trustee). The amount is then tripled by the experimenter, i.e. the trustee receives the amount $3\alpha A$. At the second stage of the game, the trustee has the option to return any part $0 \leq \beta \leq 1$ of his endowment $(A + 3\alpha A)$ to the trustor.

4 For a survey, see Güth and Kocher (2014).

The subgame perfect equilibrium of this game is easy to determine by means of backward induction. At the second stage, the trustee has no reason to give anything back to the trustor, because his dominant strategy is to choose $\beta = 0$, since any positive β reduces his payoff. The best response the trustor can have to this strategy is to choose $\alpha = 0$, i.e. not to give anything to the trustee. This means that there is no allocation in equilibrium and thus no efficient solution that can be achieved by tripling the amount of the allocation. In equilibrium, therefore, the payoff is the same for both players A. Were the trustor to send all his initial endowment to the trustee, the total payoff to both players would be $4A$. If the trustee were then to return $2A$ to the trustor, both players could double their payoffs compared to the equilibrium by showing trust (first mover) and by being trustworthy (second mover).

The sequential structure of the trust game creates scope for reciprocal behavior. This means that people react to other people acting "nicely" by being "nice" themselves and are prepared to punish people who have harmed them. The trust game experiment shows that reciprocity can certainly lead to efficiency gains.

Closely related to the trust game experiment and directly targeting reciprocal behavior is the gift-exchange game experiment, which has been investigated mainly by Fehr et al. (e.g. 1998). This experiment normally has a very special frame, with it being presented as a game between employers and employees. The subjects representing the employers make wage offers to the employees. The employees choose from the offers and then decide on the work effort they want to make in return for their wages. This effort incurs costs, i.e. the more effort workers put in for a given wage, the lower their payoff and the higher the employers' payoff. Therefore, at the second stage of the game, the employees have a dominant strategy consisting of choosing the minimum possible effort. The best response to this from employers is to offer the lowest possible wage. This, in turn, would result in an inefficient subgame perfect equilibrium, because if employers paid higher wages and workers chose higher levels of effort, both sides could improve in comparison with the equilibrium. The similarity to the trust game experiment is obvious, but the gift-exchange game experiment emphasizes even more the exchange of "non-best responses", with this exchange possibly leading to an efficiency gain.

The experimental findings from both the trust game experiment and the gift-exchange game experiment show that reciprocal behavior is chosen relatively frequently. Therefore, people are indeed in a position to achieve efficiency gains through trust and trustworthiness, but also through the exchange of gifts. However, this is not perfect. In many experiments, deviations from the maximum possible total payoff can be observed and, in particular, the behavior of the second movers in the trust game experiment is not always focused on sharing the entire payoff evenly between the two players. Frequently the returns to the trustor are designed in such a way that the trustor does not suffer any disadvantage from giving to the trustee (but he does not gain any advantage either).

A.2.5 Market Experiments

The term "market experiments" does not refer to a special experimental design, but to a whole class of experiments. The aim is to model market processes in the laboratory. In particular, the question is how pricing takes place in competitive markets and whether markets or market participants are in a position to achieve an efficient market equilibrium solely on the basis of the individual decisions of suppliers and demanders. The process that is frequently used can be roughly described as follows.

The experimenter divides the subjects into suppliers and demanders. Each supplier then receives private information about the costs incurred when selling one unit of the fictitious good, which is traded on the laboratory market. These costs are also his reservation price, since the supplier should not sell at a price below these costs. With every completed transaction his profit is equal to the *purchase price – costs*. Demanders receive information about the payment they receive when they purchase a unit of the goods, i.e. their profit is equal to the *payment – purchase price*. The payment therefore represents their maximum willingness to pay. For example, the costs and payments can be distributed in such a way that the total quantity supplied increases with the price and the total quantity demanded decreases with the price and an equilibrium, when realized, maximizes the efficiency gains that can be achieved through trading. The question then is whether there is enough private information about the respective individual reservation prices and payments to lead the market into such an equilibrium. The results of the market experiments (e.g. the experiment conducted by Smith 1962) show that market equilibrium can be achieved even with the limited information available to market participants. Whether and how quickly the equilibrium is reached, however, depends on the specific design of the markets, i.e. the way in which the offers of the suppliers and the bids of the demanders are exchanged and accepted. This is where the design of what is termed a double auction, developed by Vernon Smith, proved to be particularly robust. Vernon Smith was the first experimental economist to be awarded the Nobel Prize in 2002 for his research into the way markets function.

It should be noted at this point that experimental markets differ from completely competitive markets in an important respect. In the latter, by assumption, atomistic competition prevails among both suppliers and demanders, i.e. the actors have no scope for setting prices and therefore act as "price takers". This is noticeably different in the laboratory markets due to the limited number of subjects. Nevertheless, Vernon Smith in particular showed that prices and quantities similar to those predicted for a fully competitive market arise in in double auctions. The strategic leeway that players have in laboratory markets therefore has no effect on the allocation efficiency.

A.2.6 Lottery Choice Experiments

Decisions under risk play an important role in economic research. Risk is a very important factor in many real decision-making situations. The world is stochastic and therefore it is rarely the case that we know with certainty what consequences a decision will actually have. Risk preference, which refers to the decision-maker's attitude to risk, is very important for modeling decisions under risk. One function of lottery choice experiments is to generally analyze behavior in risky situations or to gain specific information on the risk preference of experimental subjects.

The procedure for such experiments can be simply explained by means of an experiment that can be used to determine the risk preference of a decision-maker. Not having any strategic interaction between the subjects, lottery choices are, in principle, not games. It is simply a matter of making a choice between different lotteries. In a typical experiment, for example, the subjects are presented with a lottery that realizes a payoff of X with a probability p and a payoff of zero with a probability of $(1-p)$. The expected value of this lottery is thus pX. The subject is then offered the possibility of selling this lottery for a sure payoff with the selling price varying. The sure payoff is interpreted as a lottery with a probability of 1 that this payoff is received. The risk preference can then be inferred from the subject's decision (see ▶ Sect. 2.4.1). The Becker-DeGroot-Marchak procedure is often used to determine the reservation price for the lottery directly (Becker et al. 1964). This procedure is intended to ensure that the subjects state their true reservation price for the lottery. For example, a list of prices sorted in ascending order is presented to the subject, who is asked to name the price as of which she is willing to sell the lottery. One of the listed prices is then drawn at random. If the price is above the threshold specified by the subject, the lottery is sold at the drawn price. If the price is lower, the lottery is played and the subject receives either X or zero – depending on the outcome of the lottery.

Under the rules of the Becker-DeGroot-Marchak procedure, the weakly dominant strategy is to enter the true reservation price as the limit. If this is below pX, the subject reveals that she is risk-averse, since she prefers a sure payoff to a lottery whose expected value is above this sure payoff. If the price equals pX, we speak of risk-neutral behavior and if the price is higher than pX, we speak of risk-seeking (or risk-loving) behavior. The prerequisite for the applicability of this procedure is that the subjects behave in accordance with the assumptions of expected utility theory.

However, lottery experiments are not only used to reveal risk preferences. They are also used, for example, to test basic assumptions of expected utility theory, which models behavior under risk, and its alternatives. Prominent examples of the discovery of systematic deviations from expected utility theory are the Ellsberg paradox and the Allais paradox. The results of the relevant experiments imply a violation of the independence axiom on which expected utility theory is based.

References

Becker, G. M., DeGroot, M. H., & Marschak, J. (1964). Measuring utility by a single-response sequential method. *Systems Research and Behavioral Science, 9*(3), 226–232.

Berg, J., Dickhaut, J., & McCabe, K. (1995). Trust, Reciprocity, and Social History. *Games and Economic Behavior, 10*(1), 122–142.

Brosig-Koch, J., Heinrich, T., & Helbach, C. (2014). Does truth win when teams reason strategically? *Economics Letters, 123*(1), 86–89.

Brosig-Koch, J., Heinrich, T., & Helbach, C. (2015). Exploring the capability to reason backwards: An experimental study with children, adolescents, and young adults. *European Economic Review, 74*, 286–302.

Chaudhuri, A. (2011). Sustaining Cooperation in Laboratory Public Goods Experiments: A Selective Survey of the Literature. *Experimental Economics, 14*(1), 47–83.

Fehr, E., Kirchler, E., & Gächter, A. W. u. S. (1998). When social norms overpower competition: Gift exchange in experimental labor markets. *Journal of Labor Economics, 16*(2), 324–351.

Forsythe, R., Horowitz, J. L., Savin, N. E., & Sefton, M. (1994). Fairness in simple bargaining experiments. *Games and Economic Behavior, 6*(3), 347–369.

Güth, W., & Kocher, M. G. (2014). More than thirty years of ultimatum bargaining experiments: Motives, variations, and a survey of the recent literature. *Journal of Economic Behavior & Organization, 108*, 396–409.

Güth, W., Schmittberger, R., & Schwarze, B. (1982). An experimental analysis of ultimatum bargaining. *Journal of Economic Behavior & Organization, 3*, 367–388.

Isaac, R. M., & Walker, J. M. (1988). Group Size Effects in Public Goods Provision: The Voluntary Contributions Mechanism. *Quarterly Journal of Economics, 103*(1), 179–199.

Ledyard, J. O. (1995). Public Goods: A Survey of Experimental Research. In J. H. Kagel & A. E. Roth (Eds.), *The Handbook of Experimental Economics* (pp. 111–194). Princeton: Princeton University Press.

Nash, J. F. (1950). Equilibrium points in n-person games. *Proceedings of the national academy of sciences, 36*(1), 48–49.

Selten, R. (1965). Spieltheoretische Behandlung eines Oligopolmodells mit Nachfrageträgheit. *Zeitschrift fur die gesamte Staatswissenschaft, 121*, 301–324.

Smith, V. L. (1962). An experimental study of competitive market behavior. *Journal of Political Economy, 70*, 111–137.

Index

R

S

T

U

V

W

Z